Palgrave Studies in the History of Science and Technology

Series Editors
Roger D. Launius
Auburn, USA

James Rodger Fleming
Centre Hall, USA

Designed to bridge the gap between the history of science and the history of technology, this series publishes the best new work by promising and accomplished authors in both areas. In particular, it offers historical perspectives on issues of current and ongoing concern, provides international and global perspectives on scientific issues, and encourages productive communication between historians and practicing scientists.

Janne Holmén

A History of Diagrams

Turning Points in the Illustrative Representation of Ideas

palgrave
macmillan

Janne Holmén
Department of Education
Uppsala University
Uppsala, Sweden

ISSN 2730-972X ISSN 2730-9738 (electronic)
Palgrave Studies in the History of Science and Technology
ISBN 978-3-031-99690-0 ISBN 978-3-031-99691-7 (eBook)
https://doi.org/10.1007/978-3-031-99691-7

Time series diagrams from Johann Heinrich Lambert's "Essai d'Hygrométrie ou sur la mesure de l'humidité," published in 1771 in the 1769 yearbook of the Berlin academy. The reproduced copy is from the collections of the Carolina Rediviva Library at Uppsala University

This Palgrave Macmillan imprint is published by the registered company Springer Nature Switzerland AG.
The registered company address is: Gewerbestrasse 11, 6330 Cham, Switzerland

Acknowledgments

This work is an outcome of the research project "Figuring politics," which has been conducted at Uppsala University with support from the Swedish National Research Council. Their funding also allowed me to arrange the workshop "Diagrams-Historical and interdisciplinary perspectives" in Uppsala in the autumn of 2022. At the workshop, the book project received valuable feedback from the participants Barbara Tversky, Yuri Engelhardt, Clive Richards, Ayelet Even-Ezra, Petter Hellström, Matthew D. Eddy, Malin Tväråna, and Ann Sofie Jägerskog.

Chapters of the book were also presented at the seminars for History of Education and Musicology in Uppsala, as well as at the Cognitive semiotics seminar in Lund and at the Fifth Conference of the International Association for Cognitive Semiotics (IACS-5) in Lund. Particular thanks should be directed to colleagues who have given feedback on the whole or parts of the book in different stages of its development: David Dunér, Petter Hellström, Ayelet Even-Ezra, Rikard Enberg, Johan Boberg, Michael Friendly, Iresha Gayani Wattegamaralalage Ratnayake, Ainur Elmgren, and Iva Lučić. From the publisher Palgrave Macmillan, valuable assistance has been given by Arunaa Devi and Emily Russell.

All remaining faults and errors are my own. Section "Diagrams of Heartbeats" in Chap. 6 should not be considered medical advice, and section "Stock Charts: Icons of Capitalism" in Chap. 6 is not financial advice.

Janne Holmén

Competing Interests The author has no competing interests to declare that are relevant to the content of this manuscript.

Contents

About the Author

Janne Holmén is an associate professor in the history of Education at the Department of Education, Uppsala University. In 2006 Holmén obtained a PhD in History from the Department of History, Uppsala University. He is an SHI-affiliated scholar at the Institute of Contemporary History, Södertörn University, and has also been attached to Åbo Akademi University. Much of Holmén's research has been of a comparative nature, usually within the field of education. A previous project on mental maps and historical consciousness among secondary school students around the Baltic and Mediterranean seas kindled his interest in spatial cognition, which inspired his current research on the history of diagrams.

List of Figures

CHAPTER 1

Introduction

The Catalan scholar Raymond Llull (1232–1315/16) constructed an intricate and complex system of diagrams that he hoped could be used for interreligious dialogue. His ultimate goal was that these figures, some of which included rotating disks, would help convince Jews and Muslims to convert to Christianity. However, Llull's faith in the educational power of his diagrams was shaken when he had the opportunity to present them at the University of Paris, where the learned audience had difficulty understanding them. In his autobiography Via *coaetanea* (ca. 1311), Llull wrote that "because of the weakness of human intellect," which he had witnessed in Paris, he had to simplify his diagrammatic system considerably.[1] The weakness he had observed was not limited to Paris, the medieval era, or even to academic institutions; it is a general human trait. Throughout history, our cognitive limitations have, together with other fundamental aspects of cognition, constrained and shaped the development of diagrams. Interpreted through the lens of cognitive principles, the long trends and turning points in the history of diagrams can teach us valuable lessons about how humans construct and disseminate knowledge about the world.

[1] Bonner (2007, 5). See also section "Llullism: Incomprehensible Originality" in Chap. 4.

J. Holmén, *A History of Diagrams*, Palgrave Studies in the History of Science and Technology,
https://doi.org/10.1007/978-3-031-99691-7_1

It is not difficult to understand the high expectations that Llull initially attached to his diagrammatic system. Many of us have struggled with complicated texts, only to experience an instant revelation after glancing at an accompanying diagram—or after trying to clarify the issue by constructing a diagram of our own. The concepts of the text, which were obscure from the written word alone, became clear in the illustration, where the concepts' relations to each other were represented as relations in space. As far back in history as we can trace, humans have used their cognition of space to grasp difficult thoughts and concepts and communicate them to other people. Originally, this was probably done with gestures, but later a range of cultural artifacts were applied for the same purpose. Diagrams are illustrations that resonate with the human propensity for spatial thinking. Well-known examples are evolutionary "trees of life," timelines, statistical graphics, flow charts, and the political left–right spectrum. Since they have permeated all sectors of society, such spatial models have been of central importance not only in the history of science, ideas, and education, but also in economic, social, and political history.

Today, we are beginning to understand why diagrams crafted in accordance with human cognitive abilities work wonders for our comprehension. Evidence from a variety of fields, ranging from neurology to linguistics, suggests that brain mechanisms originally developed for spatial cognition have been co-opted during evolution for the perception of time and for abstract thinking. This is done with the help of spatial analogies and metaphors, which all languages are full of: we think "back" on what happened last week and look "forward" to the coming weekend, move "up" the social ladder, feel "down" at times, and prefer our salaries to be "higher."[2]

Diagrams function as analogies where spatial relations in the figure represent relations between the concepts it analyzes. Some of these analogies are general and intuitive. For example, across almost all diagram types, closeness in space represents similarity. Other analogies rest on more specific conventions, such as that the X-axis in time series diagrams often represent time, and that in the periodic table rows represent the number of electron shells of the elements, while columns represent the number of electrons in the outer shell. These spatial analogies allow us with a glance to grasp how the population of a municipality or the earnings of a

[2] See, for example, Lakoff and Johnson (1980); and section "The Neurology of Spatial and Abstract Thinking" in Chap. 2.

company develop over time, or how heavy a chemical element is and what other elements it is likely to form molecules with.

This book surveys the history of humanity's use of diagrams, from its earliest Paleolithic roots until the present. It describes the circumstances under which new forms of diagrams have evolved, and how this development is linked to changes in the flow and transmission of knowledge, paradigmatic shifts in worldviews and understanding of space, as well as to technologies of image production. Furthermore, it analyzes how these cultural changes in particular diagrammatic forms have interacted with innate and universal principles of human spatial cognition. Hopefully, this investigation will also enable reflection on how spatial thinking affects our worldviews, and how we can best draw benefits from this ability while we avoid the pitfalls it also entails.

This book attempts to improve our understanding of the history of knowledge by analyzing it with the help of theories from the cognitive sciences. Applying theories on cognition to historical sources may shed new light on human history and suggest a way of putting the theories to a novel set of tests. We can expect that the basic mechanisms of the human cognitive apparatus have evolved little through the short timespan we call history. Therefore, the theories developed today regarding the workings of the brain should, if correct, be applicable to the output of human thought processes in the past. Ancient documents are not the kind of evidence against which, for example, neurological theories are usually tested. Typically, this takes place through controlled experiments or by observing the effects of accidental damage to the brain. However, there is significant added value in exposing theories to findings from several disparate fields of research. After all, a case is strengthened by being consilient with different and independent kinds of evidence.[3] Increased exposure to the humanities, and particularly the history of ideas and knowledge, might therefore help fine tune the theories on human cognition.[4]

The aims and approach of this study make it part of the new and small research field of cognitive history. Cognitive science has generally either focused on contemporary humans or early hominids. Cognitive history is concerned with filling the gap in the intervening time period, which we in

[3] For consilience, see Snyder (2005). The concept has been used in cognitive historiography by, for example, Ambasciano et al. (2020) and Slingerland (2014).

[4] A similar argument has been made regarding the relationship between psychology and art history. Suitner and McManus (2011, 278).

everyday language call history and contains the entire development of human culture.[5]

In an article in the first journal dedicated to cognitive history, the managing editor Leonardo Ambasciano described cognition as

> embodied (there is no mind without brain, and no brain without body), embedded (there is no cognition without extra-bodily and environmental interactions or processes), extended (physical objects are necessary to produce cognition), enactive (cognition results in the engagement of actions and behaviours), and, most importantly, encultured (the brain functions on the basis of a historically determined and learned set of beliefs, norms, practices, and worldviews).[6]

However, although this description of the scope of cognitive history is quite broad, in practice the *Journal of Cognitive Historiography*, as Ambasciano himself, has mainly published articles related to the psychology of religion. This emphasis on religion and myths can also be found in other cognitive history publications,[7] and has its foundation in a tradition of psychology of religion that predates the mid-1900s cognitive revolution by half a century.[8] This has led to a focus on the emotional aspects of cognition, which is reflected in Ambasciano's claim that humans are "emotionally primed, computationally limited, cognitively bounded, socially constrained, normatively coerced, economically naïve, and rationally misguided agents."[9] Undoubtedly, there is a lot of truth to this underwhelming account of human rationality and cognitive capacity. However, I would argue that one of the most important research problems that cognitive history has to tackle is how humans have managed to rationally address a wide range of enormously complex problems, despite the odds being stacked against them in the form of cognitive limitations. This book argues that a main strategy in this quest has been the use of spatial analogies, for example, by drawing diagrams.

Cognitive science emerged in the late 1940s, as computer scientists, neurophysiologists, and psychologists began to investigate how human cognition could be understood, a question the previously dominant

[5] Dunér (2019).

[6] Ambasciano (2017, 142).

[7] See, for example, Martin and Sørensen (2011).

[8] For example, a pioneering work was James (1902).

[9] Ambasciano (2017, 142).

behaviorist paradigm considered outside the realm of science. This development was facilitated by the fact that psychologists put their theoretical agenda aside during World War II while working on practical problems related to the war effort, such as map reading.[10]

Since the 1980s, a new turn in the cognitive sciences has emphasized the relationship between human thought and our bodies' interaction with the environment.[11] Researchers in cognitive anthropology, cognitive archaeology, cognitive history, and cognitive semiotics have found new approaches to their disciplines with the help of a set of interrelated concepts inspired by the cognitive sciences.

The embodied mind refers to humans thinking with their bodies and not just with their brains. Humans have cognitive abilities and limitations that are a function of the body and its history of interaction with the environment.[12] For example, diagrams are "crystallizations of gestures"[13] and fill a similar function in writing as gestures do in speech.

The affiliated concept *distributed cognition* refers to using an environment that enhances thinking.[14] This book presents several examples of how the spatial properties of first the natural environment, then architecture and objects, such as quipus, and finally diagrams have been utilized to enhance cognition. It is the human propensity for distributed cognition that makes cognitive history, or in fact any history writing, possible. The objects humans have created to enhance their thinking and communicate it to others constitute the sources from which we derive clues about their cognitive processes. History writing has traditionally focused on texts, but with the help of a conscious cognitive history approach, other artifacts, such as diagrams, can constitute a valuable alternative source of knowledge about the past.

Situated cognition takes the argument a step further than distributed cognition by emphasizing that human thought is not merely aided by external objects, but can in fact only function in interplay with the surrounding world, and therefore, the environment influences how we think.[15] This perspective emphasizes the contextual nature of human thought. According to this view, the history of science and ideas can only

[10] H. Gardner (1985); Holmén (2020).
[11] Dunér (2019); Clark (1997).
[12] Varela et al. (2016); Tversky (2019).
[13] Giardino (2016, 93–94).
[14] Giere and Moffatt (2003); Hutchins (2006).
[15] Roth and Jornet (2013).

be understood if we consider the environment in which the ideas have been formed. Philip Robbins and Murat Aydede use situated cognition as a higher-level concept that encompasses the embodied mind, distributed cognition, and similar concepts that describe how mental activity depends on the situation or context.[16]

Conceptual metaphors are metaphorical extensions of, for example, spatial reasoning. In order to understand a new, abstract target domain, we make use of metaphorical concepts from a well-known source domain.[17] Diagrams use space as a source domain for an abstract target domain, such as logic or political affinity. This volume frequently refers to a development of the conceptual metaphor idea, the theory of *primary metaphors*. These are a particular kind of basic metaphors that are believed to be fixated as neural connections in infancy, and constitute building blocks for more complex metaphors later in life.[18] Since a large share of the primary metaphors are spatial, they are important for our intuitive understanding of diagrams.

Another important concept is *cognitive load*, the idea that human cognition is severely restrained by limitations in working memory.[19] In fact, this constraint seems to be the reason why we, in our inability to get a direct grip on reality in all its complexity, are forced to use workarounds such as *categorization* and *externalization*. Categorization is central in the history of diagrams, since many of the advantages of distributed cognition stems from spatial visualization of categories. Most diagrams constructed through history in one way or another divide concepts or data into categories and illustrate how these categories are related to each other. Externalization refers to the unloading of our overflowing working memories into, for example, texts and diagrams, enabling the use of powerful distributed cognition.[20]

However, although the program for cognitive history sketched out by its proponents is ambitious and promising, in reality very little has been done so far in this young and small research field. The historian of science David Dunér attributes this to the fact that "[c]ognitive scientists are not

[16] Robbins and Aydede (2009, 3).

[17] Lakoff and Johnson (1980).

[18] Lakoff and Johnson (1999).

[19] See, for example, Sweller et al. (2011).

[20] For categorization, see Cohen and Lefebvre (2005). For externalization, see Clark and Chalmers (1998). They use the concept of active externalism, but research, such as Even-Ezra (2021), inspired by them use externalization instead.

trained in historical empirical research, often lack contextual knowledge and philological skills, and historians have little or no insight into what is going on in cognitive sciences and other relevant fields outside the historical disciplines."[21] What has been written about diagrams from a cognitive history perspective can therefore be summarized quite briefly.

In 1992, Nancy Nersessian used what she called cognitive-historical analysis in her investigation of the development of electromagnetism. She stressed the need for finding out how human cognitive abilities and limitations constrain scientific theories, and gave examples of how analogies and diagrams were used in theory development.[22]

Reviel Netz used the concept of cognitive history in his 1999 study of deduction in ancient Greek mathematics, where he described it as situated in the intersection between cognitive science and the history of ideas. However, he warned that cognitive scientists might view his approach as too impressionistic, and historians as over-theoretical and too eager to generalize.[23] Netz's work, which also contained a study of ancient Greek diagrams in geometry, was much closer to the history of ideas than to the cognitive sciences. In fact, it is difficult to find reference to any theory from the cognitive sciences in the book. Netz believed that cognitive historians should stay away from the biological and neural and focus solely on the historical aspects of cognition.[24] Other cognitive historians, such as Nancy Nersessian and David Dunér,[25] have rather emphasized the role of the neural in the historical processes, and that is also the position of this work. A good but rare example of historical research on diagrams through the lens of the cognitive sciences is Aylet Even-Ezra's investigation of medieval horizontal tree diagrams, which has its foundation in the theory of externalization.[26]

Cognitive anthropology and cognitive archaeology also take an interest in human cognition across time and space, and these fields are older and more developed than cognitive history. Cognitive anthropology has mainly concerned itself with language but has in the process also investigated the cognitive foundations of how humans classify and structure

[21] Dunér (2019, 7).
[22] Nersessian (1992).
[23] Netz (1999, 7).
[24] Netz (1999, 6).
[25] Nersessian (1992); Dunér (2019).
[26] Even-Ezra (2021).

knowledge hierarchically, for example, within ethnobiology.[27] This direction of research is highly relevant also for the history of diagrams, since classification is a main function of many diagram types.

Cognitive archaeology is divided into two approaches, evolutionary cognitive archaeology and what can be described as ideational cognitive archaeology. According to Thomas Wynn and Frederick Coolidge, the different theoretical foundations of these subdisciplines have meant that they have had little contact with each other. Ideational cognitive archaeology has its roots in semiotics and theories of metaphor, such as those of George Lakoff and Mark Johnson. Evolutionary cognitive archaeology, on the other hand, studies the evolution of human cognitive abilities with the help of theories from cognitive neuroscience.[28]

These two approaches are combined in this work, which utilizes concepts and ideas from semiotics, theories of metaphor, and cognitive neuroscience. Such a combination has been simplified by the emergence of intermediary disciplines such as cognitive semiotics, a development of semiotics that takes cognitive science into account,[29] and the theory of primary metaphors (also called neural theory of metaphor), a neuroscientific development of Lakoff and Johnson's theories on metaphors.[30]

Colin Renfrew warned that cognitive archaeology was "destined to tread an uneasy path between the pretentiously jargon-laden and the blindingly obvious."[31] Such a risk is of course almost unavoidable when investigating the past from the perspective of human cognition, which is at the same time always present, since we can never escape our minds, and elusively complex. However, this problem can be counteracted by referring to specific theories and empirical results from the cognitive sciences, rather than merely using wide concepts such as "the embodied mind."

Of course, the more tangible your assumptions and theories are, the greater the risk that someone can decisively prove you wrong and that time will make your analysis obsolete. For example, Renfrew's idea of using Piaget's stages of child development as a roadmap to studies of human cognitive evolution might or might not hold up to closer scrutiny.[32] However, for the general development of knowledge it is better that

[27] Blount (2011).
[28] Wynn and Coolidge (2022, 5–14).
[29] Zlatev (2012).
[30] See, for example, Lakoff and Johnson (1999).
[31] Renfrew (1982, 14).
[32] Renfrew (1982).

tangible hypotheses are put out in the open and become refuted, than that all analysis is shrouded in such general and obscure language that it eludes all criticism but is too vague to form a foundation, or even an inspiration, for further studies.

For this reason, and since the prefix "cognitive" has been attached to a variety of fields studying humanity's past and present with different aims and approaches, it is important to outline how I perceive the relationship between human cognition and cultural development. This study does not support the idea that certain diagram forms, such as matrices or tree diagrams, have a direct neurological basis. The complicated history of their development reveals that they are cultural artifacts. A tree diagram is no more biological than, for example, an axe. However, although artifacts of material culture such as axes might evolve into diverse forms for various purposes in different cultures, they are still based on fundamentals of human biology, such as length and strength of the limbs and the operation of the joints. Therefore, axes across time and space display fundamental similarities in shape and weight. As the properties of human tools for manipulating the world are founded in the biological limitations of the human body, the properties of tools for understanding the world, such as diagrams, are founded in the limitations and abilities of human cognition.

Just as you will not find an axe weighing hundreds of kilograms, you will not find a tree diagram where each node is divided into dozens of branches—and in the rare cases when you do, you have found an illustrative exception that confirms the rule. The 15 meters tall and 55 tons heavy axe in Nackawic-Millville, New Brunswick, is not designed for chopping firewood, and extremely complicated diagrams are not designed to simplify understanding.[33] Rather, they attempt to achieve an emotional effect by being overwhelming, physically and cognitively. Usually, diagrams structure knowledge in smaller and more manageable chunks ordered in hierarchical structures.[34]

Another fundamental theoretical assumption of this book is that all diagram production across time and space is based on a cognitive capacity for spatial thinking shared by all humans. In this respect, it takes a contrary position to the idea that spatial cognition has been learned through exposure to cultural artifacts such as diagrams or, in Marshall McLuhan's

[33] For complicated diagrams, see section "Labyrinthic Chartjunk: Implication Through Cognitive Overload" in Chap. 9.

[34] See folk taxonomies in section "The Shape of Life" in Chap. 5.

famous theory about typographic man, the grid-shaped spatial layout of the printed page.[35] The examples provided in the section on the prehistory of diagrams in this volume clearly illustrate that spatial cognition was a fundamental part of human culture long before the advent of print technology, and that it is oversimplified to describe non-literate societies as oral or aural, dependent on a one-dimensional narrative structure in their transmission of knowledge and culture.

At present, no comparable book exists that attempts to provide a coherent analysis of the general history of diagrams over the course of millennia. Existing academic works on diagrams are focused on particular diagram types during shorter historical periods and in particular cultural areas. Earlier books that attempt to cover longer time periods or several types of diagrams have the character of source collections, filled with high-quality reproductions of diagrams but less room for analysis.[36] Of course, a study like this, which has a global reach and a long timespan, would not be possible without previous syntheses of the many subfields it covers, such as the history of the periodic system, timelines, phylogenetic trees, and statistical graphics.[37] The European Middle Ages stands out as a period that has gained a lot of attention among researchers of diagram.[38] The use of diagrams in, for example, the Arab world, Byzantium, China, and India has received less attention, although an improvement has been seen in the last decades.[39] As a consequence, although this study attempts to outline a global history of diagrams, the majority of the examples are still from the Western tradition.

While earlier research has been invaluable, this work is not merely a summary and compilation with an overcoat of theories from the cognitive sciences. By taking a wide perspective, over disciplines as well as in time and space, it is possible to find long trends and turning points in the use of diagrams. Such an overview is impossible to achieve in focused case studies or even in syntheses limited to certain eras or areas. In addition, many of the chapters contain new and original research that highlights aspects of the history of diagrams that have not been observed in any

[35] See section "Printing and Spatial Cognition" in Chap. 5.

[36] For example, Lima (2014, 2017) and Rendgen (2019).

[37] For example, Scerri (2006); Rosenberg and Grafton (2010); Archibald (2014); and Friendly and Wainer (2021).

[38] For example, Klapisch-Zuber (2000); Kupfer et al. (2020); Even-Ezra (2021); and Verboon (2010).

[39] See, for example, Hamburger et al. (2022) and Bray et al. (2007).

previous studies. The narrative of this book travels quite rapidly along the well-trodden paths, relying on earlier research by, for example, Daniel Rosenberg and Anthony Grafton for background on the history of the timeline, or Michael Friendly and Howard Wainer for much of the development of statistical graphics.[40] Such works provide an important foundation for our understanding of the general trend, but there is no need to repeat those studies in detail here. However, when the narrative of this volume approaches important turning points, it halts and closely investigates the small leads that can help us understand how the development of diagrams is linked to cognitive principles. This means that it is not always the most well-known diagrams that are discussed and depicted in this volume, but those that help us identify turning points, support new interpretations, explore principles of spatial cognition—or those that constitute illustrative exceptions to the general rules. Hopefully, by bridging a wide range of fields, this method can provide an outline of how spatial representations of ideas and information have contributed to the development of our understanding of the world.

References

Ambasciano, Leonardo. 2017. What is cognitive historiography, anyway? Method, theory, and a cross-disciplinary decalogue. *Journal of Cognitive Historiography* 4 (2): 136–150.

Ambasciano, Leonardo, Mathias Clasen, and Darryl Jones. 2020. Horror studies between humanistic interdisciplinarity and scientific consilience: A conversation with Darryl Jones and Mathias Clasen. *Journal of Cognitive Historiography* 5 (1–2): 159–179.

Archibald, J. David. 2014. *Aristotle's ladder, Darwin's tree: The evolution of visual metaphors for biological order*. New York: Columbia University Press.

Blount, Benjamin G. 2011. A history of cognitive anthropology. In *A companion to cognitive anthropology*, ed. David B. Kronenfeld, Giovanni Bennardo, and Victor C. Munck, 11–29. Chichester: John Wiley & Sons.

Bonner, Anthony. 2007. *The art and logic of Ramon Llull: A User's Guide*. Leiden: Brill.

Bray, Francesca, Vera Dorrofeeva-Lichtmann, and Georges Métailié, eds. 2007. *Graphics and text in the production of technical knowledge in China: The Warp and the Weft*. Leiden: Brill.

[40] Rosenberg and Grafton (2010) and Friendly and Wainer (2021).

Clark, Andy. 1997. *Being there: Putting brain, body, and world together again*. Cambridge, MA: MIT Press.
Clark, Andy, and David Chalmers. 1998. The extended mind. *Analysis* 58 (1): 7–19.
Cohen, Henri, and Claire Lefebvre. 2005. Bridging the category divide. In *Handbook of categorization in cognitive science*, ed. Henri Cohen and Claire Lefebvre, 2–15. Amsterdam: Elsevier.
Dunér, David. 2019. Human mind in space and time: Prolegomena to a cognitive history. In *Cognitive history: Mind, space, and time*, ed. David Dunér and Christer Ahlberger, 3–34. Berlin: De Gruyter Oldenbourg.
Even-Ezra, Ayelet. 2021. *Lines of thought: Branching diagrams and the medieval mind*. Chicago: The University of Chicago Press.
Friendly, Michael, and Howard Wainer. 2021. *A history of data visualization and graphic communication*. Cambridge, MA: Harvard University Press.
Gardner, Howard. 1985. *The mind's new science: A history of the cognitive revolution*. New York: Basic books.
Giardino, Valeria. 2016. Behind the diagrams: Cognitive issues and open problems. In *Thinking with diagrams: The semiotic basis of human cognition*, ed. Sybille Krämer and Christina Ljungberg, 77–101. Boston: De Gruyter.
Giere, Ronald N., and Barton Moffatt. 2003. Distributed cognition: Where the cognitive and the social merge. *Social Studies of Science* 33 (2): 301–310.
Hamburger, Jefferey F., David J. Roxburgh, and Linda Safran, eds. 2022. *The diagram paradigm: Cross cultural approaches*. Washington, DC: Dumbarton Oaks.
Holmén, Janne. 2020. Time and space in time and space: Mapping the conceptual history of mental maps and historical consciousness. *Contributions to the History of Concepts* 15 (2): 105–129.
Hutchins, Edwin. 2006. The distributed cognition perspective on human interaction. In *Roots of human sociality*, ed. Nicholas J. Enfield and Stephen C. Levinson, 375–398. London: Routledge.
James, William. 1902. *The varieties of religious experience: A study in human nature*. New York: Longmans, Green, and Co.
Klapisch-Zuber, Christiane. 2000. *L'ombre des ancêtres: essai sur l'imaginaire médiéval de la parenté*. Paris: Fayard.
Kupfer, Marcia, Adam S. Cohen, and J. H. Chajes, eds. 2020. *The visualization of knowledge in medieval and early Europe*. Turnhout: Brepolis.
Lakoff, George, and Mark Johnson. 1980. *Metaphors we live by*. Chicago: University of Chicago Press.
Lakoff, George, and Mark Johnson. 1999. *Philosophy in the flesh: The embodied mind and its challenge to Western thought*. New York: Basic Books.
Lima, Manuel. 2014. *The book of trees: Visualizing branches of knowledge*. New York: Princeton Architectural Press.

Lima, Manuel. 2017. *The book of circles: Visualizing spheres of knowledge*. New York: Princeton Architectural Press.
Martin, Luther H., and Jesper Sørensen, eds. 2011. *Past minds: Studies in cognitive historiography*. London: Equinox.
Nersessian, Nancy J. 1992. How do scientists think? Capturing the dynamics of conceptual change in science. In *Cognitive models of science*, ed. Ronald N. Giere, 3–47. Minneapolis: University of Minnesota Press.
Netz, Reviel. 1999. *The shaping of deduction in Greek mathematics: A study in cognitive history*. Cambridge: Cambridge University Press.
Rendgen, Sandra, ed. 2019. *History of information graphics*. Cologne: Taschen.
Renfrew, Colin. 1982. *Towards an archaeology of mind: An inaugural lecture delivered before the University of Cambridge on 30 November 1982*. Cambridge: Cambridge University Press.
Robbins, Philip, and Murat Aydede, eds. 2009. *The Cambridge handbook of situated cognition*. Cambridge: Cambridge University Press.
Rosenberg, Daniel, and Anthony Grafton. 2010. *Cartographies of time: A history of the timeline*. New York: Princeton Architectural Press.
Roth, Wolff-Michael, and Alfredo Jornet. 2013. Situated cognition. *Wiley Interdisciplinary Reviews: Cognitive Science* 4 (5): 463–478.
Scerri, Eric R. 2006. *The periodic table: Its story and its significance*. New York: Oxford University Press.
Slingerland, Edward. 2014. Toward a second wave of consilience in the cognitive scientific study of religion. *Journal of Cognitive Historiography* 1 (1): 121–130.
Snyder, Laura J. 2005. Consilience, confirmation and realism. In *Scientific evidence: Political theories and applications*, ed. Peter Achinstein, 129–148. Baltimore, MD: Johns Hopkins University Press.
Suitner, Caterina, and Chris McManus. 2011. Aesthetic asymmetries, spatial agency, and art history: A social psychological perspective. In *Spatial Dimensions of Social Thought*, ed. Anne Maas and Thomas W. Schubert, 277–301. Berlin: De Gruyter Mouton.
Sweller, John, Paul L. Ayres, and Slava Kalyuga. 2011. *Cognitive load theory*. New York: Springer.
Tversky, Barbara. 2019. *Mind in motion: How action shapes thought*. New York: Basic.
Varela, Francisco J., Evan Thompson, and Eleanor Rosch. 2016. *The embodied mind: Cognitive science and human experience*. Revised edition. Cambridge, MA: MIT Press.
Verboon, Annemieke R. 2010. *Lines of thought: Diagrammatic representation and the scientific texts of the Arts Faculty, 1200–1500*. Dissertation. Lieden: Leiden University Scholarly Publications.
Wynn, Thomas Grant, and Frederick L. Coolidge. 2022. *An introduction to evolutionary cognitive archaeology*. New York: Routledge.
Zlatev, Jordan. 2012. Cognitive semiotics: An emerging field for the transdisciplinary study of meaning. *Public Journal of Semiotics* 4 (1): 2–24.

CHAPTER 2

The Cognitive Foundations of the History of Diagrams

This volume analyzes the history of diagrams, from their prehistoric precursors to computer-generated graphics, with the help of theories from the cognitive sciences. The second chapter of the book consists of six sections that present perspectives from the cognitive sciences to help us understand diagrams and spatial cognition. Together, the sections in Chap. 2 function as a theoretical foundation for the chronological and thematic historical analysis in Chaps. 3–9, and for the general conclusions in Chap. 10.

First, the diagram concept is defined, both through an investigation of its conceptual history and with the help of semiotic theory (section "What Is a Diagram? Conceptual History and Semiotic Definition"). This includes a description of the functional definition of diagrams that has been used in this work to decide which figures to include in the analysis. Thereafter follows section "Diagrams and the Limits of Consciousness", which discusses how human cognitive limitations contribute to making diagrams important. The section "Diagrammatic Reasoning" discusses current research on the role of diagrams in reasoning, and section "The Neurology of Spatial and Abstract Thinking" takes a closer look at the neurological foundations of spatial cognition. Section "The Evolution of Spatial Differentiation and Mythical Space" describes how human spatial cognition has evolved, and how our evolutionary history has endowed the

J. Holmén, *A History of Diagrams*, Palgrave Studies in the History of Science and Technology,
https://doi.org/10.1007/978-3-031-99691-7_2

physical dimensions of space with emotional values. Finally, section "Brain, Culture, and Diagram" summarizes how different layers of spatial cognition—innate, primary, cultural, and historical—are related to each other and discusses why this matter for the history of diagrams.

What Is a Diagram? Conceptual History and Semiotic Definition

The word "diagram" has its roots in the ancient Greek διάγραμμα (*diágramma*), meaning to mark out by lines. Incorporated into most modern languages in a variant of the Greek form, it has through the ages been used to describe a wide range of illustrations varying in composition and purpose. The same is true for similar concepts with different etymological origins, such as the Chinese *tu* (圖). Despite the challenge provided by this diversity, it is necessary to attempt a general definition of the concept before beginning to write its history. Some form of method and stringency is needed regarding which illustrations to include and which to leave out.

One possible approach is that of conceptual history: to focus on illustrations described as diagrams in the accompanying texts. Through such a study, we would find that illustrations described as diagrams are generally forms of visual communication distinct from symbolic systems, such as writing, as well as from images. Symbols have an agreed upon meaning; it is a matter of convention which sound a certain letter represents or which traffic sign represents an imperative to stop at an intersection. Images, on the other hand, function as representations by mimicking the spatial relations between the parts of the represented object. As images, diagrams also use spatial relations in order to convey meaning. However, in the case of diagrams, what they represent is neither decided by convention nor through mimicking of a represented object. Instead, diagram constructors use spatial relations in the diagram as analogies for some selected qualities or quantities of the represented object. For example, a bar chart of France's economic development has no visual similarity to France, but the heights of the bars are analogies for one particular aspect of the country, its GDP. In practice, the demarcations between these categories are seldom clear-cut. Symbols are often needed to make a diagram comprehensible, and images contain diagrammatic elements. In addition, the diagram concept is often used to describe maps,[1] blueprints, anatomical charts, and exploded diagrams. These illustrations do highlight particular aspects of their objects, but through their direct similarity to what they represent they could also be described as images.

The purpose and context of the illustration also influence whether it will be described as a diagram or not. That Leonardo da Vinci's *Vitruvian Man* is occasionally described as a diagram,[2] something that never happens to *Mona Lisa*, is not only due to the fact that *Vitruvian Man* is "marked out by lines" (Greek: *diágramma*), drawn rather than painted. More importantly, *Vitruvian Man* has an openly didactic purpose: to convey knowledge about the spatial proportions of the human body.

However, the very same types of illustrations that are commonly described as diagrams might in other texts be described as graphic presentations, charts, figures, and so forth. Since the main objective of this study is not to follow the use of the word "diagram" and its development, but the general development of figures that give spatial form to non-spatial ideas and data, it would be wrong to exclude functionally similar illustrations with other designations. The long time period of the study makes this point even more important. In ancient Greece, *diágramma* had approximate synonyms in *schêma* and *katagraphê*, and these were used by, for example, Euclid. In Latin editions of Euclid's *Elementa*, *schêma* was translated to *figura*.[3] In fact, as discussed in section "The Musical Prelude to Diagrams of Scientific Systems" in Chap. 5, until the 1600s the word "diagram" was rarely used and had a narrow meaning within the field of music. Most earlier "diagrams" were therefore called something else, commonly *figura*.

Thus, a strict conceptual history method is unsuitable for this investigation, since I want to include some illustrations not described as diagrams and exclude others that sometimes are. Geographical maps, figurative illustrations, and geometrical diagrams, such as those of Euclid, are left out. The purpose of this study is to investigate how humans have used spatial cognition in advancing and structuring their non-spatial knowledge. This is not to say that maps and geometry are unrelated to spatial

[1] The concept "map" is sometimes defined so widely, encompassing representations of non-spatial relations, that diagrams constitute a variant of maps. See, for example, Grønfeldt Winter (2020, 35–57).

[2] For example, by Strongman (2010).

[3] Safran (2020, 361).

models of non-spatial domains. On the contrary, there has been a lot of interchange between these fields. However, the history of cartography and mathematics are already two well-developed fields of research, in which spatial thinking is integrated and obvious because of the immediate relationship to geographical and geometrical space.[4] The purpose of this book is to bring together other fragmented fields of research that have touched upon diagrams of abstract relations, where the role of spatial cognition is not to think about space per se, but, through analogy, think about something else. However, cartography and geometry will be referred to when relevant to the development of diagrams in general, and in that respect, the impacts of Euclidean and analytical geometry have a particularly central role in this book.

The set of illustrations investigated here can be defined more clearly with the help of concepts from semiotics. This field, which studies communication through signs, has its roots in two different intellectual traditions. There is a linguistically inspired one, often called semiology, based on the writings of Ferdinand de Saussure, but also one with roots in logic, resting on the work of Charles Sanders Peirce. While for Saussure the relationship between the sign and the object it signifies is one of arbitrary coding, semioticity, Peirce was also interested in iconicity, the fact that some signs, which he called icons, share similarities with the objects they signify. A special kind of icon is the diagram, which has a central role in Peirce's philosophy of cognition.[5]

Peirce described aspects of signs with the help of triads (Fig. 2.1). The most well known—and the most relevant for research on diagrams—is the division into icons, indexes, and symbols. While symbols are signs in the sense of Saussure, arbitrary coding between signifiers (signs) and signified (objects), such as letters and numbers, icons bear a similarity to the objects they signify.[6] Indexes are signs that link directly to their object, such as a thermometer to changes in temperature, a speedometer to changes in velocity, or an animal track to the previous presence of an animal. In a way, all letters and figures you find in a book are indexes, since they indicate that someone has previously written and printed the book. However, the

[4] The history of cartography has been compiled in, for example, a monumental multivolume work published by the University of Chicago Press from 1987 in a still-ongoing project. For example, Harley and Woodward (1987).

[5] Stjernfelt (2007, vii–xiv, 49).

[6] Peirce's definition of an icon is very similar to how the ancient Epikurean philosophers described particular signs in their philosophy of signs. See Manetti (1993, 129).

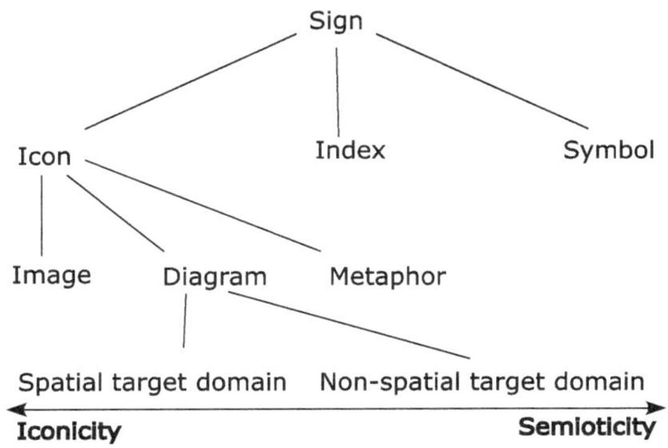

Fig. 2.1 Diagrams in Peirce's semiotics with my additions. Signs further to the left have more iconicity (are similar to their meaning), while those to the right have more semioticity (their meaning is decided through cultural conventions). This study focuses on diagrams with a non-spatial target domain, excluding, for example, maps and blueprints. Drawn by author

triad symbol, index, and icon do not consist of mutually exclusive categories, but of complementary aspects. The index aspect is particularly strong in the field of statistical graphics, as statistical diagrams are linked to datasets.

Peirce divided icons further into a triad of diagrams, metaphors, and images. His definition of these concepts was far from transparent and has been the object of different interpretations. Frederik Stjernfelt describes the difference between them as follows:

> Being an icon, the diagram is characterized by its similarity to its object—but while the image represents its object through simple qualities and the metaphor represents it through a similarity found in something else, the diagram represents it through a skeleton-like sketch of relations.[7]

The difference between images, diagrams, and metaphors might also be described as degrees of iconicity, with the image being the most iconic, the metaphor the least iconic and most semiotic, and the diagram taking a position in between.[8] While an image can be understood directly by

[7] Stjernfelt (2007, 90).
[8] Nöth (2024, 6).

similarity to its object, a metaphor rests upon a semiotic coding between parallel examples. The intermediary diagram is less similar to its object than an image, since it only represents a selection of the object's most important qualities, and its interpretation often depends on some symbolic understanding.

Peirce also emphasized the inherent vagueness of icons, and understood image, metaphor, and diagram as complementary aspects that might be present in one and the same icon.[9] Although most icons display some diagrammatic aspect, this study will focus on icons where the diagrammatic element is dominant.

From a cognitive science perspective, it is important to note that since the diagram represents a limited number of select qualities or quantities in its object, it respects the cognitive limitations of the human mind. There is no end to how complex a metaphor or an image might be, but as soon as a diagram exerts more cognitive load than we can handle, it cedes to function as a diagram and is perceived as a metaphor or an image.[10]

In addition, the endeavor of creating a hierarchical classification of signs is an attempt to make the entire universe of existing and possible signs, which contains an infinity of intermediate forms, palatable and understandable for human beings with their limited cognitive capabilities. Thus, we can also think of a continuous scale of signs from the most iconic to the most semiotic (Fig. 2.1).

For the practical purposes of this investigation, it is necessary to create yet another simple dichotomous taxonomy, based on the diagrams' function as analogies. All analogies have a source domain, from which examples are drawn, and a target domain, to which they are applied. Diagrams use space as a source domain, and some diagrams, such as maps, blueprints, and geometry problems, also have spatial target domains. However, other diagrams, such as timelines, the periodic system, and Venn diagrams, are analogies with non-spatial target domains, such as time, chemical and physical properties, or the relationship between sets. This study is limited to the latter category—diagrams where the target domain is non-spatial (Fig. 2.1).

Diagrams with a spatial target domain are more iconic and resemble images in the sense that a map can almost be said to be an image of a

[9] Stjernfelt (2007, 90).

[10] See section "Labyrinthic Chartjunk: Implication Through Cognitive Overload" in Chap. 9.

country. Diagrams with a non-spatial target domain resemble metaphors, such as the timeline being a metaphor for time. This second category of diagrams to a larger degree depends on semioticity. This means that more learning is needed in order to understand their conventions of coding.

Peirce's semiotics has been criticized for its heavy reliance on such a vague concept as similarity, which is impossible to define logically.[11] Although similarity might not be possible to define through traditional logic, it has in the last decade become easier to identify with the help of neural networks. The classic test for artificial intelligence (AI) systems of categorizing images of cats and dogs, which was an unsurmountable stumbling block for older programmed logical systems, has become child's play for modern neural networks.[12] Of course, from a strict philosophical standpoint it is possible to claim that we have merely trained the machines to mimic our own vagueness. However, this is a study of how humans actually use diagrams for thinking—not of how they should think if they were logic machines. Therefore, we will not shy away from similarity and other concepts difficult to define logically. On the contrary, similarity is of fundamental importance for categorization, which is one of the basic cognitive functions.

It could be argued that the diagrams researched in this volume are not diagrams at all in Peirce's sense, since most of his definitions described diagrams as schematic versions of images that represent the relations between their parts. He frequently used maps and plans as examples.[13] Diagrams of abstract relations, such as social pyramids, might instead, in Peirce's terminology, be classified as metaphors. However, Peirce's choice of examples was probably influenced by his own career in the United States Coast and Geodetic Survey, where he encountered diagrams of geographic space. This does not mean that his philosophical concept of diagram cannot be used for abstract spaces. In fact, Peirce's triad of image, metaphor, and diagram has even been applied to sound, since also sounds can display iconicity through different forms of similarity to what they represent.[14]

In addition, metaphor and diagram should be understood as complementary rather than exclusive dimensions, and Stjernfelt emphasizes that

[11] For example, by Nelson Goodman and the young Umberto Eco. Stjernfelt (2007, 53–66).
[12] Elngar et al. (2021).
[13] Stjernfelt (2007, 96).
[14] Enriquez (2012). See also section "Language and Gestures" in Chap. 3.

diagrammatic analysis is essential also in understanding metaphor. As an example, he pointed out that the metaphor of an ancestral tree "presupposes that the formal branching diagram is mapped from a tree onto family structure."[15] In this context, it could also be fruitful to differentiate between metaphor and analogy. Although the concepts are sometimes used synonymously, metaphors are affective and expressive, while analogies are explanatory and appeal to reason.[16] An icon, such as the social pyramid in Fig. 2.2, has aspects both of image and diagram. The pictorial elements of the image express many affective metaphorical relations, such as a bag of money for capitalism and starving children and bent backs for the oppression of the working classes. However, the diagrammatic aspect of the icon is the vertical dimension, where a scale from the most powerful on top to the most powerless at the bottom is a spatial analogy to the social stratification of society.

Many of the diagrams analyzed in this book have some aspects of images in them, although rarely to such an extent as in Fig. 2.2. However, this study does not analyze the pictorial metaphors expressed by the images, but focuses on the spatial analogies in their basic diagrammatic structure.

The most famous attempt to analyze diagrams with the help of semiotics is Jacque Bertin's *The Semiology of Graphics* (*Sémiologie Graphique*, 1967). However, this work was inspired more by the Saussurean semiotic tradition than by Peirce. Consequently, Bertin did not use the diagram concept, but described graphic displays as collections of symbols, a visual equivalent to mathematics. This was reflected in his claim that "[a] graphic can be comprehended only when the *unique* meaning of each sign has been specified (by the legend)."[17] Probably Bertin's background as a cartographer influenced his approach. However, his definition excludes many diagrams, and all that were produced before the late eighteenth century. For example, Bertin's division of visual signs into graphics, figurative imagery, and abstract imagery leaves no place for illustrations such as medieval tree diagrams for logical classification or textual analysis. Furthermore, the idea that graphics must rest upon a specified meaning does not explain why the same diagrams have frequently been reinvented independently, such as many similar variants of Venn diagrams.[18] In order

[15] Stjernfelt (2007, 92–93).
[16] Vartanian (2012).
[17] Bertin (1983, 2).
[18] Section "Circles of Inclusion" in Chap. 4.

Fig. 2.2 Pyramid of Capitalist System, published in the 1911 edition of *Industrial Worker* by the labor union Industrial Workers of the World

to understand this, we must acknowledge Peirce's idea that diagrams are icons that represent relations in their objects and combine that idea with theories on human cognition.

Diagrams and the Limits of Consciousness

It is an often-stated observation that spatial visualization of thoughts and ideas aids our cognitive processes in various ways. The psychologist Barbara Tversky has even described this as a law of cognition: "Spatial thinking is the foundation of abstract thought."[19] However, why this would be the case is far from self-evident. One of the most promising lines of explanation focuses on the severe limitations of our consciousness, and how we need effective tools and strategies to overcome them.

The cognitive neuroscientist Merlin Donald has pointed out that we can only keep a clear sensory image in our consciousness for 10 to 15 seconds. After that, the image starts to drift out of short-term memory. This has the consequence that we need to divide larger cognitive tasks into shorter thought sequences. Donald believes that we try to grasp the larger picture from our stream of consciousness with the help of verbal summaries.[20] Of course, there is also a limit to our ability to construct and remember such summaries.

Donald claims that the invention of external symbolic memory storage revolutionarily transformed conscious mental activity, since externally stored information can be made the focus of our awareness more effectively than items recalled from long-term to short-term memory. He mentions hypotheses, graphs, arguments, conclusions, and themes. According to Donald, neither spoken words nor images can be displayed long enough in working memory to allow detailed and prolonged reflection. External memory gives sharper and more durable mental representations, which allows reflection on thought itself and facilitates verification and control of the mental process.[21] Barbara Tversky has formulated this as another law of cognition: "When thought overflows the mind, the mind puts it into the world."[22]

[19] Tversky (2019, 142).
[20] Donald (2001, 15–16).
[21] Donald (2001, 308–309, 313).
[22] Tversky (2019, 190).

Donald does not differentiate between different forms of external memory, but treats them all as what in semiotics is called signs. However, Donald's argument would benefit from Peirce's distinction of diagrams as a particular kind of signs. Since diagrams represent the internal structure of the objects they signify, they facilitate reasoning.[23] We can consider diagrams a form of noise-reduced images, that, as an adaptation to human cognitive limitations, focus on a few aspects of the objects they represent that are relevant to the problem at hand.

Another perspective that links diagrammatic reasoning to the limits of human consciousness is provided by cognitive load theory. It was originally formulated in the 1980s by John Sweller as a development of George A. Miller's and Lloyd and Margaret Jean Peterson's theories about limitations to human working memory. Miller famously claimed that our working memory could not handle more than the "magical number seven" pieces of information. Later research indicates that the number of items that can comfortably be processed simultaneously is generally lower than the upper limit of seven proposed by Miller, perhaps limited to two or three. According to Sweller, human consciousness essentially performs evaluations of randomly generated combinations of items in working memory. In his view, the severe restrictions to the number of elements it can contain stem from the exponential growth of permutations of larger numbers. For example, 3 items can be combined in 6 ways, while 10 items have 3,628,800 permutations.[24]

This random-generate-and-evaluate theory is the weakest part of cognitive load theory. Although the upper limit to working memory is well established, random permutations is an unlikely explanation. To begin with, humans seem unable to generate random series of numbers. Research in this field was conducted already in the 1940s, and a meta study from 1972 found that only one study out of 15 claimed that test subjects were good randomizers.[25] A study from 2004 claimed to have found evidence that humans could randomize, but this was soon thereafter refuted in another study that replicated the experiment with better methodology.[26] John Sweller, Paul Ayres, and Slava Kalyuga recognize that problem

[23] Stjernfelt (2007, ix).

[24] Miller (1956); Peterson and Peterson (1959); Sweller (1988); Sweller et al. (2011, 40–53).

[25] Wagenaar (1972).

[26] Persaud (2005); Figurska et al. (2008).

solving through analogical reasoning might provide an alternative to random generation and evaluation, but maintain that random generation is needed when no appropriate analogy is available.[27] However, space can be used as a universal analogy applied to all kinds of problems, for example, through diagrams. Still, cognitive load theorists are correct in that the solutions reached through analogical reasoning must be tested for validity; the history of diagrams is full of false analogies. The theoretical explanation for why there is a low upper limit to working memory is of little practical consequence when we investigate the consequences of this limit for the development of diagrams. For now, it is enough to recognize that this limit exists, and we return to discussing the reasons for it in the conclusions.

The limitations in working memory capacity are alleviated by the fact that we do not always need to obtain new information from the environment. Clusters of information, schemas, can be recalled from long-term memory when cues from the environment indicate that they can be useful.[28] Diagrams can be understood as such schemas. By recalling familiar spatial schemas from long-term memory when appropriate, we can take cognitive shortcuts. This explains why some common diagrammatic forms are reused through history for different purposes.

Cognitive load theory also differentiates between intrinsic load, which is related to the fundamental structure of the information one is to learn, and extraneous load, which stems from the manner in which the information is presented.[29] Ideally, a problem should be presented in such a way that all cognitive capacity can be directed toward the intrinsic load provided by the central information, and none of it has to be reserved for dealing with, for example, the convoluted language in which the issue is presented. One of the strengths of diagrams are that they can often directly outline the central structure of a knowledge field, which provides much less extraneous load than if such a model has to be constructed from a written or oral presentation.

Cognitive load theory has inspired numerous studies on the split attention effect. They confirm that learning becomes more difficult if relevant information is separated in space or time. Sweller, Ayres, and Kalyuga exemplify this with a mathematical problem including a geometrical

[27] Sweller et al. (2011, 34).
[28] Sweller et al. (2011, 49–50).
[29] Sweller et al. (2011, 57).

diagram, which becomes much easier to solve when information about angles is located at the appropriate place in the figure than when it is found in the text below it.[30] A study that confirms this was performed in 1987 by Larkin and Simon, who compared the computational efficiency of sentential and diagrammatic representations in solving mathematics and physics problems. They found that information that needed to be calculated in the sentential representation was often explicit in the diagrammatic representations. Search for information was also expedited in diagrammatic representations, since it was organized by location. However, these advantages depend on knowledge of how to construct a good diagram.[31] An important consideration in diagram production is to display relevant information in a clear and easily understandable way. Although diagramming is based on our innate ability of spatial cognition, diagrams can often only be understood if both the constructor and the interpreter are in possession of a certain amount of cultural knowledge.

Cognitive load theory claims that we have evolved biologically to absorb certain types of knowledge. This primary knowledge is learned, but not taught, such as learning to walk or talk. Biologically secondary knowledge, on the other hand, is constituted of knowledge and skills that have appeared so recently in evolutionary terms that we have not had time to adapt to them, such as writing. Therefore, this biologically secondary knowledge has to be taught and learned, and constitutes considerable cognitive load. Culture consists of the acquisition of biologically secondary knowledge, but it is dependent on biologically primary knowledge.[32] For example, while specific forms of diagrams, such as timelines or bar charts, are secondary knowledge that has to be learned with some effort, they rest on a foundation of spatial cognition, which is an intuitively learned primary knowledge.

It can be argued that diagrams provide a higher level of aid to our cognitive processes than other forms of external memory storage. The spatial relationships presented in diagrams can be digested without overloading the limitations of our consciousness. Thereby, diagrams can instantly help us order our thoughts, both for our own comprehension and for communicating them to other people. If you try to convey the information of even a simple diagram in written form, the text will inevitably involve

[30] Sweller et al. (2011, 111–112, 116–117).
[31] Larkin and Simon (1987).
[32] Sweller et al. (2011, 5–7).

more than seven elements of information and take more than 10 to 15 seconds to read, exceeding the limits of our working memory. In order to understand the written text, we might need to go back on the page or try to recall the earlier parts from long-term memory. By that process, we can eventually construct an image of the written text that can be grasped by our consciousness, but presented as a figure the information would have been immediately available. Some diagrams, such as the periodic table, are too vast to be immediately fathomed by consciousness. In such cases, we need a process of recollection and reexamination of the different parts in order to understand the figure. However, spatial location is a very powerful cue for memory recollection, so also in such cases diagrams have clear advantages.[33]

Diagrammatic Reasoning

Based on the previous chapter, we can conclude that understanding is greatly simplified if a narrative, which takes minutes or hours to read, listen to, or experience, can be transformed into a diagram that can be perceived within an instant. Thinking that utilizes such visuospatial models is called diagrammatic reasoning. Anecdotally, several prominent scholars and scientists have had a remarkable ability to visualize problems.

Biographers of, for example, Leonardo da Vinci and Stephen Hawking have claimed that they compensated—or overcompensated—for a relative lack of proficiency in formal mathematics by visualizing problems in spatial terms.[34] Similarly, the physicist and mathematician Roger Penrose has claimed that "almost all of my mathematical thinking is performed visually and in terms of nonverbal concepts." He quoted several other researchers who had expressed similar views, including Francis Galton.[35]

Some, such as Albert Einstein and Charles Sanders Peirce, add a dimension of movement and spatial manipulation by describing their thought processes as consisting not of linguistic but of visual and muscular elements.[36] Aristotle's claim that we cannot think without an image

[33] See method of loci in section "Terrain and Architecture: Structuring Knowledge in "Oral" Cultures and Early Civilizations" in Chap. 3.

[34] Isacsson (2017, 200); Mialet (2012, 66–68).

[35] Penrose (1989, 424).

[36] Pietarinen (2016, 129); Einstein's claim was made in answer to a survey on nonverbal imagery in mathematical thought by Jacques Hadamard, who included Einstein's answer in the appendix. Hadamard (1945) 1996, 64–99, 142–143; McNeill (1992, 268–269).

probably referred to the active drawing of a diagram rather than to the passive contemplation of a picture.[37] Such examples illustrate that the spatial dimension of human cognition also has a motoric component. This is investigated in the field of embodied cognition, which is concerned with how cognitive processes are influenced by the body's morphology and sensorimotor systems.[38] The connection between bodily movement and cognition might, for example, explain why working with pen and paper can provide advantages for comprehension and memory retention.[39]

The utilization of spatial images in the thought process, by drawing, viewing, or contemplating, is not restricted to extremely gifted individuals. Rather, it has been described as typical for human reflection in general. In his book *Productive Thinking* from 1945, the Gestalt psychologist Max Wertheimer discussed the mechanisms of cognition. Although he used the examples of his friend Einstein and Galileo Galilei, both exceptional individuals in many ways, Wertheimer believed that Gestalt processes such as grouping, separating, and centering played a crucial role in all productive thinking.[40]

Charles Sanders Peirce also argued that all reasoning is conducted by drawing inferences from diagrams that we observe, and manipulate, in our minds. Thus, Peirce's model of spatial reasoning erases the dividing line between deductive and inductive reasoning, as the former is an inner version of the latter. Everybody has a small laboratory or simulator in their mind where they can perform spatial experiments. From them, we can draw conclusions not only about space but, through analogy, about abstract relationships as well.[41]

Peirce's semiotics differentiated between the object, the sign representing the object, and the representamen, the impression of the sign in the mind. Important for understanding the role of diagrams in Peirce's view of human thought is his claim that diagrams are the representamen of symbols; for example, the mind creates diagrams out of a text. Peirce likened this to how Euclid's texts were followed by explanatory diagrams.[42]

[37] Hamburger (2020, 65).

[38] Dove (2015).

[39] For the advantages of pen and paper, see Wammes et al. (2016).

[40] Wertheimer acknowledged that, particularly in the case of Galileo, his story was not a "historical reconstruction" but a "psychological hypothesis." Wertheimer (1945, 161, 189–191).

[41] Stjernfelt (2007, xiii).

[42] Stjernfelt (2007, 93).

Borrowing concepts from geometry, Peirce differentiated between corollarial and theorematic diagrammatic reasoning. Corollarial reasoning refers to conclusions that can be drawn directly from the diagram itself, while theorematic reasoning relies on additional information not present in the initial diagram.[43] Michael May has emphasized that Peirce's concept of diagrammatic reasoning was dynamic and iterative. It can be likened to a computer simulation of a complex model. We do not merely think using single diagrams but often construct a series of diagrams in a moving picture of thought. The spatial model or analogy in the diagram makes it easier to manipulate in a process of experimentation, as shown in Fig. 2.3.[44]

The shortest route, from premises via the construction and observation of a diagram to conclusion, corresponds to Peirce's corollarial reasoning, where conclusions can be drawn from the diagram itself. Longer routes, where additional loops of construction and observation are needed in order to reach a conclusion, are examples of theorematic reasoning.[45]

During most of the twentieth century, "diagrammatic reasoning" was a marginal concept that was mainly used in editions of or commentaries on Peirce's work, although it found some use in the field of psychological aptitude testing.[46] However, from the 1990s, diagrammatic reasoning was adopted by the artificial intelligence community, which brought it into wider use.[47]

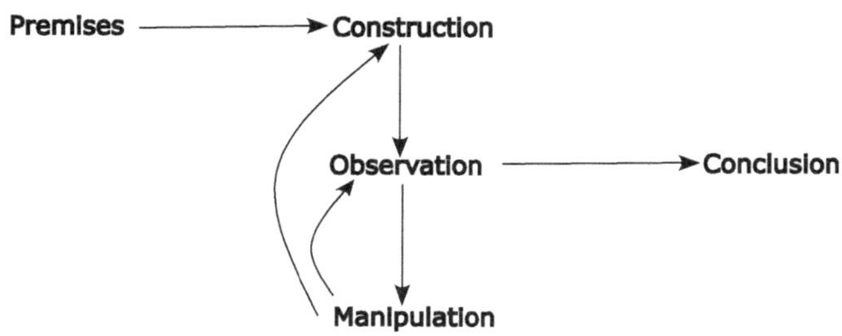

Fig. 2.3 The diagrammatic reasoning process. Drawn by author, based on May (1999, 186)

[43] Stjernfelt (2007, 107–108).
[44] May (1999, 185–186).
[45] Stjernfelt (2007, 104).
[46] For example, Holdsworth (1982, 216); Bolton (1983, 15).
[47] See, for example, Glasgow et al. (1995).

Within artificial intelligence, diagrammatic reasoning challenged the dominant idea that all cognitive processes could be broken down into simple operations of symbolic logic. The philosopher, AI researcher, and cognitive scientist Aaron Sloman likens this to Immanuel Kant's argument in *Critique of Pure Reason* that some mathematical truths cannot be derived from logic. In the AI community, a debate between the two sides has raged since the International Joint Conference on AI in London in 1971. However, according to Sloman, it is often forgotten that even in instances where diagrammatic representations are not superior to logic for the extent of what they can explain, they might have a greater heuristic power, for example, in facilitating searches for solutions. Sloman identifies two main heuristic advantages of diagrams. These have repeatedly been rediscovered by different researchers who have contemplated diagrammatic reasoning.[48]

First, diagrammatic representations restrict what is possible to express, thereby avoiding contradictions: it is possible to symbolically express that an object is both round and square shaped, but it is impossible to draw a diagram of this contradictory expression. This point is also made by, for example, Stjernfelt, who claims that since diagrams cannot be inconsistent, they can be used to weed out inconsistencies in symbolic representations.[49]

Second, diagrams provide a heuristic for search that makes it easier to find good candidates for optimal solutions. For example, on a road map, you can directly disregard all roads going off in completely wrong directions, which leave you with a limited number of options to investigate more closely.[50] This is reminiscent of human chess players, who use their visualization of the game in order to select a few possible moves for further consideration. In contrast, traditional supercomputers, such as Deep Blue that defeated Garry Kasparov in 1997, crunched through all possible moves with the help of symbolic logic and their great power of computation.[51]

However, as in other fields that have reflected on the cognitive functions of diagrams, AI researchers also seem to agree on the mutual usefulness and complementarity of the symbolic and diagrammatic approaches.

[48] Sloman (1995, 7–10).
[49] Sloman (1995, 7–10); Stjernfelt (2007, 98–99).
[50] Sloman (1995, 7–10).
[51] See, for example, Hsu (2002).

Sloman recognizes that mixed modes of representation are often the most powerful, and that the efficiency of a particular representation is dependent on the underlying machine.[52] For example, human chess players could not use the strategy of Deep Blue and vice versa, since humans lack the speed of logic operations that computers possess, and Deep Blue was not constructed for the diagrammatic reasoning that human players perform.

In 2001, the anthology *Spatial Schemas and Abstract Thought* gathered findings on spatial cognition from psychology and adjacent disciplines. According to the introduction by Merideth Gattis, spatial representations are used for three basic purposes in abstract cognition: as memorial, communicative, and logical structures. In logical reasoning, spatial structures can mark order and directionality within a dimension as well as relations between dimensions.[53] In the same volume, Barbara Tversky discussed spatial schemas in depictions, among them diagrams. She noted that in statistical graphics, certain diagrammatic forms encourage specific mental processes: bars invite comparisons, while lines evoke interpretations of trends.[54] A consciously constructed diagram might therefore nudge the viewer to draw the intended conclusions.

Diagrammatic thinking is not a panacea that can be applied to all problems without reflection, but several limitations and possible hazards have also been pointed out. In mathematics and disciplines relying heavily on it, such as economics, it has been argued that while symbols can represent the general, a diagrammatic representation of this general principle will always become a particular example. You cannot draw a line with a "general" length, since on the paper it will always become specific. Caution is therefore needed not to rely on specific properties in the diagram when drawing inferences about the general principles that it is supposed to illustrate.[55] To some extent, this problem can be counteracted through the construction of and reflection on a series of diagrams, such as Peirce suggested. General principles can be inferred from the similarities between a variety of particular diagrams that have been created in a theorematic process of experimentation.

[52] Sloman (1995, 19–24).

[53] Gattis (2001a, 2–6).

[54] Tversky (2001, 96–98).

[55] This argument was presented already by Johann Lambert in the 1700s, although it did not prevent him from pioneering diagrammatic representations of data (Bullynck 2010). For economics, see Blaug and Lloyd (2010, 6).

Even if diagrams, in contrast to symbols, cannot display contradictory arguments, diagrams that are internally consistent might, of course, still display information that is unfounded or plain wrong, in the sense that it is not supported by or contradicts external evidence. One example is the theory of the four elements (earth, fire, air, and water), which was elaborated in numerous diagrams over two millennia, only to disappear during the scientific revolution as it became clear that the theories had little connection to empirical observation. Today, diagrams describing theories in, for example, theoretical physics might be in a similar situation.[56] In addition, diagramming techniques can of course also be used to mislead intentionally, such as in political propaganda.

As all powerful tools, diagrams can be used for manipulation and control. Several scholars have noted that they might be just as misleading as verbal communication.[57] Bruno Latour has described diagrams as powerful inscriptions that corner dissenters into conformity through ever more dramatic visual effects.[58] Helen Kennedy et al. argue that four conventions of data visualization contribute to making them seem objective, transparent, and factual: two-dimensional viewpoints, clean layouts, geometric shapes and lines, and inclusion of data sources.[59] Diagrams constructed according to these principles are more likely to be taken seriously, even when they convey faulty or misleading ideas. This factor is important to remember when using diagrams as historical sources. In general, humanists and social scientists are aware of and trained in the analysis of biases in texts but less so regarding the biases inherent in diagrams.

The Neurology of Spatial and Abstract Thinking

Although our understanding of the neural foundation of spatial cognition is rapidly evolving, some aspects of this puzzle still remain to be laid, and many details will no doubt be revised. However, an overview of the field can still enlighten our understanding of diagrams and diagrammatic thinking. Some important general principles seem to be on firm ground, and these give spatial thinking a central place in human cognition.

[56] See section "String Theory: Post-Modern Platonism" in Chap. 9.
[57] Candea (2019); Bird-David (2019).
[58] Latour (1986, 18, 32).
[59] Kennedy et al. (2016).

Studies of the neurological foundation of spatial thinking did not take off until the twenty-first century, but the field has a long prehistory. Already in the 1940s, Edward C. Tolman's research with rats in mazes led him to conclude that their navigational skills could not be based on simple stimulus-response mechanisms, as was earlier believed, but required an ability to construct spatial models of the environment.[60] Initially, this cognitive map hypothesis faced staunch criticism from defenders of the then dominant behaviorist paradigm, which focused on observable behavior and considered all theories regarding mental processes unscientific. However, cognitive maps gained a neurological foundation in the early 1970s, when John O'Keefe and Jonathan Dostrovsky found that the firing rate of neurons in the hippocampus was correlated to an animal's location in space. They named these neurons "place cells," and further experiments during the 1970s confirmed their existence and investigated their functions.[61] In the 1980s and 1990s, the hippocampus was commonly equated with a cognitive map. In the early twenty-first century, a series of experiments revealed that place cells work together with grid cells, boundary cells, head direction cells, and speed cells. These are all involved in aspects of our relationship to space and are found in the hippocampus or the nearby entorhinal cortex.[62] In addition, it was revealed that place cells could also function as time cells by firing at specific times during temporally structured experiences.[63]

The connection between space and time in human cognition had been observed already earlier, based on clues from linguistics. In 1973, Kenneth E. Boulding considered it hard to believe that the

> overwhelming importance of the spatial metaphor is not related to some as yet unknown spatial pattern of the human nervous system, which enables us to perceive the four-dimensional space-time world in which we live because of some kind of corresponding structure "inside" this structure coloring all our other perception.[64]

[60] Tolman (1947). See Holmén (2020) for the history of the concepts mental and cognitive maps.

[61] O'Keefe and Dostrovsky (1971); Best and White (1999).

[62] Moser and Moser (2015); Kropff et al. (2015).

[63] Eichenbaum (2014, 742).

[64] Boulding (1973, ix).

Research in anthropological linguistics has confirmed that human cognition of time is dependent on spatial metaphors in a broad range of cultures and languages. The nature of these metaphors might vary depending on culture and context—for example, while the past is in most parts of the world considered to be behind us, some cultures perceive it to be in front of us, and in others it is even situated "downhill."[65] The fact that, despite the diversity, the basis of all metaphors is spatial indicates that our temporal cognition utilizes a mental apparatus that first evolved for orientation in space.

It is not only the perception of time that depends on spatial cognition. For example, that our understanding of numbers is based on a mental numberline is backed by both experimental data and clinical evidence. Patients with brain damage that cause them to neglect a portion of perceived space also have difficulties representing numbers located on the corresponding part of the numberline.[66] Cognitive maps have also been suggested for social relationships, color, sound, and odors, which are explained by the theory that brain mechanisms originally evolved to navigate space have been co-opted to make sense of more abstract realms.[67]

Space, time, numbers, social relations, color, and sound are phenomena humans have experienced during a long period of evolutionary history. However, we also seem to use spatial models for abstract problems that we have never encountered before. For example, when persons engage in a task that requires envisioning a two-dimensional conceptual space, their grid cells fire in the same pattern as if they would traverse physical space. This is supported by a famous experiment where subjects were confronted with images of birds whose body shapes varied by neck length and body length.[68] There was nothing spatial with the exercise per se, but the information still became spatially structured in the brains of the involved persons.

Neurology is advancing rapidly, and this is particularly true of research on spatial cognition. For a long time, the progress has been based on breakthroughs in experimental science, and few attempts have been made at constructing overarching theories. However, Jeff Hawkins and his

[65] Mitchell (1980, 542), Núñez et al. (2012), Núñez and Sweetser (2006), Núñez and Cooperrider (2013).

[66] Schultheis and Carlson (2013, 726–727).

[67] See, for example, Underwood (2016, 1378); Park et al. (2020); Ekstrom and Ranganath (2018).

[68] Constantinescu et al. (2016).

research group have attempted to formulate a general theory of intelligence that can explain all the various new findings. They have suggested that the neocortex maps the location and structure of objects in a process parallel to how the hippocampus and the entorhinal cortex locates your body in space, and that

> even abstract knowledge is stored in relation to some form of "location" and . . . much of what we consider to be "thought" is implemented by inference and behavior generating mechanisms originally evolved to move and infer with fingers and eyes.[69]

The hippocampus and the entorhinal cortex are located in the old parts in the center of the brain, while the more recently developed neocortex is a folded sheet of neurons, 2.5 millimeters thick, that covers the outside of the brain. In a human brain, the neocortex is divided into approximately 150,000 functionally differentiated columns. According to Hawkins's thousand brains theory of intelligence, each of these columns is a tiny copy of the hippocampus and entorhinal cortex, equipped with place cells and grid cells. With the help of these cells, every column—of which some are connected to visual and others to sensorimotor or auditory input—contains its own spatial model of the world. Our coherent worldview is constructed through a "referendum" among the columns.[70]

A central concept in Hawkins's theory is reference frames. The brains of our ancestors

> first evolved reference frames to learn the structure of environments so that they could move about the world. Then our brains evolved to use the same mechanism to learn the structure of physical objects so that we could recognize and manipulate them. [Then,] our brains again evolved to use the same mechanism to learn and represent the structure underlying conceptual objects, such as mathematics and democracy.[71]

Hawkins argues that thinking is a form of movement in the reference frames of abstract knowledge. Sometimes we can use a familiar reference frame for this purpose, as when we use a known place for memory storage according to the method of loci. However, often we need to construct

[69] Hawkins et al. (2017, 2019).
[70] Hawkins (2021, 59–68).
[71] Hawkins (2021, 72).

new reference frames, as in the experiment with the conceptual space of bird body types.[72]

One advantage of visual diagrams is that they might at a glance provide a person with a new abstract reference frame. It consumes much more time and effort to build up a reference frame through experience or from narrative text. However, the creation of new reference frames is more laborious than using preexisting ones. This might explain why a limited set of diagram types have come to dominate, such as tree diagrams, time series charts, and, during the Middle Ages, the square of opposition, which was borrowed from logic to physics and the four elements.[73] Once a diagrammatic form has become a commonly known schema, it is utilized to convey a range of different ideas, even if this familiar reference frame does not always fit the new domains particularly well.

An important set of building blocks for spatial reference frames of abstract ideas are the primary metaphors. Influenced by Lakoff and Johnson's work on conceptual metaphors, it was suggested in the late 1990s that we learn a common set of primary metaphors from sensorimotor experience in early childhood. For example, by observing how the surface rises in a glass when we pour water, we conflate that "More is Up." These simple metaphors become established as permanent neural connections, and form the building blocks from which we construct more complex metaphors later in life. The majority of these primary metaphors are spatial, such as "Important is Big," "Intimacy is Closeness," "More is Up," "Categories are Containers," "Linear Scales are Paths," "Organization is Physical Structure," "Control is Up," and "Similarity is Closeness."[74] Other primary metaphors, such as "Good is Bright/Bad is Dark,"[75] are also relevant for diagrams, since they can affect how different entities are contrasted to each other in figures through shading.

The primary metaphors can also be combined through conceptual blending, which can lead to new inferences and to the construction of larger complex metaphors.[76] According to Fauconnier and Turner, this blending takes place in short-term memory, combining items from

[72] Hawkins (2021, 76–81); Constantinescu et al. (2016); for method of loci, see section "Terrain and Architecture: Structuring Knowledge in "Oral" Cultures and Early Civilizations" in Chap. 3.

[73] See section "The Square of Opposition and the Wheel of Reasons" in Chap. 4.

[74] See, for example, Lakoff and Johnson (1999, 46–54).

[75] Ortiz (2011).

[76] Lakoff and Johnson (1999, 47).

long-term memory.[77] It has been pointed out that space has a general role in conceptual blending, and can be used in all domains where the concept of betweenness is applicable.[78] For example, the idea of temperature graphs might have been conceived through a conceptual blending of the domains of time and temperature—in both of which betweenness makes sense—with the domain of space. The function of betweenness in spatial organization is so important that in domains where it is not immediately given, we might have to invent it. For example, for the sound of letters and words, we have introduced the arbitrary alphabetical order, which allows us to structure words spatially, such as in bibliographies.

Julio Santiago et al. underline the spatial foundation of primary metaphors in words reminiscent of Hawkins's reference frame concept, as they suggest that:

> the capacity of holding and manipulating abstract thoughts through spatial conceptual metaphors depends crucially on labelling representations of physical objects with abstract labels and locating those "abstract" objects on the physical dimensions of space, which in turn can also hold abstract labels such as time or emotional evaluation.[79]

They believe that several conflicting spatial models can coexist in long-term memory, but that they are forced to become congruent when simultaneously recalled to short-term memory. Environmental cues might therefore nudge us toward using one of several available spatial reference frames, such as mapping the same phenomenon along horizontal or vertical axes.[80] Thus we have two ideas of what might happen if two different models are simultaneously brought to short-term memory: they might be combined through conceptual blending, essentially creating a new model, or one model might win out over the other to achieve congruence. If combination or elimination of one alternative will take place depends on whether the models are complementary or contradictory. Examples of both can be found in the history of diagrams.[81]

[77] Fauconnier and Turner (2003).
[78] Worthing (2016).
[79] Santiago et al. (2011, 78).
[80] Santiago et al. (2011, 81).
[81] See sections "The Shape of Life" in Chap. 5 and "The Inversion of the Swedish Political Scale" in Chap. 8.

It should be noted that the theory of primary metaphors does not differentiate between metaphors and analogies, and most of the primary metaphors might be described as spatial analogies. In the terminology of Peircean semiotics, these "metaphors" would instead be classified as diagrams. However, they will be referred to as primary metaphors in the following, since that clarifies their theoretical origin and helps differentiate them from other uses of the diagram concept in the text.

Other research shows that some of the associations, which according to the theory of primary metaphors are learned early in life, are found already in infants and are therefore probably innate.[82] These different explanations might be complementary rather than mutually exclusive. For example, Thomas Schubert et al. argue that innate proclivities create metaphors in interplay with the constraints of human bodies in the environment.[83]

There is one more important aspect of the neocortex's organization that we need to take into consideration: its hierarchical structure. For decades, it has been believed that sensory input is processed in one area of the neocortex before it is passed on to the following, where the procedure is repeated. For each level in the hierarchy, the information becomes more abstract. While the first layers are devoted to simple pattern recognition, the highest chisel out general theories and worldviews. However, contrary to what was earlier believed, this is far from a one-way process. In fact, there are more synapses going from the higher levels to the lower than in the other direction. The neocortex is now understood to function as what in computer science is described as an auto-associative memory. When our senses perceive part of a cat, a word, or the trajectory of a thrown ball and send this input up the cortical hierarchy, memories of cats, words, or projectile trajectories are sent down the hierarchy. This means that our perception of the world is to a large degree influenced by our memories and preexisting worldviews. We see patterns in the world that are not found directly in the perceptory input but are derived from our expectations of what should be there.[84]

Of course, the observation that our cognitive processes are "filling in the blanks" where we lack data has been made in other disciplines long ago. For example, the Gestalt principles of perception, developed by psychologists in the first half of the 1900s, describe how we through

[82] For example, association between line length and number, Spelke (2011, 290).

[83] Schubert et al. (2011, 176).

[84] Hawkins (2005, 44–47, 73–75); Hawkins (2021, 30–32).

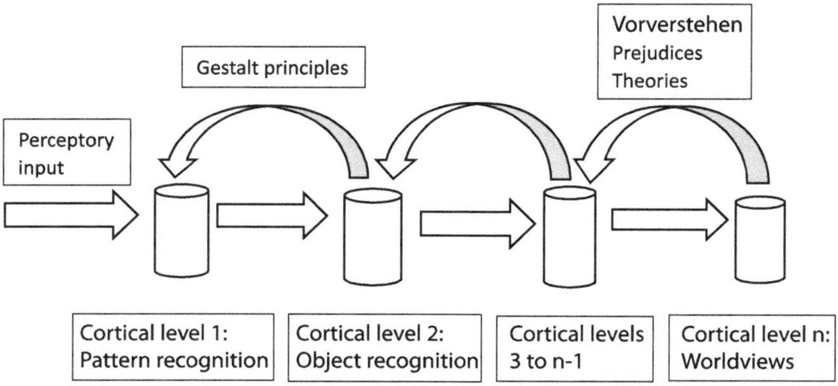

Fig. 2.4 The process of feedback from higher to lower levels in different parts of the cortical hierarchy has been described by a range of concepts in psychology, philosophy, and the humanities. Drawn by author

processes such as reification, multistability, invariance, and grouping infer spatial information that is not explicit in the perceptory input.[85] While psychological phenomena such as the Gestalt principles illustrate multidirectional signals in the lower layers of the cerebral hierarchy, the humanities and social sciences have focused on problems caused by similar mechanisms in its higher levels (Fig. 2.4).

For example, in the philosophical tradition of hermeneutics, which is concerned with (primarily but not exclusively textual) interpretation, the concept *Vorverstehen* or preunderstanding describes how we, when we encounter new information, do not start with a blank slate, but have a prejudice or pre-judgment based on our knowledge of the context.[86] The Gestalt psychologist Max Wertheimer extended the Gestalt concept to describe such higher-level worldviews, talking about the strong *Gestalt* of pre-Einsteinian physics, similarly to how Thomas Kuhn would later use the term paradigm.[87] This indicates that Wertheimer considered the same mechanisms to be at work at different levels of human cognition.

Concerns about the role of prejudices, norms, and discourses in premature judgment of fellow human beings have inspired attempts to

[85] Leeuwen (1998).
[86] Palmer (1969, 24–25, 51, 121, 134–136, 181–184).
[87] Wertheimer (1945, 188); Kuhn (1962).

counteract their effect, such as deconstructionism and norm criticism. By becoming conscious of our prejudices, it is hoped that we can avoid stereotypical oversimplifications. However, we cannot escape the fact that our brains are constantly making judgments and predictions based on interpretations of sensory data that are heavily influenced by our higher-level hypotheses of how the world is constituted. Without this process, the world would be an incomprehensible mess of unprocessed sensory information.

The hierarchical schema in Fig. 2.4 is highly simplified, as the connections between brain regions are more reminiscent of a complex network than a tidy hierarchy.[88] However, the main point that the simplification is intended to convey is that our abstract worldviews are shaped in the same process of interaction between perception and previously learned models as our mental maps of the physical surroundings. Since the cortical apparatus originally evolved for creating a spatial model of the world and keeping track of the location and orientation of oneself and objects in it, the cortical areas related to high-level worldviews are also to their fundamental nature spatial cognitive maps.

According to Hawkins, we have these models of the world in our brains because they help us make predictions.[89] Making a prediction about the future is analogous to filling in the blanks of our perception in a spatial frame. For example, according to the Gestalt principle of continuity, we assume that objects continue behind other objects that are concealing them from our view. If our cognition of time is indeed based on the same spatial mental framework of diagrammatic reasoning, we make predictions by assuming that current developments will continue behind the future's cloak of uncertainty. Fundamentally, this is explained by the brain's properties as an auto-associative memory: when the beginning of a development is relayed up the cortical hierarchy, it activates complete memories of similar developments in the past, including their endings. The predictions we make are firmly rooted in the diagrammatic models of the world that we have created based on our experiences.

Lakoff and Johnson have emphasized that the hierarchical structure of the brain forces us into categorization, which is thereby an "inescapable

[88] See, for example, the diagram of the monkey visual cortex in Felleman and Van Essen (1991).

[89] Hawkins (2005, 87–89).

consequence of our biological makeup." For example, the eyes have 100 million light sensing cells, but only 1 million fibers lead to the brain, which means that each image must be reduced in complexity by a factor of 100. Similar simplification through categorization is needed in the higher levels of the cortical hierarchy. Lakoff and Johnson therefore conclude that even when we think that we are deliberately forming new categories, we are affected by our unconscious categorization.[90] For each higher cortical level, the image received by our eyes becomes more of a diagram that preserves only the most central aspects of the observed objects.

An interesting question is whether the spatial foundation of human cognition is a unique outcome of our particular path of evolutionary history, or if it is a necessary feature in all cognitive systems. Some observations related to the rapid development of artificial intelligence seem to strengthen the latter hypothesis. A 2023 paper by Wes Gurnee and Max Tegmark claimed that space and time neurons were spontaneously formed during the training of the large language model Llama.[91] This occurred despite the fact that such models are only trained to predict the next symbol in a series, based on reading large amounts of texts. Although the article has attracted criticism for overstating its point,[92] it is not unreasonable to assume that when artificial intelligence systems begin to display human-like abilities, their neural networks will acquire structures for spatial cognition similar to those of the human brain. It is likely that this development will become more accentuated with the proliferation of real-world "embodied" AI systems. These might move in and interact with the environment, thereby improving their spatial cognition and ability to reason by drawing analogies from space to abstract domains.[93] However, as will be discussed in the next section, the sense of space acquired through embodied interaction with the environment, both inherited from evolutionary history and learned from personal experience, is fundamentally different from the neutral space we encounter in geometry textbooks.

[90] Lakoff and Johnson (1999, 18).
[91] Gurnee and Tegmark (2023).
[92] See, for example, Marcus (2023).
[93] For a discussion of this, see Zador et al. (2023).

The Evolution of Spatial Differentiation and Mythical Space

The idea of a universal continuous space, in which the three dimensions are interchangeable with identical properties, emerged along with the coordinate system in the early modern period.[94] We humans stem from a series of organisms shaped by the spatial experience of living on Earth. That experience is far from the neutral ideal of analytical geometry, as the dimensions here are charged with different values and feelings. The earliest lifeforms were self-replicating molecules, for which the three dimensions of the world they inhabited were of minor importance. Still today, many single-celled organisms have little or no spatial differentiation, consisting of a nucleus surrounded by a membrane that is more or less uniform in all directions. However, several circumstances have contributed in creating organisms that are spatially differentiated. The sun, from which most forms of life derive their energy, is located up. For multicellular organisms, that means that the uppermost layers of cells should be protected against radiation, and organs using photosynthesis should maximize their surface area facing up.[95] In addition, gravity pulls you down. This means that, for organisms not floating freely in water or air, some kind of supportive structure has to be created underneath them, such as feet or roots. It is also easier to move in the horizontal plane than in the vertical dimension. As Barbara Tversky has phrased it, "[g]oing against gravity takes time, force, strength, energy, even money."[96] This has consequences for how we perceive up and down.

Mobile organisms have also developed a differentiation between forward and back. In general, animals have a preferred direction of motion, in which it is much easier for them to move than in the other. This is partially dictated by the dynamics of moving through water or air, but primarily it depends on the specialization of joints, muscles, and tendons, which are optimized for movement in one direction. A knee that can bend both ways is not as efficient as a specialized one; simply making a turn before moving in the other direction is a more cost-effective solution.

[94] This kind of space is often called "Euclidean space," although alien to the geometry of Euclid, which only recognized space within geometrical forms, without placing these forms in a grid pattern of infinite extension. Boberg (2020, 112–113). See also section "Statistical Graphics" in Chap. 6.

[95] Rothschild (2003).

[96] Tversky (2011, 19), see also Morey-Holton (2003).

Consequently, walking, swimming, or flying backward is considerably more difficult than moving forward. This means that motion primarily occurs in one direction, and sensory and manipulative organs such as eyes, trunks, and hands are therefore placed facing forward.

However, while differentiation between up and down as well as forward and back developed relatively early and is shared by most species, left–right specialization is less prominent. In fact, you see very little difference between the left and right sides in the physical appearance of most animals, as both sides of the body are almost identical copies of each other. The fact that most organisms are symmetrical has contributed to us perceiving symmetrical shapes as beautiful and pleasing, which also affect our construction and perception of diagrams.[97]

In animals where a left–right differentiation exists, it is usually confined to the neurological area.[98] Animals who use their extremities to manipulate their environment can reach a higher level of "dexterity" by specializing in the use of one side of their body. In cases where there is a physical difference between left and right, such as in male fiddler crabs that have one giant claw for fighting and attracting mates and one small for feeding, it stems from such a lateral specialization in manipulation of the environment. It is not a coincidence that crabs are also among the few animals that regularly move in all four directions, and have joints better adapted for sideways than for forward motion. When making a sideways rush to attack an adversary with its giant claw, the crab's weapon is effectively facing "forward" in relation to its direction of motion.[99]

Thus, the different properties of the dimensions have been etched into our bodies and minds during our evolutionary history. In addition, our everyday perception confirms the difference between the dimensions. According to the philosopher Ernst Cassirer "physiological space differs from metric space in that here right and left, before and behind, below and above are not interchangeable, since motion in any of these directions involves specific organic sensations." He concluded that this has resulted in the dimensions carrying specific mythical feeling values.[100]

[97] For symmetry, see Hodgson (2009), and Fig. 4.5 and section "String Theory: Post-Modern Platonism" in Chap. 9, in this volume.

[98] Güntürkün et al. (2020).

[99] Perez et al. (2015).

[100] Cassirer (1955, 84–85).

Graphic displays often only use vertical and horizontal orientations. According to Tversky, the vertical is defined by gravity, and the horizontal by the horizon, and vision is particularly sensitive along these axes. Memory is also better for these orientations, and slightly tilted lines are remembered as more vertical or horizontal than they actually are. The vertical has precedence over the horizontal. This might be due to the fact that in the vertical, the sky dominates over the Earth, while no such asymmetry is prevalent on the horizontal axis. In linguistic metaphors, up is consequently associated with positive concepts such as better, more, and stronger, and in diagrams, ideals are placed on top.[101] In three dimensions, in front and behind display a similar asymmetry as top and down.[102] Lakoff and Johnson have pointed out that abstract concepts, such as love, happiness, and social status, are made understandable through orientational metaphors based on the perceivable physical reality: we are "in love," "depressed," or "rise" on the social ladder.[103]

Although far less distinct than the up–down and front–back dimensions, the left–right axis is also charged with some value, stemming from the fact that most humans are right-handed. Growing up in a literary culture where writing generally takes place in a set direction also affects how left and right is perceived.[104] For example, in cultures where you write from left to right, more powerful and agentic actors are often placed to the left in images and diagrams, and movement takes place from left to right. This left–right polarity, called spatial agency bias, has for decades been observed by art historians.[105] In a convincing confirmation of this, Robert Couzin has compared medieval illustrations from left to right Greek and Latin areas with those from right to left Hebrew and Syriac areas.[106]

However, under this cultural variability there seems to be an innate aesthetic preference for activity from left to right, which is believed to have its foundation in the lateralization of the brain.[107] Therefore, non-literate cultures, which are not affected by any direction of writing, tend to

[101] Tversky (2001, 99–102).

[102] Gattis (2001b, 228–229).

[103] Lakoff and Johnson (1980, 14–17).

[104] This was suggested already in Weyl (1952, 24).

[105] Maass and Schubert (2011, 4); Chatterjee (2011); section "The Rise of the Left–Right Political Scale" in Chap. 8.

[106] Couzin (2021, 192–210).

[107] Friedrich et al. (2014).

compose images where people and animals are facing right.[108] This might help explain why the spatial agency bias is stronger in cultures writing from left to right than in those writing from right to left. It is not merely a case of right-to-left writers being more exposed to left-to-right writing than the opposite.[109] In their case, the spatial agency bias caused by writing direction is also counteracted by brain lateralization.

Thus, in addition to their neutral, rational properties, the spatial dimensions also hold mythical and metaphorical values. In diagrams, these aspects of the spatial dimensions are combined in ways that are undeclared and often also unconscious. The values conveyed by spatial metaphors might become easier to accept when dressed in the superficially neutral costume of a diagram, a fact that has been utilized in propaganda.[110] Therefore, it is worth reflecting on the spatial choices that have been made in the construction of a diagram, such as what has been placed on top or to the left.

Brain, Culture, and Diagram

The preceding chapters gave an overview of current research in the cognitive sciences relevant for understanding diagrams and their development. However, the general principles put forward there are not always new. For example, the conceptual metaphor theory of Lakoff and Johnson has been criticized for its lack of historicity and neglect for the rich tradition of similar views. Ulrike Schröder has even claimed that for this theory,

> the majority of the basic hypotheses and even the main part of the examples have already been discussed since the beginning of the seventeenth century, especially in philosophical treatises (Clauberg, Vico, Locke, Leibniz, Lambert, Kant, Nietzsche, Mauthner, Vaihinger, Richards, Blumenberg), but also from anthropological (Herder, Boas, Sapir, Whorf, Gehlen), psychological (Stählin, Bühler) and linguistic (Paul, Wegener, Gerber, Biese, Lady Welby, Black, Weinrich) perspectives. All these contributions already imply a profound awareness of the cognitive-epistemological function of metaphor, its ubiquity in everyday life, as well as its affection both on a conceptual and linguistic level.[111]

[108] Boiteau et al. (2020).

[109] This interpretation has been used to explain differences between children with English, Hebrew, and Arabic as their first language. Tversky et al. (1991).

[110] Sections "Revolutionary Diagrams: Breaking the Barrier" in Chap. 8 and "Labyrinthic Chartjunk: Implication Through Cognitive Overload" in Chap. 9.

[111] Schröder (2014, 80).

Likewise, the idea that our perception is not only conditioned by sensory input but also through previous experience was mentioned already by Aristotle (as the third kind of perception described in *De anima*), and by the Buddhist philosopher Dignaga (480–540 CE).[112]

The existence of predecessors does, however, not make modern cognitive science less relevant as a theoretical basis for this investigation. Rather, according to the principle of consilience, the credence of these theories is elevated by the fact that the issue has throughout history repeatedly been understood similarly by individuals approaching it from different disciplinary angles. In addition, some theories from the cognitive sciences go well beyond earlier attempts in the level of detail in which they describe how these cognitive principles have developed. For example, even if we would accept that the conceptual metaphor theory mostly reproduces ideas presented earlier, the theory of primary metaphors is no doubt a significant development in several respects. Most importantly, concepts such as the primary metaphors are well suited for the analysis of an empirical material such as historical diagrams.

Given the multidisciplinary, heterogeneous, and fragmented nature of the theories presented in sections "Diagrams and the Limits of Consciousness", "Diagrammatic Reasoning", "The Neurology of Spatial and Abstract Thinking" and "The Evolution of Spatial Differentiation and Mythical Space", it is necessary to here give a collected and structured account of the layers of cognitive factors that influence how humans construct and understand diagrams. Using a simple categorization that respects the cognitive limitations of the author and the readership, we can identify four different levels of spatial cognition and diagrammatic thinking:

1. **Innate** There is an innate layer of perception, which, for example, predisposes us to use horizontal and vertical lines more than tilted ones or to quickly detect symmetrical shapes and consider them aesthetic. These fundamentals of perception nudge diagram construction in certain directions, such as favoring horizontal and vertical lines, and, in combination, grid-shaped patterns, as well as symmetrical shapes. The limits of human cognition, such as working memory capacity, are also innate, and they restrict how diagrams can be constructed. In order for a diagram to be comprehensible, the levels

[112] Gregoric and Fink (2022); Hayes (1988, 135).

of simplification, granularity, categorization, and hierarchization must adhere to Miller's magical number seven and cognitive load.

2. **Primary** On top of this innate layer of perception and cognitive restrictions, there is a set of learned but common to all principles, which have their roots in shared experiences of growing up in the spatial conditions of Earth. These are the primary metaphors, such as "More is Up," which creates the foundation for diagrammatic forms such as bar charts. Different opinions exist regarding, for example, if the association between line length and number is innate or a learned primary metaphor. However, since all persons who have ever constructed a diagram have been in possession of both innate and primary repertoires of spatial cognition, the exact distinction between the two are of minor importance for analyzing the history of diagrams. In addition, we might expect a process where weak innate tendencies are reinforced during infancy by learning primary metaphors.[113]
3. **Cultural** Yet another layer is constituted by principles which are not entirely universal but particular to large cultural areas. The most well researched of them is the spatial agency bias determined by the direction of writing, which influences how we perceive direction and dominance in the horizontal dimension.
4. **Historical** On top of that, there is a set of learned conventions, which are needed in order to understand particular kinds of diagrams. These might even vary between academic disciplines and professional fields. Such conventions also entice constructors of new diagrams to mold them into preexisting schemas in order to achieve comprehensibility.

The remainder of the book will investigate how such learned conventions of diagrammatic communication have evolved throughout human history. This will help us understand how new diagrammatic forms have developed under the influence of changes in worldviews, technology and the production and spread of knowledge, and in interplay with preexisting historical conventions. Through a cognitive history approach, we will also attempt to investigate how this historical development of diagrams is conditioned by primary and innate understandings of space.

[113] See, for example, Scholl (2005).

References

Bertin, Jacques. 1983. *Semiology of graphics: Diagrams, networks, maps.* Madison: University of Wisconsin Press.

Best, Phillip J., and Aaron M. White. 1999. Placing hippocampal single-unit studies in a historical context. *Hippocampus* 9 (4): 346–351.

Bird-David, Nurit. 2019. Dis/working with diagrams: How genealogies and maps obscure nanoscale worlds (a hunter-gatherer case). *Social Analysis* 63 (4): 43–62.

Blaug, Mark, and Peter Lloyd. 2010. Introduction. In *Famous figures and diagrams in economics*, ed. Mark Blaug and Peter Lloyd, 1–23. Cheltenham: Edward Elgar.

Boberg, Johan. 2020. *Scientifically minded: Science, the subject and Kant's critical philosophy.* Uppsala: Uppsala University.

Boiteau, Timothy W., Cameron Smith, and Amit Almor. 2020. Rightward directional bias in art produced by cultures without a written language. *Laterality* 25 (2): 165–176.

Bolton, G. M. 1983. *Testing in selection decisions.* London: NFER-Nelson.

Boulding, Kenneth. 1973. Foreword. In *Image and environment: Cognitive mapping and spatial behavior*, ed. Roger M. Downs and David Stea, vii–xi. Chicago: Aldine.

Bullynck, Maarten. 2010. Johann Lambert's scientific tool kit: Exemplified by the measurement of humidity. *Science in Context* 23 (1): 65–89.

Candea, Matei. 2019. On visual coherence and visual excess: Writing, diagrams, and anthropological form. *Social Analysis* 63 (4): 63–88.

Cassirer, Ernst. 1955. *The philosophy of symbolic forms, vol. 2: Mythical thought.* New Haven, CT: Yale University Press.

Chatterjee, Anjan. 2011. Directional asymmetries in cognition: What is left to write about? In *Spatial Dimensions of Social Thought*, ed. Anne Maas and Thomas W. Schubert, 189–210. Berlin: De Gruyter Mouton.

Constantinescu, Alexandra O., Jill X. O'Reilly, and Timothy E. J. Behrens. 2016. Organizing conceptual knowledge in humans with a gridlike code. *Science* 352 (6292): 1464–1468.

Couzin, Robert. 2021. *Right and left in early Christian and medieval art.* Leiden: Brill.

Donald, Merlin. 2001. *A mind so rare: The evolution of human consciousness.* New York: Norton.

Dove, Guy. 2015. How to go beyond the body: An introduction. *Frontiers in Psychology* 6:660.

Eichenbaum, Howard. 2014. Time cells in the hippocampus: A new dimension for mapping memories. *Nature Reviews Neuroscience* 15 (11): 732–744.

Ekstrom, Arne D., and Charan Ranganath. 2018. Space, time, and episodic memory: The hippocampus is all over the cognitive map. *Hippocampus* 28 (9): 680–687.

Elngar, Ahmed A., Mohamed Arafa, Amar Fathy, Basma Moustafa, Omar Mahmoud, Mohamed Shaban, and Nehal Fawzy. 2021. Image classification based on CNN: A survey. *Journal of Cybersecurity and Information Management* 6 (1): 18–50.

Enriquez, León. 2012. A Peircean model for music and sound-based art: A pragmatist approach to experiences in the artistic use of sound. In *Proceedings of the electroacoustic music studies network conference 'meaning and meaningfulness in electroacoustic music.'* Stockholm: Electroacoustic Music Studies Network.

Fauconnier, Gilles, and Mark Turner. 2003. Conceptual blending: Form and meaning. *Recherches en Communication* 19:57–86.

Felleman, Daniel J., and David C. Van Essen. 1991. Distributed hierarchical processing in the primate cerebral cortex. *Cerebral cortex* 1 (1): 1–47.

Figurska, Małgorzata, Maciej Stańczyk, and Kamil Kulesza. 2008. Humans cannot consciously generate random numbers sequences: Polemic study. *Medical hypotheses* 70 (1): 182–185.

Friedrich, Trista E., Victoria L. Harms, and Lorin J. Elias. 2014. Dynamic stimuli: Accentuating aesthetic preference biases. *Laterality* 19 (5): 549–559.

Gattis, Merideth. 2001a. Space as a basis for abstract thought. In *Spatial schemas and abstract thought*, ed. Merideth Gattis, 1–12. Cambridge, MA: MIT Press.

Gattis, Merideth. 2001b. Reading pictures. In *Spatial schemas and abstract thought*, ed. Merideth Gattis, 223–245. Cambridge, MA: MIT Press.

Glasgow, Janice, N. Hari Narayanan, and B. Chandrasekaran, eds. 1995. *Diagrammatic reasoning: Cognitive and computational perspectives.* Menlo Park, MA: MIT Press.

Gregoric, Pavel, and Jakob Leth Fink. 2022. Introduction: Sense perception in Aristotle and the Aristotelian tradition. In *Forms of representation in the Aristotelian tradition, vol. 1: Sense perception*, ed. Juhana Toivanen, 15–39. Leiden: Brill.

Grønfeldt Winter, Rasmus. 2020. *When maps become the world.* Chicago: University of Chicago Press.

Güntürkün, Onur, Felix Ströckens, and Sebastian Ocklenburg. 2020. Brain lateralization: A comparative perspective. *Physiological Reviews* 100 (3): 1019–1063.

Gurnee, Wes, and Max Tegmark. 2023. Language models represent space and time. https://arxiv.org/abs/2310.02207. Accessed 29 July 2024.

Hadamard, Jacques. (1945) 1996. *The mathematician's mind: The psychology of invention in the mathematical field.* Princeton, NJ: Princeton University Press.

Hamburger, Jeffrey F. 2020. Mindmapping: The diagram paradigm in medieval art—And beyond. In *The visualization of knowledge in medieval and early*

Europe, ed. Marcia Kupfer, Adam S. Cohen, and J. H. Chajes, 61–86. Turnhout: Brepolis.

Harley, John Brian, and David Woodward. 1987. *The history of cartography, vol. 1: Cartography in prehistoric, ancient, and medieval Europe and the Mediterranean*. Chicago: University of Chicago Press.

Hawkins, Jeff. 2005. *On intelligence*. New York: Henry Holt.

Hawkins, Jeff. 2021. *A thousand brains: A new theory of intelligence*. New York: Basic Books.

Hawkins, Jeff, Subutai Ahmad, and Yuwei Cui. 2017. A theory of how columns in the neocortex enable learning the structure of the world. *Frontiers in Neural Circuits* 11:81.

Hawkins, Jeff, Marcus Lewis, Mirko Klukas, Scott Purdy, and Subutai Ahmad. 2019. A framework for intelligence and cortical function based on grid cells in the neocortex. *Frontiers in Neural Circuits* 12:121.

Hayes, Richard P. 1988. *Dignaga on the interpretation of signs*. Dordrecht: Kluwer.

Hodgson, Derek. 2009. Evolution of the visual cortex and the emergence of symmetry in the Acheulean techno-complex. *Comptes Rendus Palevol* 8 (1): 93–97.

Holdsworth, Ruth. 1982. *Psychology for careers counseling*. London: Macmillan.

Holmén, Janne. 2020. Time and space in time and space: Mapping the conceptual history of mental maps and historical consciousness. *Contributions to the History of Concepts* 15 (2): 105–129.

Hsu, Feng-Hsiung. 2002. *Behind Deep Blue: Building the computer that defeated the world chess champion*. Princeton, NJ: Princeton University Press.

Isacsson, Walter. 2017. *Leonardo da Vinci*. New York: Simon and Schuster.

Kennedy, Helen, Rosemary Lucy Hill, Giorgia Aiello, and William Allen. 2016. The work that visualisation conventions do. *Information, Communication & Society* 19 (6): 715–735.

Kropff, Emilio, James E. Carmichael, May-Britt Moser, and Edvard I. Moser. 2015. Speed cells in the medial entorhinal cortex. *Nature* 523 (7561): 419–424.

Kuhn, Thomas. 1962. *The structure of scientific revolutions*. Chicago: University of Chicago Press.

Lakoff, George, and Mark Johnson. 1980. *Metaphors we live by*. Chicago: University of Chicago Press.

Lakoff, George, and Mark Johnson. 1999. *Philosophy in the flesh: The embodied mind and its challenge to Western thought*. New York: Basic Books.

Larkin, Jill H., and Herbert A. Simon. 1987. Why a diagram is (sometimes) worth ten thousand words. *Cognitive Science* 1 (1): 65–100.

Latour, Bruno. 1986. Visualization and cognition: Thinking with eyes and hands. In *Knowledge and society: Studies in the sociology of science, past and present, vol. 6*, ed. Henrika Kuklick and Elizabeth Long, 1–40. New York: JAI Press.

Leeuwen, Cees van. 1998. Perception. In *A companion to cognitive science*, ed. William Bechtel and George Graham, 265–281. Malden, MA: Blackwell.

Manetti, Giovanni. 1993. *Theories of the sign in classical antiquity*. Bloomington: Indiana University Press.

Marcus, Gary. 2023. Muddles about models: Why it's a stretch to say that large language models represent "literal world models." https://garymarcus.substack.com/p/muddles-about-models. Accessed 31 December 2023.

May, Michael. 1999. Diagrammatic reasoning and levels of schematization. In *Iconicity: A fundamental problem in semiotics*, ed. Troels Degn Johansson, Martin Skov, and Bertil Brogaard, 175–194. Århus: NSU Press.

McNeill, David. 1992. *Hand and mind: What gestures reveal about thought*. Chicago: University of Chicago Press.

Mialet, Hélène. 2012. *Hawking incorporated: Stephen Hawking and the anthropology of the knowing subject*. Chicago: University of Chicago Press.

Miller, George A. 1956. The magical number seven, plus or minus two: Some limits on our capacity for processing information. *Psychological Review* 63 (2): 81–97.

Mitchell, William John Thomas. 1980. Spatial form in literature: Toward a general theory. *Critical Inquiry* 6 (3): 539–567.

Morey-Holton, Emily R. 2003. The impact of gravity on life. In *Evolution on planet Earth*, ed. Lynn J. Rothschild and Adrian M. Lister, 143–159. Amsterdam: Academic Press.

Moser, May-Britt, and Edvard I. Moser. 2015. Understanding the cortex through grid cells. In *The future of the brain: Essays by the world's leading neuroscientists*, ed. Gary Marcus and Jeremy Freeman, 67–77. Princeton, NJ: Princeton University Press.

Nöth, Winfried. 2024. Peirce's iconicity and his image-diagram-metaphor triad revisited: Complements to Stjernfelt's sheets, diagrams, and realism. *Semiotica* 258:143–167.

Núñez, Rafael, and Kensy Cooperrider. 2013. The tangle of space and time in human cognition. *Trends in Cognitive Sciences* 17 (5): 220–229.

Núñez, Rafael, and Eve Sweetser. 2006. With the future behind them: Convergent evidence from Aymara language and gesture in the crosslinguistic comparison of spatial construals of time. *Cognitive Science* 30 (3): 401–450.

Núñez, Rafael, D. Kensy Cooperrider, and Doan, and Jürg Wassmann. 2012. Contours of time: Topographic construals of past, present, and future in the Yupno Valley of Papua New Guinea. *Cognition* 124 (1): 25–35.

O'Keefe, John, and Jonathan Dostrovsky. 1971. The hippocampus as a spatial map: Preliminary evidence from unit activity in the freely moving rat. *Brain Research* 34 (1): 171–175.

Ortiz, Maria J. 2011. Primary metaphors and monomodal visual metaphors. *Journal of Pragmatics* 43 (6): 1568–1580.

Palmer, Richard E. 1969. *Hermeneutics: Interpretation theory in Schleiermacher, Dilthey, Heidegger, and Gadamer*. Evanston, IL: Northwestern University Press.

Park, A., Douglas S. Miller, Hamed Nili, Charan Ranganath, and Erie D. Boorman. 2020. Map making: Constructing, combining, and inferring on abstract cognitive maps. *Neuron* 107 (6): 1226–1238.

Penrose, Roger. 1989. *The emperor's new mind: Concerning computers, minds and the laws of physics*. London: Vintage.

Perez, Daniela M., Siobhan J. Heatwole, Lesley J. Morrell, and Patricia R. Y. Backwell. 2015. Handedness in fiddler crab fights. *Animal Behaviour* 110:99–104.

Persaud, Navindra. 2005. Humans can consciously generate random number sequences: A possible test for artificial intelligence. *Medical Hypotheses* 65 (2): 211–214.

Peterson, Lloyd, and Margaret Jean Peterson. 1959. Short-term retention of individual verbal items. *Journal of Experimental Psychology* 58 (3): 193–198.

Pietarinen, Ahti-Veikko. 2016. Is there a general diagram concept? In *Thinking with diagrams: The semiotic basis of human cognition*, ed. Sybille Krämer and Christina Ljungberg, 121–137. Boston: De Gruyter.

Rothschild, Lynn J. 2003. The sun: The impetus for life. In *Evolution on planet Earth*, ed. Lynn J. Rothschild and Adrian M. Lister, 87–107. Amsterdam: Academic Press.

Safran, Linda. 2020. A prolegomenon to Byzantine diagrams. In *The visualization of knowledge in medieval and early Europe*, ed. Marcia Kupfer, Adam S. Cohen, and J. H. Chajes, 361–382. Turnhout: Brepolis.

Santiago, Julio, Antonio Román, and Marc Ouellet. 2011. Flexible foundations of abstract thought. In *Spatial dimensions of social thought*, ed. Anne Maas and Thomas W. Schubert, 39–108. Berlin: De Gruyter Mouton.

Scholl, Brian J. 2005. Innateness and (Bayesian) visual perception: Reconciling nativism and development. In *The innate mind: Structure and contents*, ed. Peter Carruthers, Stephen Laurence, and Stephen Stich. New York: Oxford Academic.

Schröder, Ulrike. 2014. Rediscovering the cognitive-semiotic and cognitive-pragmatic approaches to metaphor in the work of Johann Heinrich Lambert and Philipp Wegener. *Metaphoric.de* 25:79–102.

Schubert, Thomas W., Sven Waldzus, and Beate Seibt. 2011. More than a metaphor: How the understanding of power is grounded in experience. In *Spatial dimensions of social thought*, ed. Anne Maas and Thomas W. Schubert, 153–185. Berlin: De Gruyter Mouton.

Schultheis, Holger, and Laura A. Carlson. 2013. Spatial reasoning. In *The Oxford handbook of cognitive psychology*, ed. Daniel Reisberg. Oxford: Oxford University Press.

Sloman, Aaron. 1995. Musings on the roles of logical and nonlogical representations in intelligence. In *Diagrammatic reasoning: Cognitive and computational perspectives*, ed. Janice N. Glasgow, Hari Narayanan, and B. Chandrasekaran, 7–32. Menlo Park: MIT Press.

Spelke, Elizabeth S. 2011. Natural number and natural geometry. In *Space, time and number in the brain: Searching for the foundations of mathematical though*, ed. Stanislas Dehaene and Elizabeth Brannon, 287–317. London: Academic Press.

Stjernfelt, Frederik. 2007. *Diagrammatology: An investigation on the borderlines of phenomenology, ontology, and semiotics*. London: Springer.

Strongman, Luke. 2010. Force field: Vitruvian Man and the physics of sensory perception. *International Journal of Arts and Sciences* 3 (9): 218–226.

Sweller, John. 1988. Cognitive load during problem solving: Effects on learning. *Cognitive Science*. 12 (2): 257–285.

Sweller, John, Paul L. Ayres, and Slava Kalyuga. 2011. *Cognitive load theory*. New York: Springer.

Tolman, Edward C. 1947. Cognitive maps in rats and men. *Psychological Review* 55 (4): 189–208.

Tversky, Barbara. 2001. Spatial schemas in depictions. In *Spatial schemas and abstract thought*, ed. Merideth Gattis, 79–112. Cambridge, MA: MIT Press.

Tversky, Barbara. 2011. Spatial thought, social thought. In *Spatial dimensions of social thought*, ed. Anne Maas and Thomas W. Schubert, 17–38. Berlin: De Gruyter Mouton.

Tversky, Barbara. 2019. *Mind in motion: How action shapes thought*. New York: Basic.

Tversky, Barbara, Sol Kugelmass, and Atalia Winter. 1991. Cross-cultural and developmental trends in graphic productions. *Cognitive Psychology* 23 (4): 515–557.

Underwood, Emily. 2016. Expanding our mental maps. *Science* 352 (6292): 1378.

Vartanian, Oshin. 2012. Dissociable neural systems for analogy and metaphor: Implications for the neuroscience of creativity. *The British Journal of Psychology* 103 (3): 302–316.

Wagenaar, Willem A. 1972. Generation of random sequences by human subjects: A critical survey of literature. *Psychological Bulletin* 77 (1): 65–72.

Wammes, Jeffrey D., Melissa E. Meade, and Myra A. Fernandes. 2016. The drawing effect: Evidence for reliable and robust memory benefits in free recall. *Quarterly Journal of Experimental Psychology* 69 (9): 1752–1776.

Wertheimer, Max. 1945. *Productive thinking*. New York: Harper.

Weyl, Hermann. 1952. *Symmetry*. Princeton, NJ: Princeton University Press.

Worthing, Daniel. 2016. *Computational conceptual blending as a model of spatial transfer in natural language*. Thesis. Edinburgh: University of Edinburgh. https://project-archive.inf.ed.ac.uk/msc/20162033/msc_proj.pdf. Accessed 30 July 2024.

Zador, Anthony, Sean Escola, Blake Richards, Bence Ölveczky, Yoshua Bengio, Kwabena Boahen, Matthew Botvinick, et al. 2023. Catalyzing next-generation artificial intelligence through neuroai. *Nature Communications* 14 (1): 1597.

CHAPTER 3

The Prehistory of Diagrams

When diagrams first appeared in human history, they did not arrive out of the blue. Humanity had an evolutionary history that produced the abilities for spatial cognition that form the foundation of diagrammatic thinking. Of course, it is extremely difficult to follow such a cognitive development in the anthropological and archaeological record. However, evolutionary cognitive archaeologists interpret the emergence of symmetrically shaped hand axes around 500,000 years ago as a sign that many aspects of spatial cognition were developed already. The knapping of such tools required coordination between symmetry sensitive visual cell groups and cell groups for motor activity. Compared to our ancestors and relatives among the apes, this required developments in the processing of visuospatial working memory.[1]

Between that time and the emergence of fully developed diagrams some thousand years ago, there was a prehistory of diagrams during which humans utilized various spatial aids in order to enhance their cognition. This chapter of the book begins with discussing the role of gestures in speech, which is similar to that of diagrams in text (section "Language and Gestures"). In addition, it examines the role of physical locations in terrain and architecture as a form of memory loci in cultures without a written

[1]Wynn and Coolidge (2022, 53–55).

J. Holmén, *A History of Diagrams*, Palgrave Studies in the History of Science and Technology,
https://doi.org/10.1007/978-3-031-99691-7_3

language (section "Terrain and Architecture: Structuring Knowledge in "Oral" Cultures and Early Civilizations"). Memories could also be stored with the help of portable objects, such as Incan quipus, which are discussed in section "Quipu: Inca Infographics". With the invention of writing, text became spatially structured in ways that aided cognition, for example, in lists and tables (section "Writing, Lists, and Tables").

Language and Gestures

It is likely that the first human use of language was spatial both in form and content: body language that conveyed a message about location. One aspect of human language that separates it from the sounds and signals of most animals is displacement, which enables a speaker to inform a listener about events neither of them can see since they are distant from them in spatial, temporal, or even imaginary dimensions. Derek Bickerton and Eörs Szathmáry have investigated the roots of language by contemplating during which situations Paleolithic hominins might have needed the ability of displacement. They have suggested that confrontational scavengers needed to communicate location to their kin in order to recruit their help in competing with large predators over valuable carcasses. According to Bickerton and Szathmáry, this probably occurred among *Homo erectus*, who evolved 2 million years ago.[2]

Some philosophers, linguists, and psychologists argue that spoken language was probably preceded by a period when gestures dominated human communication.[3] In gesturing, humans use their bodies to perform motions in space. Peter Woelert has described gestures as a "cognitively potent part of the process of mobilizing space for the constitution, comprehension, and communication of abstract structures of meaning."[4] The human face is also, through its many small muscles, uniquely well equipped to perform spatial operations, for example, raising the eyebrows or the corners of the mouth. When spoken language arrived, gestures continued to complement speech. In fact, it has been convincingly argued that many spoken words are formed by imitating gestures with tongue and palate. For example, "you" is pronounced by pointing at someone with your lips and tongue as you would point at them with your finger. When we say that something is "tiny," we make our oral cavity small as when we hold our

[2] Bickerton and Szathmáry (2011).
[3] Woelert (2011, 118–119).
[4] Woelert (2011, 129).

thumb and index finger close together. Similar patterns occur across a vast range of languages. Experiments also indicate that, because of such principles, humans from all language groups understand nonsense words similarly, for example, when asked to decide whether "mil" or "mal" represent a larger object.[5] Early spoken language was probably a mixture of gestures and words, which both to a large degree consisted of iconic spatial diagrams.

However, when language became codified through writing, bodily gestures disappeared. Today, when written language has for the first time become utilized for instant person-to-person communication on a large scale, the gestures have reappeared in the form of emojis, such as raised thumbs, clapping hands, or smiling faces.

A laboratory study of 19 participants at the University of Chicago revealed that gestures are frequently used in explanations, as a form of spatial analogies. These are often coherent over a long explanation of several minutes, forming a spatial analogous system (Fig. 3.1). In the experiment, which involved an explanation of positive and negative feedback systems, spatial expressions were used four times more often in gesture than in speech. Thus, the researchers concluded that spatial gestures are not merely a manual echo of spatial analogies in speech. Instead, we use spatial analogies for reasoning about abstract systems, also when this is not

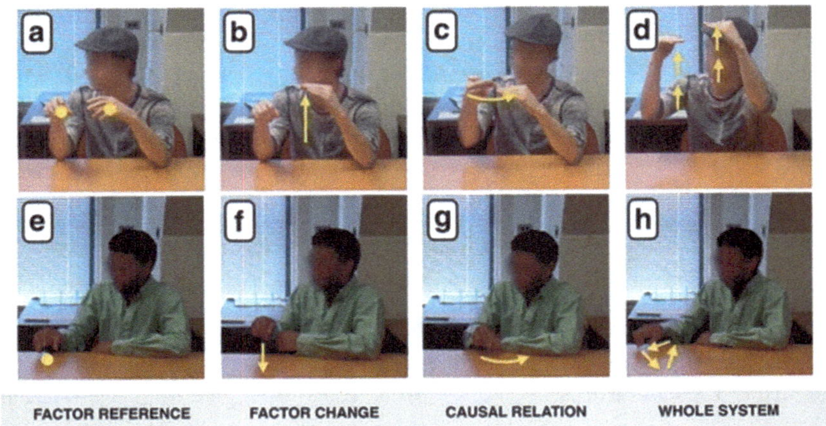

Fig. 3.1 Gestures convey spatial analogies in complex relational reasoning. Cooperrider et al. (2016)

[5] Mithen (2024, 124–153).

revealed in spoken language. Although the researchers acknowledge that familiarity with diagrams might influence use of gestures and vice versa, they are of the opinion that both have their foundation in an urge to spatialize relational structures.[6]

In the relatively new research field of embodied mathematics, abstract mathematical concepts are considered to be rooted in embodied activities. In mathematical jargon, natural numbers "grow" indefinitely, points "lay" on a line, and functions "move" to zero. Gestures are better than spoken language at expressing continuity, time, and movement. When communicating with pen on paper, diagrams possess the same advantages.[7]

That the written word and diagrams complement each other in written text in the same way speech and bodily gestures do in person-to-person communication was expressed already in 1161 CE by the Song dynasty scholar Zheng Qiao (1104–1162): "To see the writing without the image is like hearing a voice without seeing the form; to see the image without the writing is like seeing a person but not hearing his words." Therefore, Zheng believed that technical information was best conveyed with the help of diagrams paired with explanatory text.[8]

The link between gesturing and diagrams is also manifested in mnemonic and cognitive systems based on body parts, such as the Guidonian hand of music theory or several similar examples in mathematics. Historically, drawings of body parts were common illustrations in situations where we would today use more stylized diagrams. For example, before arrows appeared in diagrams in the 1700s, pointing fingers were used for marking directions or drawing attention to a particular spot.[9] However, it was not only one's own body that was used as a spatial reference frame for knowledge in pre-literary societies, but the entire landscape.

Terrain and Architecture: Structuring Knowledge in "Oral" Cultures and Early Civilizations

Cultures without writing have relied on speech and memorization for the survival of their cultural heritage. However, although it is convenient to describe these cultures as "oral," this does not give justice to their

[6] Cooperrider et al. (2016).

[7] Giardino (2016, 93–94).

[8] Bray (2007, 1); Châtelet (2000, 10).

[9] Heeffer (2010); Berger (1981); Gombrich (1990, 28); section "Flow Charts: From Liquids to Paper and Back Again" in Chap. 7.

multimodal systems of knowledge transmission. According to Lynne Kelly, most oral cultures use various forms of spatial mnemonic aids. For the Australian Aborigines, locations in the landscape were tied to different myths and narratives. The heavenly constellations were used in a similar fashion, and sedentary cultures across the world have erected megalithic structures used for the same purpose. These monuments were often circular, as they could be connected to the annual cycle of rituals through astronomical alignment. Different kinds of portable mnemonic devices were also used.[10]

Contrary to a common misunderstanding, learning in oral cultures was not informal and spontaneous. With the help of learning aids, oral culture was transmitted through formal teaching processes structured as several levels of initiation. During the initiation processes, initiates were secluded from the rest of society for intensive bouts of studies. These often incorporated the learning of songs and dances, which could help in the memorization of, for example, spatial, mythological, biological, environmental, or calendric information. In Australian Aboriginal cultures, it could take 40 to 50 years to learn all the songs and dances, and the sites associated with them. According to Lynne Kelly, in such oral cultures various aspects of knowledge were integrated and blended, since rigorous classification of the fields of knowledge only arrived with the logical structure provided by writing.[11]

For example, the Yanyuwa in Australia used dreaming tracks or songlines. Each location along these tracks was associated with a particular song, documenting stone tool making, place names, peoples, winds, seasonal events, hunting and foraging methods, cooking, toolmaking, or land use rights. At an ancient site for stone tool making, one of the elders could in 1992 still sing verses describing these technologies, although they had not been used for almost a century, and he had not visited the site for 50 years.[12] Thus, songlines index the landscape and transform it into a knowledge space, facilitating memorization. The same mnemonic structure survived in the songs of sedentary cultures. For example, the cognitive psychologist David Rubin has pointed out that classical epics such as

[10] Kelly (2015, 8, 9); for portable objects, see section "Quipu: Inca Infographics" in Chap. 3.

[11] Kelly (2015, 30–33).

[12] Bradley (2008).

The Iliad have "a clear spatial path that is tied to the story line," which together with the graphic imagery aids memorization.[13]

This is reminiscent of the method of loci that modern day memory champions use to memorize huge swaths of information. They visualize moving along a sequence of locations they associate with the sequence of data they are trying to memorize.[14] The oldest preserved descriptions of this technique are from ancient Greek and Roman texts on rhetoric. The Roman *Ad Herennium* (ca. 86–82 BCE) stressed that the mental image at each location should be as striking and vibrant as possible in order to be the most memorable through, for example, beauty, ugliness, comic effect, heroism, disasters, and great feats. Following the same principle, the dreaming tracks of oral cultures connected memory loci that were striking natural formations or rock painting locations.[15]

There are many examples of Paleolithic art similar to art from later periods that we know have functioned as memory loci. Since the context is lost, it is difficult to know whether Paleolithic rock paintings had the same function. However, rock paintings from Europe in the period from 40,000 to 10,000 BCE displayed a remarkable continuity regarding the geometric signs that were painted alongside the figurative motives.[16] These signs are often interpreted as symbols, but might also have functioned as diagrams, on their own or in combination with their environment. However, as will be discussed in the next section, it is difficult to attempt clear distinctions between symbols, diagrams, and images before the appearance of fully developed writing systems.

Kelly suggests that the construction of Neolithic megalithic monuments, such as Stonehenge, might have been a way for no longer nomadic populations to create memory loci for the annual cycle, which replaced their earlier migrations through the landscape.[17] The temple architecture of early civilizations might have had a similar function. For example, Wu Hung claims that early Chinese ritual structures such as the Mingtang (Hall of Light) were a "microcosmic architectural representation of the entire cosmic order." Another example was the grave of the first Qin emperor, which was a model of the world complete with rivers and oceans.

[13] Rubin (1995, 39).
[14] Bellezza (1996).
[15] Kelly (2015, 69–71).
[16] Von Petzinger (2017).
[17] Kelly (2015, 204).

In words reminiscent of Peirce's distinction between diagram and image, Wu Hung describes these buildings as two ways of representing the universe: the Mingtang constitutes a diagram of it while the Qin tomb pictures it. The earliest Hall of Light was a temple for state ceremonies during the Western Zhou dynasty (1046–771 BCE), but it was reinvented by the succeeding Eastern Zhou, that lasted until replaced by the Qin dynasty around 250 BCE. According to Wu Hung, this reinvention was heavily influenced by traditions of diagramming from several scholarly fields. The Eastern Zhou was a period of political turmoil and erosion of royal power, during which the Western Zhou was looked back upon as a political ideal. Thus, rebuilding the Hall of Light became synonymous with creating an ideal political order for a unified country.[18]

The ancient Egyptian Heb-sed festival had a similar role in unifying the country through rituals in a spatial setting that metaphorically represented the realm. During the festival, which was held for the first time when a pharaoh had reigned for 30 years, the monarch ran around a ritual course symbolizing the limits of the realm, alternately wearing the crowns of Upper and Lower Egypt. The ritual is mainly interpreted as a way of demonstrating or rejuvenating the pharaoh's strength, but has also been seen as a symbolic re-enactment of the unification of Egypt. The northern part of the complex where the rites were performed represented Lower Egypt (the Nile delta), while the southern represented Upper Egypt.[19]

It would be tempting to suggest a chain of development from nomadic oral cultures, where the entire universe is used as a diagram to which knowledge can be attached, via megalithic monuments and temple architecture to drawn diagrams. However, such a simple narrative of progressively reduced size and increased systematization is contradicted by Wu Hung's abovementioned claim that drawn diagrams also affected the construction of temples. Another example of this is Hindu temples, which are constructed based on the Vastu-purusha Mandala, a fractal diagram of the structure of the cosmos.[20]

Rather than viewing diagrams as pure cultural artifacts that have developed from a single origin, it is more appropriate to consider them manifestations of an innate human propensity for spatial thinking, which has resurfaced in one form or another in many human cultural expressions

[18] Hung (2007, 191).
[19] See, for example, Gohary (1992, 9–25).
[20] Trivedi (1989).

throughout history. In an example of situated cognition, this tendency for spatialization has manifested itself differently in oral cultures, early civilizations, and the modern world, since the cognitive demands and the technologies for externalization developed to meet them have varied.

Quipu: Inca Infographics

Oral cultures have also used a range of portable objects as aids for memory recollection. Native American groups used items such as winter counts (for example of Buffalo hide), song boards, and birchbark scrolls. In the Luba Kingdom in the Kongo, "men of memory" used *lukasa*, objects adorned with pearls and shells in order to recite genealogies, king lists, migration stories, political practices, etiquette, hunting, smelting, blacksmithing, and other technologies.[21]

However, the most commonly known of these objects are the quipus (Fig. 3.2), which were made from knotted strings. In fact, the quipus are often described as unique, although they are but one example of a class of

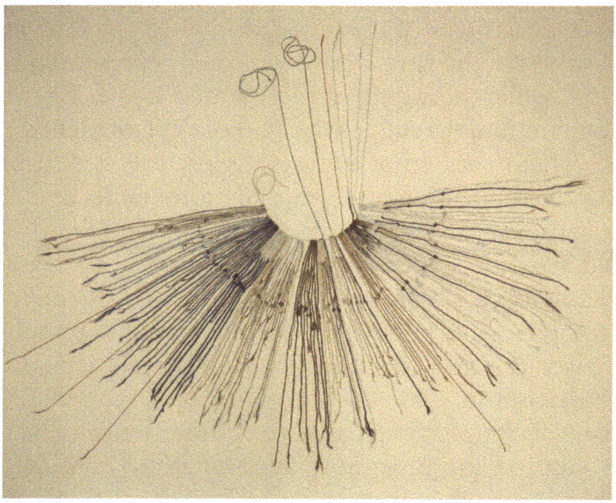

Fig. 3.2 Quipu from Inca period Peru, fifteenth to sixteenth centuries. To the right, brown cords are branching from white cords, forming a tree structure. Metropolitan museum of arts, accession number 1940.469

[21] Kelly (2015, 75–80).

objects found all over the world. One reason for the quipus' fame is perhaps that they cannot be mistaken for stylized or abstract art, as many of the other objects. Their system of memory storage was also more formalized and complex than in most other cases, bordering on a written language.

Although the Andean Inca quipus are the most well known, record keeping with knotted strings is known from many other cultures. Relatively recent ethnological observations describe the use of quipus on Polynesian islands, for example, Hawaii and the Marquesas. Ancient Chinese and Tibetan texts also describe a time before writing was invented when record keeping was conducted with knotted strings—in Taoist texts, quipus are associated with the simple virtues of the old days. The use cases were the same as among the Inca: accounting, taxation, genealogy, and history writing.[22] On Okinawa and the surrounding islands, *warazan*, a similar system of straws instead of strings was used for calculations and taxation. They disappeared in the early 1900s, when Meiji period educational reforms made literacy widespread.[23] The use of knotted cords as stores of information has also been recorded among the Salish in the American Pacific Northwest, various Nigerian tribes, the Li of Hainan, the Sonthals of Bengal, and ancient Semitic peoples.[24]

Although we know that quipus have been in use in many places around the world, actual preserved specimens are rare outside the Inca Empire, where the dry conditions favor their preservation. Woolen strings perish easily, and we have evidence of their existence only where conditions for preservation are good, where they have been used so recently that we have anthropological records or collections, or where clues are found in ancient texts written when their use was still remembered. Perhaps the tales of the Moirai and Norns, who spun the life threads of mortals in Greek and Norse mythology, fall back on memories of quipu-like record keeping in the western parts of Eurasia.

The quipu can be understood as a diagram outlined with the help of strings, instead of drawn with pen on paper or on the computer screen. The backbone of the Inca quipu is called the primary cord, and attached to it is a number of pendant cords. The pendant cords might in turn have subsidiary cords attached to them, which might themselves carry

[22] Jacobsen (1983).
[23] Rosa (2016).
[24] Kelly (2012, 120).

subsidiaries. Up to six levels of subsidiaries have been found in quipus.[25] Thus, the quipus have the branching structure of a tree diagram.

Around two-thirds of quipus can be interpreted as numerical. The Incas used a decimal system, where different kinds of knots represented numbers from 1 to 9 and the place value of the number was decided by distance to the primary cord. However, one-third of quipus follow some other not yet readable system. These non-decimal or narrative quipus are supposed to have been used for recollection of stories and songs.[26]

While some aspects of the quipus are well understood, such as the decimal numerical system, others are shrouded in secrecy, since the art of reading quipus has disappeared. However, a similar system was still used in a rebellion in Chile in 1792.[27] Locally, quipus survived even longer for some specialized use cases. In 1897, the archaeologist Max Uhle found that quipus were still used in the interior of Peru and Bolivia, for example, for accounting herds of sheep. Female animals were always listed on the border of the quipu, while the males were found in the center. Secondary cords attached to the main cords denoted numbers that should be subtracted from the total, such as sheep eaten by the shepherd.[28]

Although quipus, especially those with several levels of subsidiary cords, are reminiscent of tree diagrams, the uncertainty around their use makes it difficult to evaluate if and how they functioned as diagrams. Gary Urton believe that quipus should not be understood in isolation. For example, he suggests that seven quipus found together in a burial are part of a three-tiered hierarchical system of accounting. In another example, Urton proposes that a pair of quipus might represent the lower and upper districts of the capital Cuzco. The main cords of the two quipus should be arranged in a circle, with the pendant cords directed outward, forming a wedge diagram. Supposedly, the numbers on the cords represent census data from Cuzco, where districts were arranged according to their alignments (*ceques*) from the main temple in the city center.[29] The *ceques* were a system of roadways radiating out from Cuzco, and they had similar functions as ritual roads connected to cultural knowledge as the dreaming tracks

[25] Urton (2017, 46).
[26] Urton (2017, 47–49).
[27] Nadaillac and Dall (2005, 458).
[28] Benson (1975, 19–22).
[29] Urton (2017, 91–95, 143–153). For a later wedge diagram, see Fig. 6.6.

Fig. 3.3 Proto-cuneiform tablet with administrative account of barley distribution, 3100–2900 BCE. Image, symbols, or diagram? These categories were not separated until the development of complete writing systems. Metropolitan Museum of Art, Accession Number: 1988.433.1

mentioned in section "Terrain and Architecture: Structuring Knowledge in "Oral" Cultures and Early Civilizations".[30]

Frank Salomon has likened quipus to early writing systems that were not connected to the structure of languages, such as proto-cuneiform and proto-elamite, both written on clay tablets (Fig. 3.3). Common to these systems is that they lack syntax and phonetic coding and are used in restricted contexts of application. At least proto-elamite shared a diagrammatic aspect with quipus, in that the horizontal axis represented continuation and the vertical represented summation. Solomon claims that these systems of proto-writing "are more akin to data graphics as Tufte and Bertin conceive them, or to computer 'infographics,' than to its eventual mutant offspring, 'true writing.'" He suggests we should call these systems "data writing," in contrast to later systems of "speech-sound writing."[31]

[30] Kelly (2012, 93, 322).
[31] Salomon (2004, 280–281).

Thus, in proto-writing systems such as quipus, the diagrammatic element initially dominated and symbolism was restricted. However, at least in the case of proto-cuneiform, the symbolic dimensions grew in importance until the medium could reproduce spoken language. In this process, the diagrammatic and symbolic aspects were separated, which paved the way for the development of illustrations that we today unambiguously would identify as diagrams.

Writing, Lists, and Tables

The introduction of spoken language was no doubt an important initiative for the development of human culture. Through language, thoughts, memories, and skills could be passed on from one individual to another, allowing an accumulation of knowledge far surpassing what single individuals could amass in own experience during their lifetimes. However, the invention of writing greatly enforced the information carrying capacity of language. By allowing storage of knowledge, it enabled language to traverse greater distances in space and time. The introduction of written language was facilitated by the human inclination for spatial metaphors of time. In spoken language, sounds form words and sentences that follow each other in time. In written language, this flow of time in the narrative is represented by spatial order: you read from left to right, right to left or from the top down. The sounds are represented by signs, but the order of the sounds is spatially represented. When you reach the end of the row or column, the text continues at the beginning of the next row or column. All writing systems invented by humanity have been laid out in such a grid pattern. Examples of meandering or spiral writing can be found, for example, on rune stones, but those are of limited length. There are also examples of diagonal writing and diagonal tables, for example, in Arab and Persian manuscripts, which seem to have been devised to save space and achieve aesthetic effect. However, even in diagonal tables the actual cells are arranged in a vertical-horizontal grid pattern.[32]

It took time for writing to develop into a system that could express all the nuances of spoken language. Before written language had acquired grammar, the Sumerian city-states used lists to collect and analyze information. According to Merlin Donald, "[l]ist arrangement can

[32] MacMurdie (2022, 338–345); Khoddari Naeeni and Akbari (2019); grid patterns are discussed in section "Printing and Spatial Cognition" in Chap. 5.

facilitate the sorting, summarizing, and classifying of items and can reveal patterns otherwise not discernible."[33]

These classification systems became more advanced with the full development of writing, as it allowed more complex systematization and categorization. One of the earliest systems of writing was the cuneiform, which was developed by the Sumerians, living at the mouth of the Euphrates and Tigris Rivers, by the thirty-first-century BCE. Around 90% of cuneiform inscriptions are administrative texts. Usually they are list-like documents containing inventories, ration lists, receipts, or work rosters. Often, they contained complex columns of subtotals accounting for the intersection of, for example, workers and rations. Columns organized as intersections between categories and subcategories also remained a dominant form of scholarly production for two and a half millennia. A common form of early lists was grammatical texts used for the education of scribes, which contained complex hierarchies of subcategories. Other areas of knowledge, such as astrology, divination, and medicine, were systematized using similar lists.[34]

In the first half of the 1900s, Assyriologists commonly saw the hierarchical structure of Sumerian lists as an attempt by the scribes to classify the world around them in a search for order. Everything was assigned to a conceptual category represented by a character, and items within each category were ordered in comprehensive lists. A classic example of this "Sumerian list science" is the Standard Professions List from Uruk. It was described as a hierarchical representation of Uruk society, beginning with the king, followed by heads of departments, and everyone else in order. However, more recent scholarship has questioned the search for order in the world as a motivation behind the hierarchical structure of the lists and instead suggests that this structure merely served to facilitate memorization.[35] Spatial organization can no doubt be used both to describe hierarchical order and as a mnemonic aid, and it is entirely possible that the Sumerian lists performed both functions in parallel.

A science of lists and tables appeared in other centers of civilization as well. In China, the Han historian Sima Qian claimed to have invented the chronological table (*bian*) ca. 90 BCE. However, in 1977, bamboo strips with chronological tables that were about a century older were found in the Anhui Province. Each strip of bamboo constituted a column in the

[33] Donald (1991, 288).
[34] Gragg (1995, 19).
[35] Taylor (2013, 295–296).

table.[36] In this case, the strip-shaped medium facilitated the tabular form of representation.

There are competing theories of how the human propensity for classification has come about. For example, in 1903, Émile Durkheim and Marcel Mauss argued that humans have no innate capacity for classification, but the habit of dividing the world into classes grew from the social practice of dividing humans into tribes. However, Rodney Needham pointed out some flaws in their argument, methodology, and use of sources in the introduction to the 1963 English edition.[37] In contrast, Henri Cohen and Claire Lefebvre have claimed that categorization is the foundation for the construction of our knowledge of the world, and thereby the most basic function of cognition.[38] The cognitive neuroscientist Stevan Harnad goes one step further by claiming that cognition is categorization.[39] Today, there is agreement in the cognitive sciences that we have an innate predisposition for categorization, although only some go as far as suggesting that there are also innate categories.[40]

From the perspective of cognitive load, the reason we tend to construct hierarchies seems quite obvious. Already Eleanor Rosch, one of the pioneers in the anthropological study of categorization, suggested that it was a question of cognitive economy: category systems provide maximum information with minimum cognitive effort.[41] Thus, the reason we classify phenomena in the world is our inability to handle too many items of information without arranging them in hierarchical tree structures of classification with a limited number of branches at each node. Lists were only a first attempt to externalize this logical structure of categorization spatially. For the last two millennia, an increasing number of diagrams have been produced for this purpose.

References

Bellezza, Francis S. 1996. Mnemonic methods to enhance storage and retrieval. In *Memory: Handbook of perception and cognition*, ed. Robert A. Bjork and Elisabet Ligon Bjork, 345–380. San Diego: Academic Press.

[36] Vankeerberghen (2007).
[37] Needham (1963).
[38] Cohen and Lefebvre (2005).
[39] Harnad (2005).
[40] Cohen and Lefebvre (2005).
[41] Rosch (1978).

Benson, Elizabeth P. 1975. The quipu: 'Written' texts in ancient Peru. *The Princeton University Library Chronicle* 37 (1): 11–23.
Berger, Carol. 1981. The hand and the art of memory. *Musica disciplina* 35:87–120.
Bickerton, Derek, and Eörs Szathmáry. 2011. Confrontational scavenging as a possible source for language and cooperation. *BMC Evolutionary Biology* 11 (1): 261.
Bradley, John J. 2008. When a stone tool is a dingo: Country and relatedness in Australian Aboriginal notions of landscape. In *Handbook of landscape archaeology*, ed. Bruno David and Julian Thomas, 633–637. Walnut Creek, CA: Left Coast Press.
Bray, Francesca, Vera Dorrofeeva-Lichtmann, and Georges Métailié, eds. 2007. *Graphics and text in the production of technical knowledge in China: The Warp and the Weft*. Leiden: Brill.
Châtelet, Gilles. 2000. *Figuring space: Philosophy, mathematics, and physics*. Dordrecht: Kluwer.
Cohen, Henri, and Claire Lefebvre. 2005. Bridging the category divide. In *Handbook of categorization in cognitive science*, ed. Henri Cohen and Claire Lefebvre, 2–15. Amsterdam: Elsevier.
Cooperrider, Kensy, Dedre Gentner, and Susan Goldin-Meadow. 2016. Spatial analogies pervade complex relational reasoning: Evidence from spontaneous gestures. *Cognitive Research: Principles and Implications* 1 (1): 1–17.
Donald, Merlin. 1991. *Origins of the modern mind: Three stages in the evolution of culture and cognition*. Cambridge, MA: Harvard University Press.
Giardino, Valeria. 2016. Behind the diagrams: Cognitive issues and open problems. In *Thinking with diagrams: The semiotic basis of human cognition*, ed. Sybille Krämer and Christina Ljungberg, 77–101. Boston: De Gruyter.
Gohary, Jocelyn. 1992. *Akhenaten's Sed-festival at Karnak*. London: Kegan Paul.
Gombrich, Ernst. 1990. Pictorial instructions. In *Images and understanding: Thoughts about images. Ideas about understanding*, ed. Horace Barlow, Colin Blakemore, and Miranda Weston-Smith, 26–45. Cambridge: Cambridge University Press.
Gragg, Gene B. 1995. Babylonian grammatical texts. In *Concise history of the language sciences from the Sumerians to the cognitivists*, ed. Ernst Frideryk Konrad Koerner and Ronald E. Asher, 19–21. Oxford: Pergamon.
Harnad, Stevan. 2005. To cognize is to categorize: Cognition is categorization. In *Handbook of categorization in cognitive science*, ed. Henri Cohen and Claire Lefebvre, 19–43. Amsterdam: Elsevier.
Heeffer, Albrecht. 2010. The body in Renaissance arithmetic: From mnemonics to embodied cognition. In *AISB 2010 convention*, 21–22. The Society for the Study of Artificial Intelligence and the Simulation of Behaviour.

Hung, Wu. 2007. Picturing or diagramming the universe. In *Graphics and text in the production of technical knowledge in China: The warp and the weft*, ed. Francesca Bray, Vera Dorrofeeva-Lichtmann, and Georges Métailié, 191–214. Leiden: Brill.

Jacobsen, Lyle E. 1983. Use of knotted string accounting records in old Hawaii and ancient China. *The Accounting Historians Journal* 10 (2): 53–61.

Kelly, Lynne. 2012. *When knowledge was power*. Dissertation, LaTrobe University. https://opal.latrobe.edu.au/articles/thesis/When_knowledge_was_power/21858909.

Kelly, Lynne. 2015. *Knowledge and power in prehistoric societies: Orality, memory and the transmission of culture*. Cambridge: Cambridge University Press.

Khoddari Naeeni, Saeed, and Elham Akbari. 2019. A study on diagonal writing in Islamic manuscript layout from 11th–15th centuries. *Honar-Ha-Ye-Ziba: Honar-Ha-Ye-Tajassomi* 24 (3): 65–76.

MacMurdie, Meekyung. 2022. Proven recipes: Geometry and the art of Arabic medicine. In *The visualization of knowledge in medieval and early Europe*, ed. Marcia Kupfer, Adam S. Cohen, and J. H. Chajes, 331–358. Turnhout: Brepolis.

Mithen, Steven. 2024. *The language puzzle: How we talked our way out of the stone age*. London: Profile books.

Nadaillac, Jean-François-Albert du Pouget, and W. H. Dall. 2005. *Prehistoric America*. Tuscaloosa: University of Alabama Press.

Needham, Rodney. 1963. Introduction. In *Primitive classification,* ed. Émile Durkheim and Marcel Mauss, vii–xliii. Chicago: University of Chicago Press.

Rosa, Mark. 2016. *Native writing systems in the Okinawan islands.* Dissertation. Tokyo: University of Tokyo. https://www.l.u-tokyo.ac.jp/postgraduate/database/2015/165.html.

Rosch, Eleanor. 1978. Principles of categorization. In *Cognition and categorization*, ed. Eleanor Rosch and Barbara B. Lloyd, 27–48. Hillsdale: Lawrence Erlbaum.

Rubin, David C. 1995. *Memory in oral traditions: The cognitive psychology of epic, ballads, and counting-out rhymes.* Oxford: Oxford University Press.

Salomon, Frank. 2004. *The cord keepers: Khipus and cultural life in a Peruvian village*. Durham, NC: Duke University Press.

Taylor, Jon. 2013. Administrators and scholars: The first scribes. In *The Sumerian world*, ed. Harriet Crawford, 290–304. Abingdon: Routledge.

Urton, Gary. 2017. *Inka history in knots: Reading khipus as primary sources.* Austin: University of Texas Press.

Vankeerberghen, Griet. 2007. The tables (*biao*) in Sima Qian's *shiji:* Rhetoric and remembrance. In *Graphics and text in the production of technical knowledge in China: The warp and the weft*, ed. Francesca Bray, Vera Dorrofeeva-Lichtmann, and Georges Métailié, 295–311. Leiden: Brill.

Von Petzinger, Genevieve. 2017. *The first signs: Unlocking the mysteries of the world's oldest symbols*. New York: Atria.
Woelert, Peter. 2011. Human cognition, space, and the sedimentation of meaning. *Phenomenology and the Cognitive Sciences* 10 (1): 113–137.
Wynn, Thomas Grant, and Frederick L. Coolidge. 2022. *An introduction to evolutionary cognitive archaeology*. New York: Routledge.

CHAPTER 4

Diagrams for the Structuring and Classification of Knowledge

For most of human history, knowledge of the world has been collected by direct observations through the senses. Few exact measurement techniques were developed until the scientific revolution in the seventeenth to eighteenth centuries. Thus, the older world was largely devoid of "data" in the modern sense. There are exceptions that confirm the rule, primarily the field of astronomy, where data was collected from an early age. The planets' movements relative to the constellations are slow enough to be tracked with an almanac, while their position could be fixed quite exactly with simple instruments for angular measurement and knowledge of geometry. This bias toward astronomy in early data collection might help explain why ancient geometry seems so advanced compared to the state of the sciences and mathematics in general at the time—and the advanced state of geometry made it an ideal for other intellectual fields.[1]

Beside exceptions such as astronomy, observations and ideas were not expressed in data and numbers but in language, and language became the object of logical analysis.[2] It is mainly in this logical analysis of language and texts that diagrams played a role in the premodern world. Aristotle drew parallels between analyzing arguments and diagrams in his writings.[3]

[1] For ancient astronomy and geometry, see Hoskin (1997).
[2] See, for example, the discussion in Gontier (2011).
[3] Byrne (1997).

J. Holmén, *A History of Diagrams*, Palgrave Studies in the History of Science and Technology,
https://doi.org/10.1007/978-3-031-99691-7_4

The relationships between logic and spatial diagrammatic thinking, and the extent to which spatiality has influenced the language for discussing logic, were well expressed by Walter Ong (1912–2003):

> Description of mental activity in terms of "composition" (putting together), or "implication" (folding up or falling back upon), or "definition" (setting bounds or limits), or "division," or even "description" (drawing a line around or sketching) all exploit, more or less evidently, an analogy between the field of intellectual activity and a field which involves local motion and is sensorily apprehended in terms of sight. Formal logic is fundamentally an attempt to deal with the activity of the mind in terms of these and related visualist analogies. It is interested in the "structure" of our intellectual activity – a notion which cannot even be conceived except by analogy with some sort of spatial diagram.[4]

For millennia, the overwhelming majority of diagrams were constructed in order to illustrate such logical operations, relations, and distinctions. As the deductive methods of geometry became an ideal for philosophy in general, the diagrams of Euclidean geometry became models for diagrams in all fields. They remained so until the 1700s, when revolutions in data collection together with a transformation of geometry heralded a new age in the history of diagrams.

This chapter of the book begins with discussing the emergence of the earliest diagrams in different cultural areas, and the philosophical contexts in which they were situated (section "Ancient Diagrams and Their Philosophical Foundations"). Thereafter follows presentations of some common diagrammatic forms, such as the square of opposition (section "The Square of Opposition and the Wheel of Reasons") and, arguably the most common of all, different forms of tree diagrams (section "Tree Diagrams"). Section "The Scholastic Diagram Habitus" is dedicated to the explosion of new diagrams that took place in twelfth-century Europe. It discusses the causes of this development and the question of whether a common scholastic habitus of subdivisions permeated the texts, diagrams, and architecture of the age. Section "Similarities and Differences Between Scholastic, Byzantine, Arabic, and Chinese Tree Diagrams" compares tree diagrams in Western Europe, Byzantium, the Arab world, and China in an attempt to understand why they become more common in some civilizations than in others. Section "Genealogical Trees and Diagrammatic

[4] Ong (1958, 107).

Heraldry" explores the diagrammatic principles of genealogy and heraldry. Thereafter, section "Llullism: Incomprehensible Originality" discusses why the advanced diagrammatic system of Raymond Llull, despite revisions, ultimately failed to become widely adopted. This chapter concludes with section "Circles of Inclusion", which analyzes diagrams illustrating intersections, such as Venn and Euler diagrams, and their predecessors that go back several millennia.

Ancient Diagrams and Their Philosophical Foundations

Many of the diagrammatic forms described in this volume are found in similar iterations in several cultural areas. In many respects, Europe, Byzantium, and the Arab and wider Islamic world are to be considered one area, since they to a large extent produced similar diagrams that were used to illustrate, comment on, and develop the same Greek heritage.[5] Eastern civilizations, China and India, constitute a separate case through their relative isolation from the West. Based on similarities and differences between these cultural areas, it is at least possible to attempt a discussion of the universal contra the culture specific in the processes of spatial cognition behind common diagram types. Such a discussion is most relevant for the oldest historical periods, since cultural interchange between these areas increased with time, and in modern history they can no longer be considered separate cases.

The first diagrams in the sense intended in this book, spatial representations of non-spatial relations, appear in the first millennium BCE. Diagrams were essential to ancient Greek mathematics and science. It is believed that in the time of Aristotle, lectures were illustrated by diagrams drawn on *leukoma*, literally "whiteboard." The leukoma were probably painted wooden boards that, in contrast to modern whiteboards, were not easily erasable.[6] Ample references to diagrams by Aristotle, for example, reveal that they were in use. Aristotle apparently used lettered diagrams, as he refers to the letters in his texts.[7]

[5] However, there are also some important differences between them, discussed in section "Similarities and Differences Between Scholastic, Byzantine, Arabic, and Chinese Tree Diagrams".

[6] Netz (1999, 16).

[7] Natali and Hutchinson (2013, 113–116).

Few diagrams from ancient Greece have survived until today. It has generally been believed that the copyists who transferred the ancient heritage to medieval codices focused on the texts, and ignored or distorted the diagrams. However, the diagrams on surviving papyri fragments of, for example, Euclid are very similar to those found in early medieval codices. This has led David Creese to claim that diagrams were not more fragile to hand copying then the texts themselves, and that medieval codices might therefore give us an insight into the world of ancient diagrams.[8] Although this is certainly possible, Euclidean "texts" might be an exception in this regard, since in them the diagrams were at the center of the presentation while the text took second stage. It is possible that copying of diagrams in papyri on other subjects, where diagrams were mere illustrations to the texts, was less conscientiously executed.

In Chinese, the character *tu* (圖) refers to technical diagrams, distinct from pictures and paintings (*hua*) and technical texts (*shu*). Examples of *tu* are maps, anatomical charts, illustrations of edible plants in "famine manuals," geometrical diagrams, and tree diagrams. The character *tu* has been found on bronze plates from 1000–800 BCE, when it seems to have had the connotations of "to order" and "to position in space." The earliest example of an illustration captioned *tu* is from 310 BCE.[9] Also in China, few really old diagrams are known, in part because of the destruction of older documents by the emperor Qin Shi Huang in 213 BCE.

Although diagrams were known in all major ancient cultural areas, there are differences in their purposes, which have their roots in the role of science and philosophy in these societies. In contrast to the originators of Western logic, Plato and Aristotle, the roughly contemporary Mohist Chinese logicians during the warring states period (ca. 479–221 BCE) did not try to establish generalized, abstract principles, but worked with the help of examples. Chinese philosophy did not distrust perception, and therefore did not partake in the search for truths beyond the senses that characterized Platonism.[10] This might be linked to the concern of Chinese geometry with practical problems, such as measuring land, and the absence

[8] Creese (2010, 63).

[9] Bray (2007, 1–14).

[10] Willman (2021). For a discussion of the role of proofs and axioms in European, Chinese, and Indian mathematics, see Katz (1998, 328).

of engagement in theoretical proofs from axioms, which was at the core of Greek geometry.[11]

Plato's thoughts about a world of ideas truer than the empirical world were inspired by geometry, as was his insistence that all philosophers needed to be trained in mathematics. His theory of the "Platonic solids" as the building blocks of the elements even put geometrical shapes as the foundation of reality (Fig. 4.1). For Plato, the main benefit of geometry was not its practical application, but that it drew the soul to the truth. This was in line with a change in the nature of mathematics around 600 BCE, which separated Greek civilization from older civilizations in, for example, Egypt and Babylonia. There, mathematics had been a practical tool for the central administration of empires and large kingdoms. The small Greek city-states lacked central administration of comparable size. However, rule of law, and in some cases democracy, forced citizens to learn skills of argument and debate. According to Victor Katz, this atmosphere encouraged the invention of proofs in mathematics.[12] Geoffrey Lloyd has instead convincingly argued that the idea of proofs entered Greek philosophy through Plato's and Aristotle's *criticism* of democracy; by irrefutably demonstrating some important truths, they could place them out of reach of the demagogical rhetoric that characterized Athenian democracy.[13]

Regardless, the influence between philosophy and geometry in ancient Greece undoubtedly went both ways. Advances by early Greek mathematicians were systematized by Euclid, who probably was a generation younger than Aristotle and worked at the library in Alexandria. He depended on mathematical concepts developed by the Athenian philosophers, primarily Aristotle's distinction between number and magnitude.[14] Euclid's geometry textbook *Elementa* was in use for two millennia, and in large part because of its influence, geometry came to be seen as an ideal for other branches of science and philosophy.[15]

Netz has claimed that not only Euclid's diagrams but also the hierarchical structure of his accompanying texts had an important influence on how science was to be given spatial form. Euclid's formulae had a branched

[11] Lay Yong (1994); Siu (1993).

[12] Katz (1998, 47); Emmer (1982).

[13] Lloyd (2014, 6–29).

[14] Katz (1998, 58–59).

[15] Dunér (2019) discusses the influence of this tradition with a focus on the scientific revolution but draws its roots back to Euclid.

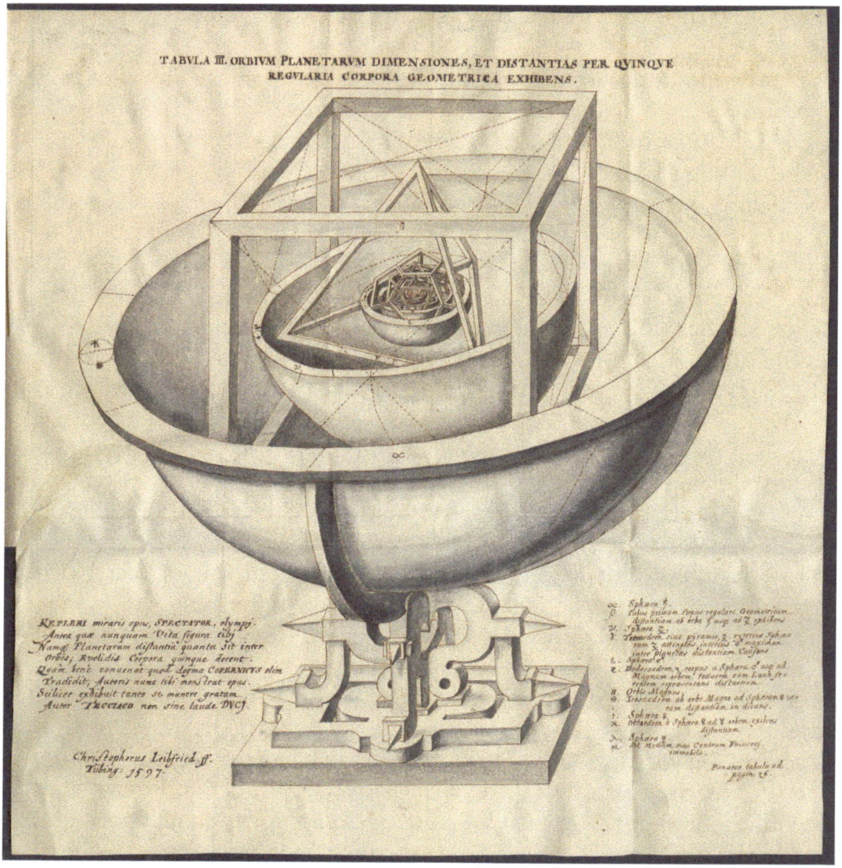

Fig. 4.1 The platonic solids, three-dimensional bodies constructed of uniform two-dimensional surfaces, continued to be of significance in the sciences for millennia. For example, Johannes Kepler believed that they explained the distances between the planetary orbits. Kepler (1597)

structure with usually no more than two or three branches at each node.[16] This was a format that would characterize scientific endeavors for millennia. However, this is not necessarily a result of influence from Euclidean

[16] Netz (1999, 146).

geometry or Greek science. Since this pattern is so common, found literally everywhere in all eras,[17] regardless of Greek influence, it might also simply reflect that Euclid was adapting his presentation to the same cognitive limitations of the audiences that have challenged scholars in all epochs and cultures.

The Square of Opposition and the Wheel of Reasons

The square of opposition (Fig. 4.2) has, at least since the second century CE, been used to illustrate the relationships between the four basic categorical propositions in traditional logic. Although it is generally believed that the diagram itself was first drawn by the Roman Apuleius (124–170 CE), it was based on a development of Aristotle's logic.[18] Aristotle also described a structurally identical diagram in his *Nichomachean Ethics*, although it was used to discuss economic exchange.[19]

The square of opposition became so common and well known in the Middle Ages that the same diagrammatic form was used for other

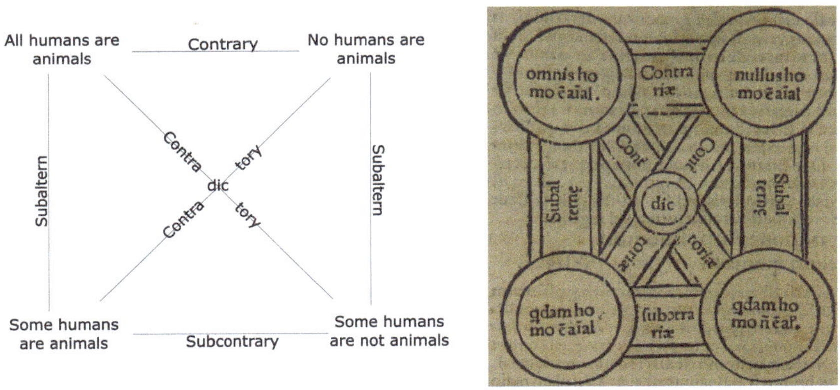

Fig. 4.2 The square of opposition was used to illustrate how the four logical propositions were related to each other. Translation to the left by author, original to the right from Reisch (1503)

[17] See folk taxonomies in section "The Shape of Life" in Chap. 5.

[18] Londey and Johanson (1984).

[19] Section "Diagrams in Economic Theory" in Chap. 6.

purposes. For example, it could describe the four basic intervals of musical theory or, in the ophthalmological handbook by Ali ibn Isa of Mawsil, the consequences of greater and smaller quantity as well as thickness and thinness of fluid in the eye. A common application was the relationships between the four elements and their opposing qualities: dryness, wetness, hotness, and coldness. However, this was as a forced fitting. While the concept of contradictory and perhaps contrary qualities could be meaningfully applied to Aristotelian physics, subaltern and subcontrary relations could not. Still, all these concepts were sometimes borrowed from the logical square of opposition to the square of elemental qualities.[20] This illustrates how attempts to reduce cognitive load by reusing familiar spatial schemas for new purposes might lead to unwanted confusion, if it is done without reflection on the appropriateness of the analogies underlying the diagram.

The connection between these similar diagrams is illustrated by graffiti on a church wall in Bro parish on the island of Gotland in the Baltic Sea, which depicts the square of opposition and a diagram of the four elements side by side. The cuttings in the plaster are probably from the thirteenth century, and the handwriting suggests they were performed by the same person or at least persons with a common education.[21] Apparently produced as teaching aids, these cuttings are examples of Aristotelian philosophy and the associated diagrams being known and used in smaller educational institutions outside the larger university environments.

When used for its original purpose, the square of opposition was no doubt an effective pedagogical tool that simplified comprehension through its spatial layout. However, attempts were made to visualize logic relationships that required up to 12 nodes to be connected to each other, and they were far from simple and accessible (Fig. 4.3).[22] The cognitive load that such diagrams exerted far exceeded the limitations for what humans can comfortably handle.[23]

A diagram similar to the square of opposition, although in three dimensions, was Isidorus of Seville's *Cybus Elementorum*. In this cube, the opposites blunt–sharp, mobile–immobile, and thin–thick constituted opposing

[20] J.E. Murdoch (1984, 67–71, 353); Verboon (2010, 143–149).
[21] O'Meadhra (2012).
[22] J.E. Murdoch (1984, 64–65).
[23] Similar diagrams that cause a cognitive overload are discussed in section "Labyrinthic Chartjunk: Implication Through Cognitive Overload" in Chap. 9.

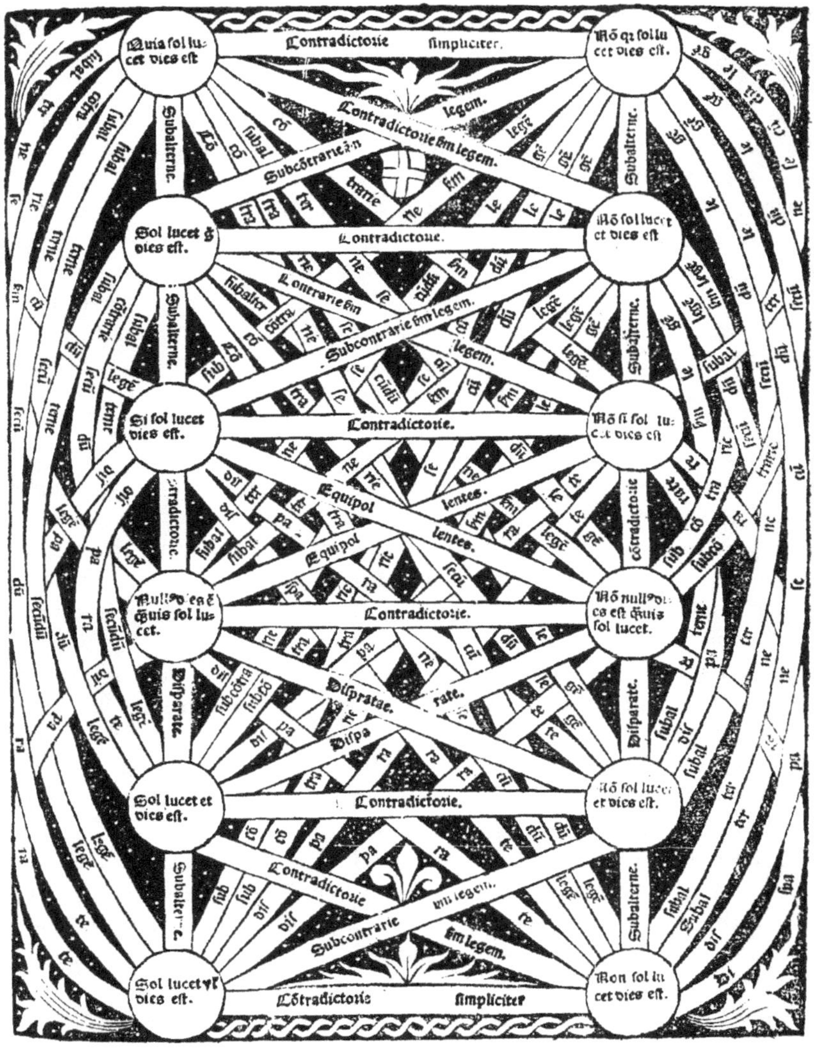

Fig. 4.3 A complicated diagram of logical relationships that exceeds the cognitive load that humans can comfortably handle. Celaya (1525)

sides, while the four elements were placed on four of the points. However, the comprehensibility of this relatively simple model was distorted when translated into drawn diagrams, since the illustrators had difficulty representing three dimensions on a codex page.[24]

Around the time the oldest remaining copies of the square of opposition were drawn in Europe, a structurally similar diagram appeared in Indian Buddhist logic, The Wheel of Reasons (Hetucakra). The text describing the Hetucakra was first written by Dignaga (ca. 480–540 CE), although the only preserved complete version is a much later Tibetan print from the eighteenth century. That text is an unillustrated 12-verse poem with a complicated translation history, which has made it open to different interpretations.[25] However, we know that Dignaga was concerned with the rules for sound inference. In a square-shaped table with three cells by three cells, he calculated all possible combinations of similar (*sapakṣa*) and dissimilar (*vipakṣa*) instances.[26]

Dignaga was trying to find out under which circumstances it was possible to determine whether an individual was in possession of a certain property by observing that the individual possessed another property.[27] Dignaga's own examples were quite complicated, comparing the qualities of clay pots, sound, and the ether, but Fig. 4.4 contains a modern simplified exemplification of the Hetucakra three-by-three table. It outlines under which circumstances it is possible to infer that a tea is Chinese by tasting that it is bitter, provided that Indian and Chinese are the only types of tea in the world. This is possible in the top middle location, if all Chinese and no Indian teas are bitter, and in the bottom middle, where some Chinese and no Indian teas are bitter. However, if all Chinese and all Indian teas are bitter, such as in the top left, or some from each country, such as in the bottom right, no inference can be made about whether a tea is Chinese from the observation that it tastes bitter.

Later scholars criticized Dignaga's argument that sound inference had to be based on both positive and negative supporting examples. Therefore, Uddyotakara (sixth century) expanded the wheel into a four-by-four

[24] J.E. Murdoch (1984, 280–281).

[25] He and van der Kuijp (2016, 268).

[26] A more detailed description is found in Stcherbatsky (1984, 320–324). A slightly different interpretation is found in Ganeri (2004, 344–350).

[27] Hayes (1988, 114).

1: + sapakṣa, + vipakṣa **All Chinese tea is bitter** **All Indian tea is bitter**	**2: + sapakṣa, – vipakṣa** **All Chinese tea is bitter** **No Indian tea is bitter**	**3: + sapakṣa, ± vipakṣa** **All Chinese tea is bitter** **Some Indian tea is bitter**
4: – sapakṣa, + vipakṣa **No Chinese tea is bitter** **All Indian tea is bitter**	**5: – sapakṣa, – vipakṣa** **No Chinese tea is bitter** **No Indian tea is bitter**	**6: – sapakṣa, ± vipakṣa** **No Chinese tea is bitter** **Some Indian tea is bitter**
7: ± sapakṣa, + vipakṣa **Some Chinese tea is bitter** **All Indian tea is bitter**	**8: ± sapakṣa, – vipakṣa** **Some Chinese tea is bitter** **No Indian tea is bitter**	**9: ± sapakṣa, ± vipakṣa** **Some Chinese tea is bitter** **Some Indian tea is bitter**

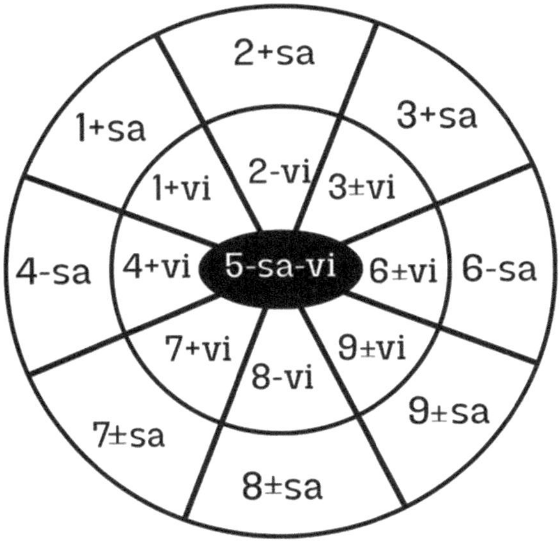

Fig. 4.4 An example of the Hetucakra applied to under which circumstances it is possible to infer that a tea is Chinese based on the observation that it is bitter (if all tea is either Chinese or Indian) (+ = all, ± = some, – = none). Modification of Jha (2020, 474). The diagram below is adapted from circular interpretation of the Hetucakra by He and van der Kuijp (2016). Drawn by author and Jacob Holmén

square grid that also included what he called universally negative and universally positive inferences.[28]

That the Hetucakra was called a wheel (cakra) although it is square shaped was probably influenced by many prominent Indian diagrams

[28] Ganeri (2004, 349–350).

being round, such as the Wheel of Life or Wheel of Suffering (Bhavachakra). Earlier research on the Hetucakra has pointed out that cakra might also be translated as cycle, group, set, or multitude. However, He and van der Kuijp have, by carefully comparing different Tibetan translations, presented an interpretation of the Hetucakra that is similar to the eight-spoked Dharmachakra wheel of Buddhism. In that model, the *vipaksa* and *sapaksa* are placed in separate concentric circles, and combinations of them can be produced by rotating the wheels, as in Raymond Llull's volvelles.[29]

That the most important Indian diagram was concerned with inference from sense perception reflects a difference between Eastern and Western philosophical traditions. The professor of Buddhist philosophy Richard P. Hayes has made a distinction between the epistemology of the Western monotheistic religions, inspired by Platonism, and the Eastern religions emphasizing the search for Nirvana. While the Western tradition searched for truths behind the senses, the Eastern tradition viewed such a search as a cause of suffering that distracted from Nirvana. Hayes argues that the Buddhist tradition recognized the limitations of humanity's ability to understand the world, and likens the Nirvana concept to the philosophy of the Greek skeptics.[30] Thus, while the Western tradition compiled its knowledge into diagrams that was supposed to reflect a truer level of reality, the Eastern tradition instead contemplated the limitations to our possibility of reaching any sure knowledge. For this purpose, they could use Dignaga's Hetucakra, which illustrated how difficult correct inference was. In the words of Hayes, for Dignaga logic was not "intended to serve as a means of adding to our knowledge but rather as a means of subtracting from our opinions."[31]

However, despite these philosophical differences, the pedagogical trajectories of the two dominating logical diagrams in these traditions were still similar. As the square of opposition was created by later scholars to illustrate principles from Aristotle's philosophy, the Wheel of Reasons was invented to teach principles earlier developed by Vasubandhu (fourth to fifth century CE).[32] These examples illustrate that diagrams are often developed as pedagogical tools for wider dissemination of ideas that are

[29] He and van der Kuijp (2016). For Lull, see section "Llullism: Incomprehensible Originality".

[30] Hayes (1988, 32–35).

[31] Hayes (1988, 35).

[32] Anacker (2005, 34).

already known inside an intellectual elite. The need to explain ideas for a broader audience encourages the use of diagrams, since space is a frame of reference that the initiated and uninitiated have in common.

Tree Diagrams

A common diagram type is a line or a similar connecting element that is successively subdivided into more and more lines. Because of this branching structure, in the Arab and European Middle Ages such diagrams were called tree diagrams, *tashjir* or *arbores*. In Europe they were often, but far from always, portrayed as organic trees with leaves, roots, and so forth. However, similar diagrams have also been used in other cultural spheres, and earlier in European history but under different names. All diagrams that display a branching structure will in the following be referred to as tree diagrams, regardless of whether they are portrayed as physical trees or not, or whether they are labeled tree diagrams.

Umberto Eco differentiates between tree diagrams concerned with definitions, such as logical diagrams, and those concerned with classification, such as the trees of knowledge that have been used to present the branches of learning. Eco calls these two types dictionaries and encyclopedias.[33] The dictionaries are developed from general principles through a deductive process, while the encyclopedias are formed through induction by a classification of observations. Eco's categorization is not complete, since it does not describe all tree diagrams. For example, in probability theory trees are used as aids in calculations.[34] However, the early history of tree diagrams is firmly rooted in definitions and classification, and it is trees for those purposes that will be the focus of the following chapters.

Jean Laponce has argued that the tree structure, as the matrix, is a powerful tool for ordering and generating phenomena and ideas, but that the tree has the added advantage of being easy to ascend or descend toward greater or lesser generality. Based on Miller's suggestion that human cognitive abilities decline sharply beyond seven items, Laponce claimed that it is difficult for the human mind to handle more than a limited number of categories without grouping them into local clusters. Classifications into four categories are more common than five, while three categories are even more abundant and dichotomies are the most prevalent of all. That

[33] Eco (2014, 3–33).
[34] S. Evans (2008).

we by preference limit ourselves to grouping information into clusters of two or three is confirmed by later research in the footsteps of Miller.[35] An early example of a dichotomous tree diagram was the Pythagorean Ypsilon (Y), where the two arms represented the choice between good and evil. This spatial model of a fork in the road as an analogy for moral choice was later adopted by the early Christian church.[36]

Plato's method of defining a concept through diaresis, a series of dichotomous distinctions, was an inspiration for ancient Greek and Roman classificatory systems, for example, in zoology, botany, and medicine. No tree diagrams from this period are preserved. However, although we do not know whether Plato illustrated his texts or used diagrams at his Academia, Sybille Krämer has claimed that Plato's texts are still examples of diagrammatic reasoning.[37] Plato's disciple Aristotle used the concepts *genera* and *species* to describe the different levels of a classification. The terms were relative, as species became genera on the next level where they were in turn subdivided into species.[38]

A common form of tree diagram is the evolutionary chart. There are two schools of thought regarding its epistemological status: the first claims that tree thinking has developed into a necessary form of "biological literacy" and the second that the tree of life reflects a natural reality, independent of culture.[39] These constructivist and realist positions can be applied to tree diagrams in general; are they a mere convention of scholarly presentation, or do they reflect an underlying reality in the world? However, a third alternative is that the tree structure is a mental scaffolding enforced upon us by our cognitive limitations. We construct tree diagrams not because they perfectly reflect nature, but because successive divisions into a manageable number of branches are a representation of reality simple enough to be grasped by our limited consciousness. Only through a series of divisions and subdivisions can we fathom such diversity as displayed by, for example, biological species, languages, and branches of knowledge. Although not perfect, such simplified systems of classification are often sufficiently adequate models of reality to prove advantageous; that is why the propensity to construct them have developed and survived through the evolutionary history of the brain.

[35] Laponce (1981, 14–21); Miller (1956); Gobet and Clarkson (2004).
[36] M. Evans (1980); Couzin (2021, 179–191).
[37] Krämer (2016).
[38] Granger (1980, 37).
[39] See discussion on this in Hellström (2019, 18).

However, although categorizations through divisions and subdivisions might be a universal trait of human cognition, that does not mean that tree diagrams are used in all times and places. In many cultures we do not find them at all, and in others they are extremely common. In order to explain this, we need to investigate the historical context in which the diagrams have been constructed. The European Middle Ages was a period well endowed with tree diagrams.[40] By the thirteenth century, tree diagrams could be applied to trees of Love, Virtues, Vices, Sciences, Life, Knowledge, Wisdom, Genealogy, Jesse, the Ten Commandments, the Ages of Man, Affinity and Consanguinity, and Heresy.[41] Probably the oldest and also one of the most common was the tree of Porphyry, which was used to illustrate Porphyry's (ca. 234–305) introduction, *Isagogue,* to Aristotle's logic. It was associated with the old logic known in Europe even before the rediscovery of additional Aristotelian texts in the twelfth century, and every medieval scholar was familiar with it, since it formed an important part of the instruction in logic at the faculties of art, that were the basis of further studies at the European universities that appeared around this time.[42]

No early manuscripts of Porphyry's writings contain the tree named after him. Other diagrams were used to illustrate the same principle, but they were not branched but resembled a descending ladder, as did the actual description in Porphyry's text. The earliest diagrams of a tree of Porphyry, although the name was not yet used, are from a ninth-century manuscript of Boetius's (ca. 480–524) commentary on the *Isagogue*. Since the manuscript where the tree is found might be the oldest known copy of Boethius's work, Annemieke Verboon suggests that Boetius originated the tree of Porphyry.[43]

In the 600s, Pophyry's *Isagogue* was translated from Greek to Syriac, and contained *pulaga*, a Syriac version of what would later be known in Europe as the tree of Porphyry.[44] In the thirteenth century, Peter of Spain[45] wrote an introduction to logic for beginners, *Tractatus* or

[40] See, for example, Klapisch-Zuber (2000). Although focused on genealogical trees, it gives an overview of other types of medieval tree diagrams that contributed to their development.

[41] Verboon (2014, 107–1080).

[42] Verboon (2010, 37).

[43] Verboon (2010, 46–50).

[44] King (2013); Vat. Syr. 158.

[45] His exact identity is disputed, but he is usually identified with Pope John XXI (ca. 1215–1277), and stemmed from today's Portugal rather than Spain, as the Latin *Hispania* encompassed the whole peninsula.

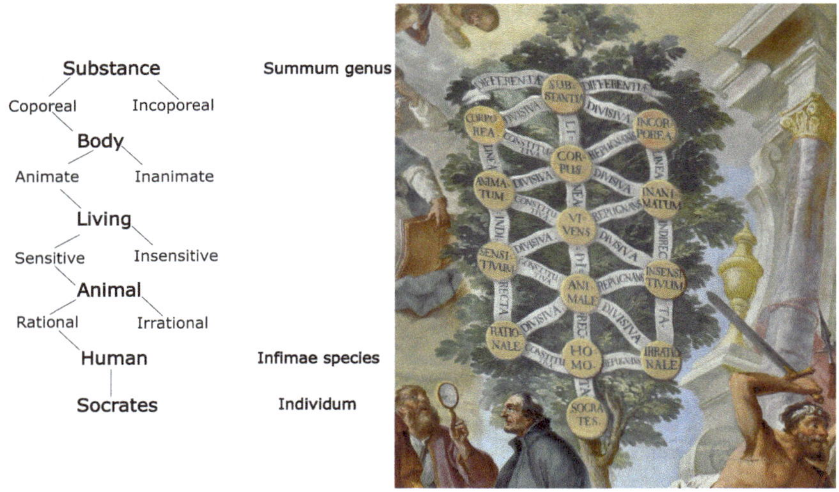

Fig. 4.5 Tree of Porphyry. Logically, only one side of the diagram should be developed (left), but illustrators and artists often preferred more symmetrical shapes (right). This illustration is decorated as a tree, but has lost its tree structure. Left: Drawn by author. Right: Fresco by Franz Georg Hermann (1692–1768), Library of Schussenried monastery. Photo: Andreas Praefcke

Summulae logicae, which avoided difficult philosophical questions and complicated terminology. In the name of pedagogical simplification, it also included a diagram of how genera were divided into species. For the first time, this diagram was now referred to as the tree of Porphyry. In the thirteenth century, it became common to ornament the diagrams, which had until then been simple and stylized, into physical trees with leaves and branches.[46] Verboon has pointed out that the adoption of this tree metaphor could lead the logical analogy astray. For example, medieval illustrators aimed for symmetry in their diagrams, while from a logical standpoint only one side of the diagram should be developed. In the early modern period, symmetry was often combined with logical consistency by writing "repugnans" (contradictory) on the connectors from the side that was not developed, such as in Fig. 4.5.[47]

[46] Hacking (2007, 242–245); Verboon (2010, 59–63).

[47] Verboon (2014, 96–105). For symmetry see also section "String Theory: Post-Modern Platonism" in Chap. 9.

In the tree of Porphyry, genera were divided into species with the help of differentiators such as inanimate–animate or irrational–rational (Fig. 4.5). Species become genera at the next level. The first genus, that is not the species of any higher-level genera, was called *summum genus*, and the species that was not the genus of any lower-level species was referred to as *infimae species*. Examples of this species were called *individum*. Typically, a total of five to seven hierarchical levels were displayed in the tree of Porphyry.[48] This is similar to the five-level folk taxonomies that cognitive anthropologists have found in traditional classificatory systems all over the world. The upper limit of the levels in these systems is probably related to limitations in short-term memory.[49]

Umberto Eco has claimed that the scholastic tradition at medieval European universities lacked a deeper analysis of metaphors. He linked this to the tree of Porphyry, which suggested that there could be a single objective classification of all things. According to Eco, scholasticism refused to theorize metaphors, since that would have suggested that classifications and trees were mere pragmatic tools. However, Ayelet Even-Ezra has identified several medieval horizontal trees that were used for detailed analysis of metaphors. The rich variation of different classificatory trees produced during the period also contradicts Eco's thesis that medieval scholasticism was bound by a notion of one true classification of all things.[50] Even so, why this variation and richness of diagrams appeared in the High Middle Ages is still in need of explanation.

The Scholastic Diagram Habitus

There are several competing theories of why diagrams became more common in Europe from the twelfth century. The most established explanation is that it was connected to a revival of ancient memory culture. Greek and Roman treatises on rhetoric emphasized the importance of memory techniques based on spatial visualization. For example, Cicero claimed that

> the best aid to clearness of memory consists in orderly arrangement. . . . [P]ersons desiring to train this faculty must select localities and form mental images of the facts they wish to remember and store those images in the

[48] Wilkins (2009, 29–32).

[49] For folk taxonomies, see section "The Shape of Life" in Chap. 5.

[50] Even-Ezra (2021, 100–105).

> localities, with the result that the arrangement of the localities will preserve the order of the facts, and the images of the facts will designate the facts themselves, and we shall employ the localities and images respectively as a wax writing tablet and the letters written on it.[51]

This tradition, called method of loci, was continued during the early Middle Ages. The impression of images and visualizations was considered important in order to heighten readers' awareness and enhance their memory process. Occasionally, actual images and diagrams were drawn in codices, but more often a written description of an image would suffice, and they could also be referred to as *picturae*. However, during the twelfth century drawn illustrations become much more common. It seems as if this coincided with attempts to reach less learned audiences, such as novices at cathedral schools.[52]

The Casanatensis manuscript, which contains copies of common medieval diagrammatic illustrations, gives a hint of the relative frequency of different diagram types in medieval Europe. It displays two diagrams in the form of angels, one in the form of a building, one conceived as a man with branches sprouting from his body, some simple linear diagrams like concentric circles or ladders, but no less than 26 tree diagrams. Most of them were used to illustrate ethical questions.[53] Verboon connects this proliferation of tree diagrams to the common use of "arborization" as a mnemonic technique for orators.[54] Inspired by analogies of good and bad trees in the New Testament (Gal. 5:22; Matt. 7:17), tree diagrams were often used to illustrate how vices sprouted out of the root of Superbia while Humilitas gave fruit to virtues. The most popular tree diagram of the Middle Ages was Saint Bonaventure's (1221–1274) *lignum vitae*, tree of life. It illustrated the history of dogma by comparing approved and heretical interpretations of the Bible.[55]

Some of Aristotle's work was known in Europe already in the early Middle Ages through Latin translations made by, for example, Boethius in the late Roman period. However, the bulk of Aristotle's writings were unknown in the West until translated from Arabic and Greek in the second half of the twelfth century. The reading of these rediscovered texts,

[51] Cicero et al. (2014, 467).
[52] Carruthers (2008, 274–275, 291–293, 307–308, 329).
[53] Saxl (1942).
[54] Verboon (2014, 107–108).
[55] Saxl (1942, 108).

collectively referred to as the New Aristotle, became the backbone of teaching at the faculties of Arts at the recently formed universities, such as Paris. To facilitate learning of these texts, a heterogeneous range of manuals and study-aids were created, especially in the thirteenth century. These textbook genres—abbrevationes, compendia, conclusiones, flores, auctoritates, tabule, notule, and more—were particularly well endowed with diagrams.[56]

Diagrams were not only drawn in books. The theologian Peter of Poitiers (ca. 1130–1215) created visual aids for classroom use in the form of charts on parchment scrolls. They displayed biblical genealogy using lines and circles arranged in tree structures.[57] Peter of Poitiers was part of a scholastic environment at the Cathedral School of Paris, where Cicero's and Quintilian's ideas about the use of rhetorical images for mnemonic purposes were translated into visual images.[58] Mary Caruthers has shown that Peter was heavily influenced by Hugh of St. Victor (1096–1141) and his ideas about memory. She also sees Peter of Poitiers's genealogical trees as representative of medieval diagrams intended to aid memory, building on methods used in ancient rhetoric.[59]

Thus, a strong and relatively old research tradition on medieval scholarly diagrams has emphasized their function as mnemonic aids. According to this tradition, books were rare in an overwhelmingly oral culture, and the main purpose of the text was to help the reader memorize its content. Illustrations helped by personifying abstract concepts, or by the spatial subdivision of the content in diagrams. These practices were connected to ancient memory techniques, such as the method of loci.[60]

However, Ayelet Even-Ezra has pointed out that despite the emphasis on mnemonic techniques in earlier research it is still unclear to what extent they were actually practiced. According to her, the long tradition of memory research in medieval studies has resulted in a tendency to interpret all diagrams as mnemonic devices, which has obscured other uses. Although some simple tree diagrams would have been useful for memorization, the more complex ones were not. She is of the opinion that in many cases, diagrams were constructed in order to facilitate learning and

[56] Verboon (2010, 2, 17–18).

[57] Rosenberg and Grafton (2010, 31–32, 35). See also Fig. 4.10.

[58] Murat (2015).

[59] Carruthers (2008, 328–329).

[60] Pioneering works in this tradition were Yates (1966) 2014and Carruthers (1990) 2008.

comprehension, either by the author or by the reader as marginal notations. Out of a sample of 71 manuscripts on Aristotelian logic, Even-Ezra found that 52 contained horizontal tree diagrams in the margins.[61] Indeed, one purpose of the diagrams might have been to unload the memory of facts in order to achieve a better comprehension through externalization:

> [E]xternalization operates in a twofold manner: it takes on the load of short-term and long-term memory; and it encourages selective attention by separating information and specific operations from the chaos in our minds.[62]

This externalization was of course dependent on the availability of manuscripts in which you could make marginal annotations. Contrary to proponents of mnemonic interpretations of medieval diagrams, which has emphasized the scarcity of books, Even-Ezra claims that the growing demand for manuscripts as well as the beginning of commercial production and standardization allowed more people to own copies and hold them in front of their eyes.[63]

Whether the purpose was memorization, externalization, or both, the tree with divisions and subdivisions was a common scholastic diagrammatic structure. Parallels have been drawn to the figures adorning medieval churches, and to gothic architecture with its divisions and subdivisions of vaults and columns. Although this connection between the scholastic structure of knowledge and gothic architecture, first suggested by Erwin Panofsky,[64] has been put into question, the link is often made explicit in manuscripts where illustrations of columns and vaults function as dividers and brackets structuring categories in diagrams (Fig. 4.6).[65] Such medieval architectural drawings for illustrating abstract principles have been most thoroughly studied by Karl Kinsella. He concluded that they were used not only for mnemonic purposes, but also to organize and index abstract information. The architectural illustrations constituted tangible models that helped the understanding of difficult concepts, and they were often referred to with spatial terminology in the texts. Architectural diagrams were common in arithmetic, music, and geometry, but not in astronomy. Kinsella's explanation for this is that astronomical knowledge already had

[61] Even-Ezra (2021, 8, 47–48, 67).
[62] Even-Ezra (2021, 37).
[63] Even-Ezra (2021, 52–53).
[64] Panofsky (1951).
[65] For examples of the diagrams, see J.E. Murdoch (1984, 33).

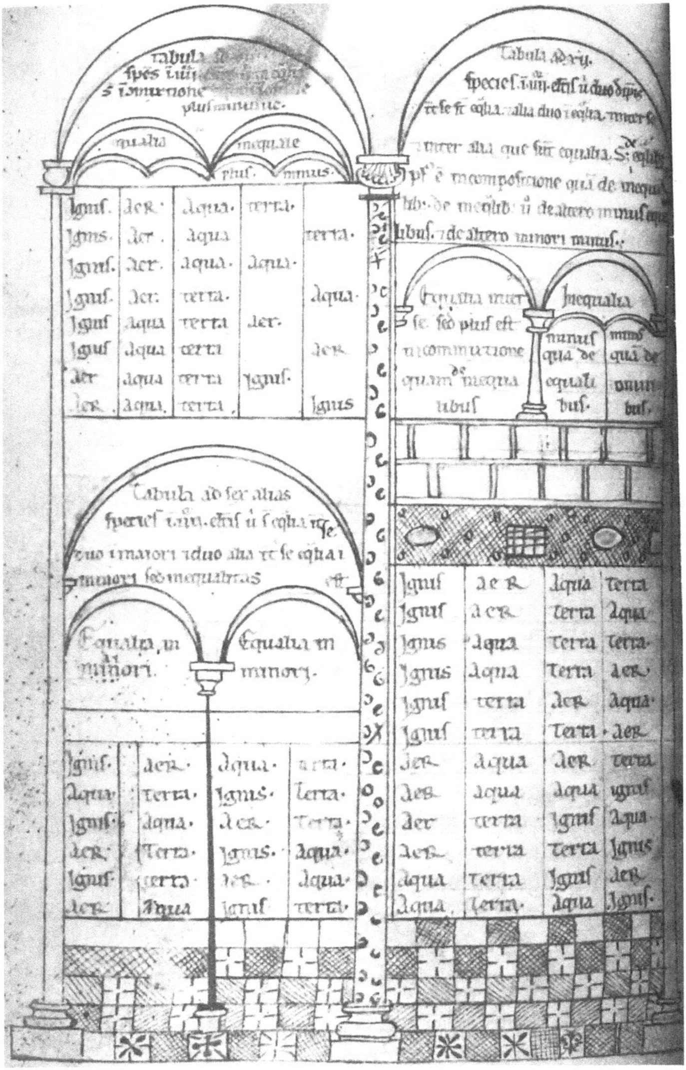

Fig. 4.6 Diagrammatic representation of possible mixtures of air, fire, earth, and water, with divisions and subdivisions in the form of architectural vaults. Twelfth century. British Library, Harley MS4348, fol. 26v

a spatial structure, and therefore did not need to borrow a tangible form from architecture.[66]

Panofsky explained the similarity between architecture and the logical structure of scholasticism with "mental habits" formed in the eleventh century in the area of Paris, which influenced the scholastic structure of knowledge as well as the architecture. In his commentary on Panofsky's work, the French sociologist Pierre Bourdieu adopted the Latin concept habitus, which had been used frequently by medieval scholars such as Thomas of Aquino. Ultimately, it was a translation of Aristotle's *hexis*, which denoted learned habits that, together with *ethos*, controls an individual's actions. Although Panofsky and Bourdieu did not link mental habits/habitus to moral philosophy, their basic understanding of the concept was the same as it had been for Aristotle. Bourdieu described habitus as a system of internalized figures that enables the development of thoughts, perceptions, and actions and that are particular to the culture of a certain group.[67]

However, neither Bourdieu nor Panofsky studied actual diagrams in manuscripts in their discussion of scholasticism, but focused on the structure of the written text. They were writing in the 1950s and 1960s, before the wave of medieval diagram studies that took off in the 1980s. For example, Even-Ezra's research illustrates how university students could acquire a scholastic habitus of divisions and subdivisions not only by contemplating diagrams in manuscripts, but by drawing their own in order to structure their thoughts (Fig. 4.7).

In summary, the High Middle Ages saw an increase in the use of diagrams and in the variety of diagram types. This development was spurred by the growing number of people involved in teaching and learning at cathedral schools and universities, and the number of books, manuals, and pedagogical aids needed in the process of education. In addition, the influx of freshly translated works of Aristotle heightened the need for diagrams that could help scholars understand and develop novel ideas and transmit them to their students. As the subdivided vaults of Gothic cathedrals distributed the load of the roofs, the tree diagrams divided the cognitive load of scholastic learning into manageable pieces. The diagramming practices in medieval Europe, and the conditions that gave rise to them,

[66] Kinsella (2016).

[67] Lyhne (2002); Bourdieu (2002).

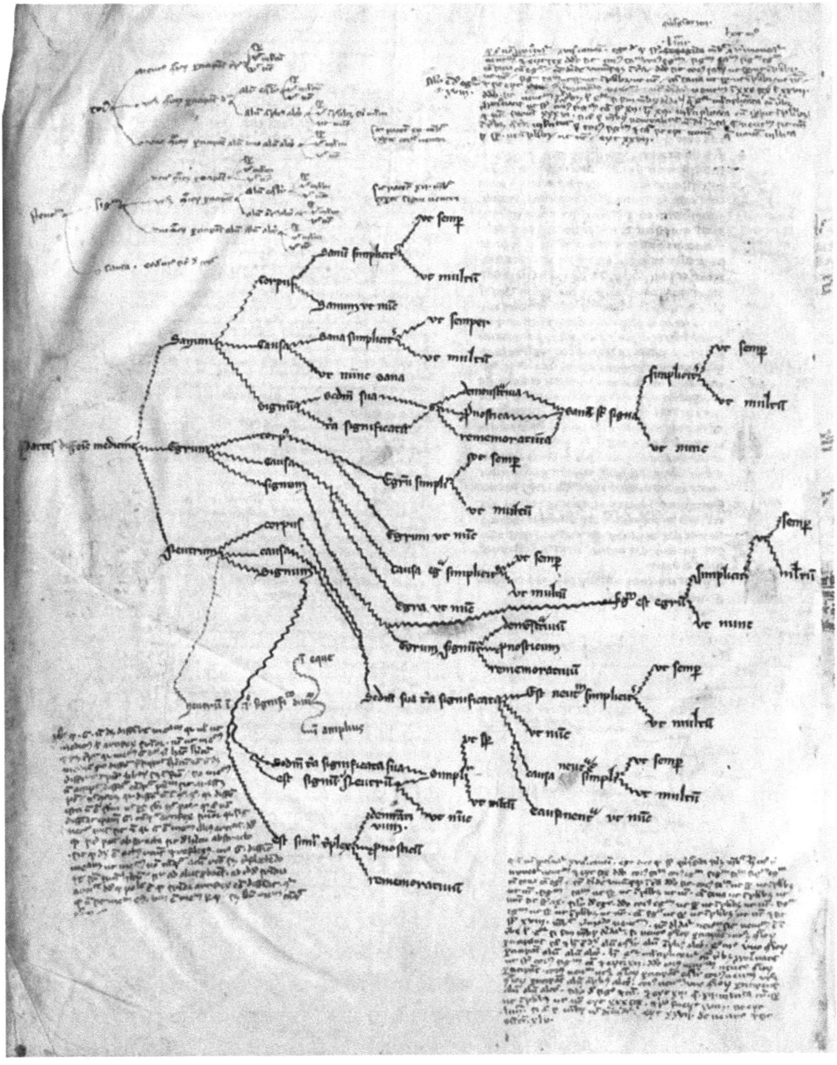

Fig. 4.7 Manuscript from Paris, according to Even-Ezra (2021, 41-42) used by a scribe to order his thoughts on a subject through externalization. Bibliothèque nationale de France, Manuscript Lat. 16174, folio 116v

can be better understood through a comparison with other cultural areas around the same time.

Similarities and Differences Between Scholastic, Byzantine, Arabic, and Chinese Tree Diagrams

Tree diagrams were used both in Western Europe and in the Byzantine Empire, but several Western forms were lacking in the Byzantine culture, such as some types of tree diagrams and genealogical diagrams. Other Western types, such as the consanguinity diagrams, were only found in peripheral areas of the Byzantine empire which were affected by Latin culture. Greek *diaereses*, trees of genera and species, were common in the Byzantine world from at least the ninth century. However, these Byzantine "tree" diagrams were never described as trees, and were much simpler than their Western counterparts, often consisting of a single dichotomy or tripartite division. Linda Safran attributes this to the fact that the Roman treatises on memory were not known in the East. Aristotle's *On Memory* was preserved, but the idea of memory loci was a development by his Roman successors, and a Greek translation of the Roman *Ad Herennium* was not completed until 1460. According to Safran, this late Byzantine interest in Latin memory culture might have contributed to the surge in the number of diagrams observed in Byzantine manuscripts at the time.[68] Thus, the division of the Roman Empire into Latin and Greek halves constituted a kind of natural experiment, which seems to confirm the hypothesis that there is a strong connection between Latin memory culture and the proliferation of tree diagrams.

However, this hypothesis appears less likely when we consider that in addition to Western Europe and Byzantium, the Arabic civilization also built on the heritage of the classic Mediterranean civilization. Like Byzantium, they relied on Greek literature and had no knowledge of the Latin *Ad Herennium*, but they still produced tree diagrams, as did the Latin scholars of Europe. When the Arabic civilization spread to lands in the Eastern Mediterranean in the seventh and eighth centuries, a process of translation and compilation of Greek texts began. At least from the ninth century, Arabic texts contained tree diagrams. For example, the

[68] Genealogy and consanguinity charts, see section "Genealogical Trees and Diagrammatic Heraldry"; Safran (2020, 376); Even-Ezra (2021, 24); Arethas of Caesarea and Michael Share (1994).

tenth-century writer Ibn Fariun, from the Emirate of Chaghaniyan in present-day Tajikistan, used them in his classification of the sciences (Fig. 4.8).[69]

Tree diagrams were particularly common in Arabic translations and compilations of Galen's medical works. One compiler described his reasons for using diagrams as follows:

> I found seven chapters of the book, and so I decided to arrange them in branch-diagrams . . . and tables . . . so that the book would be a better abridgment, more worthy of being taken seriously, more attractive to see, easier to read, and not so boring.[70]

These diagrams were, according to Emilie Savage-Smith, probably intended to simplify memorization of medical terms. She believes that the Galenic tree diagrams were of Greek Alexandrian origin, although no such manuscripts remain, and that they were later adapted by the Arab scholars who translated Galen's works. This kind of tree diagram later spread to the Latin west, where they became important in the scholastic tradition.[71]

After a period when the Arab world had gone through a process of intense translation, compilation, and commentary of Ancient Greek works, Europe entered a similar phase of translation from Greek and Arab manuscripts. In both these processes, diagrams and particularly tree structures were used to systematize vast amounts of new knowledge.

If Savage-Smith is correct, tree diagrams would have been invented by Greek scholars in Alexandria, whereafter they were adopted by Arabs and through them by Western Europeans. However, similar tree diagrams seem to have developed independently in China. From the fourth century CE, Buddhist *kewen* diagrams were used for teaching purposes. They abbreviated texts in a tree structure, providing a diagrammatic presentation of headings and paragraphs (Fig. 4.9).[72] These tree diagrams appeared in China in the first centuries after Buddhism had arrived from India. This is another example of how, when a large amount of knowledge is presented to a new audience, it creates a cognitive overload that encourages the creation of spatial diagrams for systematization and externalization.

[69] Ibn Fariun (1985).
[70] Savage-Smith (2002, 122).
[71] Savage-Smith (2002).
[72] Lackner (2007).

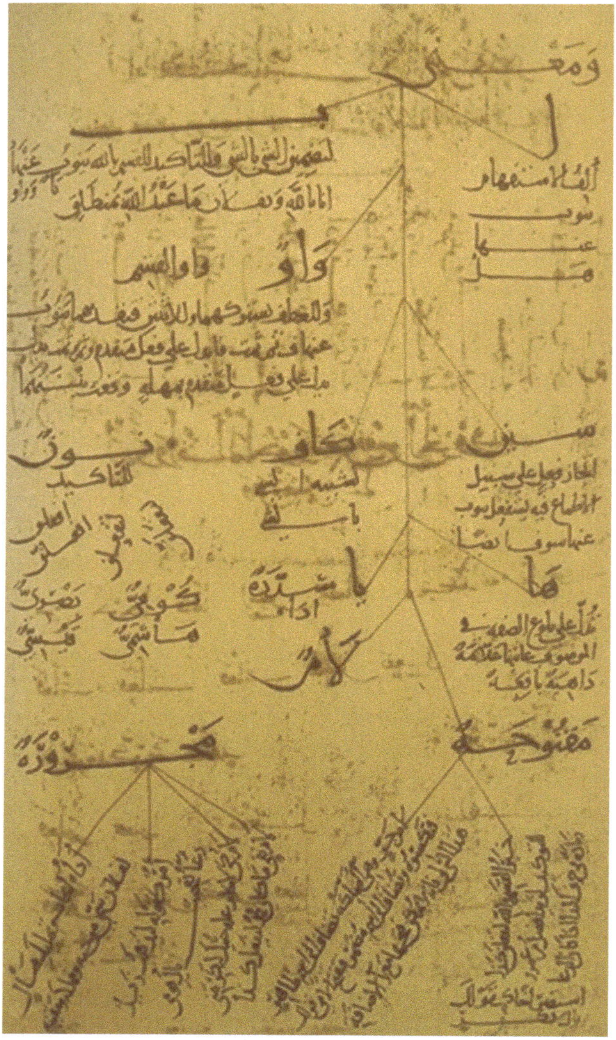

Fig. 4.8 Detail from tree of knowledge from tenth-century Central Asia. The whole manuscript is dedicated to trees of different branches of knowledge. Ibn Fariun (1985, 12)

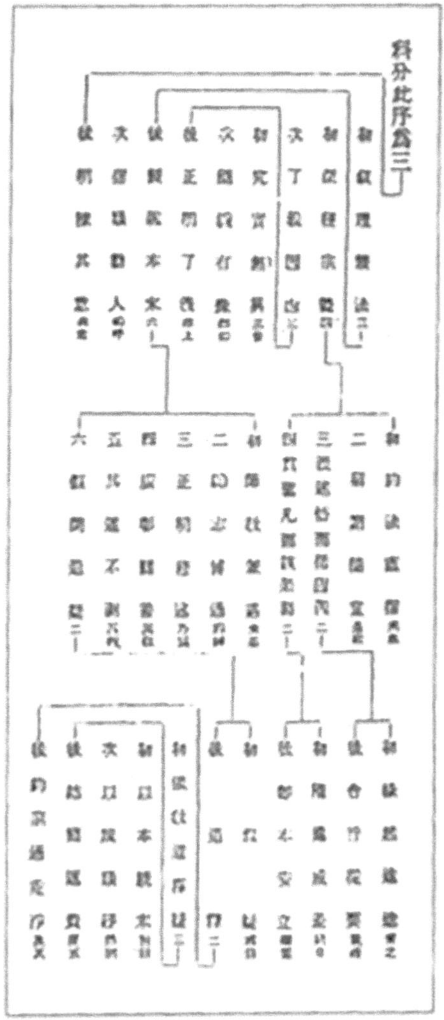

Fig. 4.9 Buddhist *kewen* ascribed to the Tang dynasty scholar Zongmi (780–841). Lackner (2007, 358)

Although these diagrams were introduced in China in the fourth century, they reached their apex parallel to the proliferation of tree diagrams among the Arabic and European civilizations. Michael Lackner claims that although most Chinese "diagrams" would have been classified as images by Charles Sanders Peirce, from the mid-Song to the Yuan dynasty (twelfth to fourteenth century) a tendency toward rationalization drove Chinese *tu* close to diagrams in the sense of Peirce: iconic representations of structural aspects of the signified. During this period, a form of tree diagrams was used for analysis of the Classics and the writings of the Neo-Confucian schools. Lackner compares these to the tree diagrams used in Europe at the time. In both cases, the tree structure was used to analyze topics into parts and sub-parts, and the didactic aspect was important. However, while Chinese *tu* generally analyzed specific textual passages, European diagrams more often dissected general concepts, such as the "division of sciences" and "division of virtues." Lackner believes that it was the unambiguous interpretations conveyed by diagrams that led to their early decay in China. Some liberty of interpretation was warranted in the teaching of classical texts.[73]

Although Europe had general diagrams of knowledge or vices such as Lackner mentions, there was also a European diagram type similar to the Chinese diagrams for analysis of particular texts: those that Ayelet Even-Ezra has labeled "horizontal trees." As they were also used for textual analysis, Even-Ezra finds it natural that they flowed horizontally from left to right as writing, and not vertically as other tree diagrams, such as genealogies or the tree of Porphyry.[74] The comparable Chinese diagrams flowed vertically from top to bottom, following the direction of classical Chinese writing.

While the introduction of *kewen* diagrams in the fourth century was associated with the influx of new Buddhist texts, the diagrams during the Song and Yuan dynasties were mostly found in Neo-Confucian texts. During this period, Confucian learning, which was the basis of the imperial examinations that selected China's leading bureaucracy, reached many more persons than before. Literacy rates increased, and a large market was established for printed texts. Thus, the old Confucian texts now reached new groups, which heightened the need for diagrams that simplified learning.[75]

A later offshoot of the Chinese textual tree diagrams is found among the *Ten Diagrams* by the Korean Neo-Confucian Yi Hwang (1501–1570). Published in 1568, the ten diagrams were intended as an instruction in

[73] Lackner (2007).

[74] Even-Ezra (2021, 36).

[75] For the increase in literacy during this period, see, for example, He (2020, 15–28).

Confucianism for the young ruler. Condensed in order to be written on ten plates, the diagrams would do the teaching that the old and ailing Yi Hwang was no longer capable to perform in person.[76] Thus, in this case diagrams were not only mandated by a restriction in the capacity of the learner, but also of the teacher.

Although there were several similarities between the use of diagrams in Western Europe and other cultural areas, there were also important differences. While used for textual analysis, tree diagrams were of no importance in East Asian logic. Chinese logicians did not try to establish generalized, abstract principles, but worked with the help of examples.[77] In addition, Chinese logic did not classify things into a treelike hierarchical structure, as the genera and species of Aristotelian logic.[78] These differences might help explain why abstract diagrams, and particularly trees, became much more widely used in European philosophy and science than in Chinese, although they were certainly known in both cultures. After the Warring States period, logic even disappeared as a separate field of study in China and was not revived until under Western influence from the sixteenth century.[79] The majority of the rich Buddhist logic tradition that developed in India never even gained a foothold in China.[80]

In summary, it is unlikely that the richer multitude and complexity of tree diagrams in Latin Europe compared to Byzantium could be explained by the tradition of Roman memory culture, considering the global reach of similar diagrams. A probably more important explanation for the relative abundance of diagrams in general, and particularly tree diagrams, in medieval Europe, is the translation explosion that forced scholars to process vast amounts of new texts. In the Arab world, a similar intense period of translations began slightly earlier, and China experienced a period with *kewen* tree diagrams after the influx of Indian Buddhist texts. However, the Byzantine Greek world never lost access to the body of Classical Greek texts and did not experience comparable periods of cognitive overload, through massive translations or spread of knowledge to broader groups, which would have necessitated a wide deployment of categorizing diagrams.

[76] Yi (1988, 24–25).
[77] Willman (2021, 328).
[78] Tiles and Jinmei (2004, 137–149).
[79] Willman (2021).
[80] Lusthaus (2003, 363).

Genealogical Trees and Diagrammatic Heraldry

Many non-literary cultures have artifacts that help them keep track of genealogy. It might be staffs with knobs, one for each generation, as on Rarotonga or among the Maori, or stools with carved representations of ancestors, as among the Iatmul in New Guinea. Aided by these stools, learned men memorize around 20,000 personal names, which function as an index to the Iatmul's mythological songs.[81] Today, genealogy is often represented in tree diagrams. The abovementioned mnemonic staffs have some similarities with trees, and if some of the today undecipherable quipus were indeed genealogical, they might also have had a tree structure. The type of genealogical trees we have today developed in the Middle Ages (Fig. 4.10), but had predecessors in antiquity. According to Pliny the Elder and Seneca, wealthy Roman families often displayed their heritage by hanging wax masks of their deceased ancestors on their walls, where the direct patrilineage was marked with ribbons, *stemmata,* from the oldest ancestor on top. Sometimes the *stemmata* had branches, but it is debated to what degree they resembled later genealogical trees.[82]

Another diagrammatic form with Roman roots was the consanguinity chart. It was used in legal texts to illustrate concepts describing family relationships, such as great grandparents and nephews. The consanguinity charts consisted of cells, where the ego was placed in the middle, with descendants below and ancestors above. When both the maternal and paternal sides were depicted, and only direct descendants of the ego were included, the outline of the diagram was similar to a tree with stem. From 800 CE, the charts were sometimes described as trees, and from the twelfth century they were called *arbores consanguinitatis.*[83] These "trees" were depicting family relationships in general, and not the roots of any particular family. Neither did they always resemble trees, and when they did so it was because of their contours and decoration rather than through a branching structure. However, the consanguinity trees might have functioned as inspiration for more treelike diagrams of particular families.

There is evidence that a genealogical tree of the Carolingians was produced already in 908–911, and a small number of other trees of the Carolingian and Ottonian imperial dynasties are known from the tenth

[81] Kelly (2012, 115–116).
[82] Archibald (2014, 33–34).
[83] Holladay (2019, 11–14).

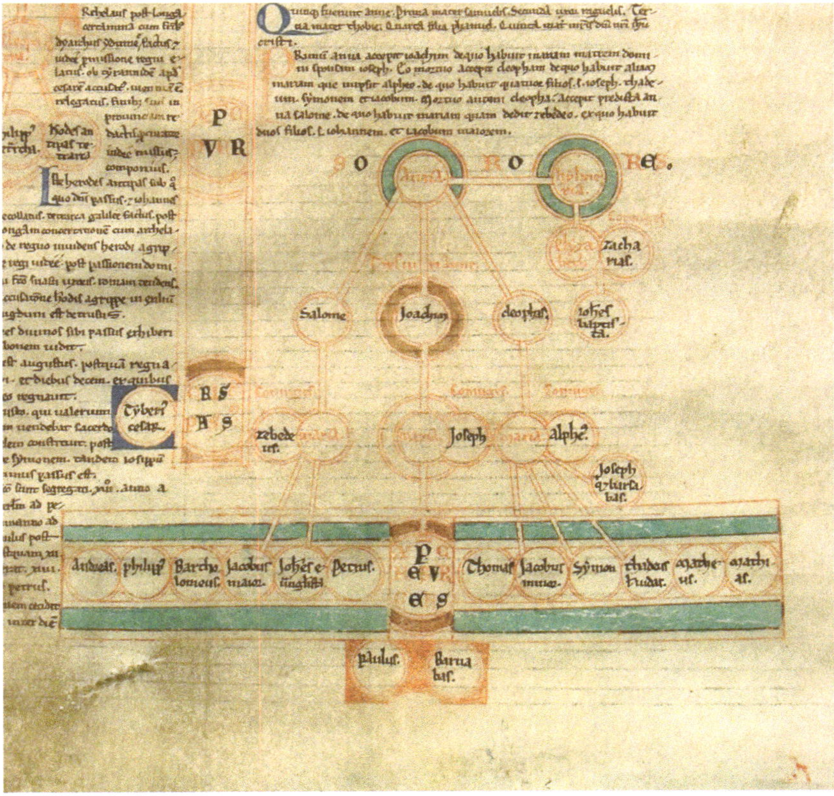

Fig. 4.10 This fragment of a British copy of *Compendium historiale in genealogia Christi* by Peter of Poitiers from ca. 1230 depicts the genealogy of Christ. Metropolitan Museum 2002.433

and eleventh centuries. A tradition of genealogical literature can be traced back to the 950s and the monasteries that functioned as dynastic necropolises for the counts of Anjou and Flanders. This literature became widespread from the twelfth century, and with time the genealogies were accompanied by charts.[84] From the early family trees of emperors and high nobility, the practice trickled down to ordinary nobility.

[84] Holladay (2019, 14–18).

Although it is clear that such genealogical trees were part of an attempt to legitimize power and possession of land, there are several theories regarding why they appeared at this time. John H. Holladay has linked it to the twelfth-century Renaissance, which sought ancient precedents in a range of cultural and scholarly fields. Practices such as the Roman *stemmata* should, according to this theory, have inspired medieval followers.[85] Others have pointed to the increased efforts of the church to regulate consanguinity in marriage in the High Middle Ages, which increased the importance of documented ancestry.[86]

The interest in family genealogy in the twelfth century had a parallel in depictions of the Holy Family around the same time. According to an interpretation of Isaiah 11:1, the Virgin Mary was a "rod out of the stem of Jesse." Inspired by this, medieval artists developed visual representations of Jesus's ancestry. The oldest preserved Jesse Tree is found in the Vysehrad Codex from 1085. It has been claimed that it did not, as later Jesse trees, refer to the genealogy of Mary, but to "the royal Christological genealogy to which all Christian monarchs claim to belong." Thus, the codex, which was produced for the coronation of Duke Vratislav as king of Bohemia, attempted to legitimize the new monarch's claims to the throne. In the almost equally old Exeter manuscripts, the reference to the rod out of the stem of Jesse is more obvious, as Mary is seated between Jerome and Isaiah, carrying a sprig.[87]

Some of the early, simple Jesse trees are similar to illustrations of a passage in the Sanskrit epic *Mahabharata*, where a divinity is born from a lotus flower sprouting from the navel of the Eternal Being. In the late nineteenth and early twentieth centuries, it was suggested that the Jesse trees were inspired by Asian predecessors. However, the distance in space and time—the Asian images are considerably older—makes it difficult to establish a direct link.[88]

In the twelfth century, Cistercian monks used the Jesse Tree to emphasize the humanness of Jesus's mother Mary, in order to counteract the Cathari heretics who claimed that Jesus was of entirely divine origin. The Cistercian experiments with Jesse trees ended when Bernard of Clairvaux—probably in 1152—managed to convince the order that artwork itself was

[85] Holladay (2019, 10–18).
[86] Crisp (2003).
[87] Dodwell (1993, 193–195); Williams (2000, quote on 20).
[88] Watson (1934, 58–66).

heretical. However, the Cistercians left a lasting impression on the genre; for example, they invented the iconography of a tree shooting out from the loins of Jesse.[89] Since the Jesse Tree and noble genealogical charts developed at the same time, it is likely that the traditions co-evolved and influenced each other, rather than that one of them can be seen as the predecessor of the other.

There are several parallels between the development of genealogical charts in Europe and in China. Today, Chinese families have genealogical books tracing their lineages many generations back. These originated with Ouyang Xu, who after his mother's funeral in 1053 CE raised a stele at her grave with a chart of the family's ancestry. The charting method had been developed by Ouyang Xu's colleague Lü Xiaqing, who had compiled an official version of the Tang dynasty's history. Until then, genealogy was generally presented in essays that maintained a single line of descent, but the charts could trace multiple lines. Ouyuang Xu's charts were tables with five generations presented per page. With time, the practice of compiling genealogical charts spread from the wealthiest families to the lower strata of society. This was linked to the expansion of literacy during the Song dynasty. The charts became compiled to books, where the main line was presented first and branching lines toward the end. Soon, diagrams appeared where descent was represented by lines, in what we would call tree diagrams.[90] Such genealogical tree diagrams were also used to analyze the transmission of books and philosophical schools (Fig. 4.11).[91]

Thus, in Europe as well as in China, genealogical charts appeared in the eleventh to twelfth centuries. In both cases, they were preceded by written genealogies. Another similarity is that they appeared first among the highest strata in society, the families of emperors, whereafter they gradually found their way to broader circles. Yet another parallel is that this happened during a period of weakening central government, where strengthened local dignitaries emulated the genealogical practices of imperial historians and genealogists. Both in Song China and in twelfth-century Europe, the spread of literacy and book culture was a facilitating factor.

In Europe, genealogy was from the High Middle Ages closely related to heraldry, the construction of arms that displayed the ancestry of nobility. Since these arms are symbolic representations of families, heraldry has

[89] Dodwell (1993, 211–215).
[90] He (2020, 63–67).
[91] Lackner (2007, 343, 360).

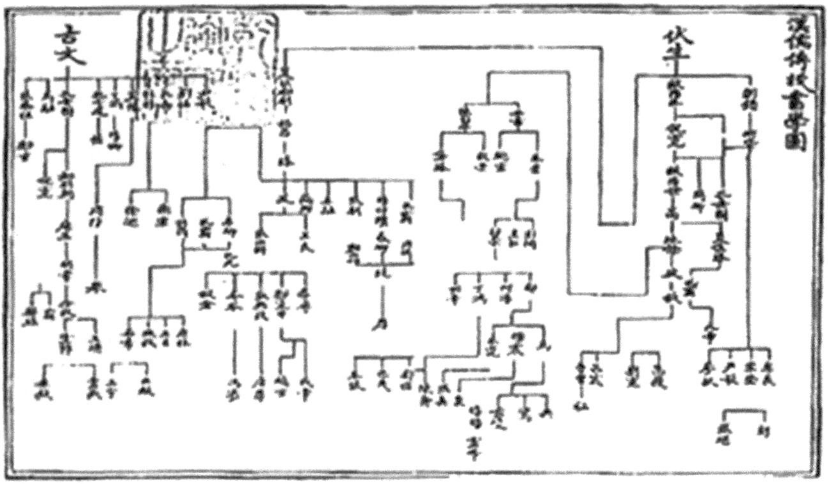

Fig. 4.11 Diagram of the transmission of the *Book of documents*, twelfth century. From *Wujing tu*, "Illustrations of the five classics." *Tu* might also be translated as diagrams. Lackner (2007, 360)

been described as a symbolic language.[92] However, heraldry might also describe descent by arranging the symbols of different families according to a diagrammatic spatial schema. Heraldic arms could through quarterings become intricately complex, displaying multi-generational information normally found in trees. Compound shields for the following generations were created through a number of complicated rules, which depended on whether the women who married into the families were heraldic heiresses or not (Fig. 4.12).

The "simple quartering" of the shield into four or more fields could be followed by a number of "compound quarterings," in which each of the primary quarters was in turn subdivided.[93] In this way, a hierarchical system of quarterings could be constructed, similar to a tree diagram with a limited number of branches at each node, following the principles of folk taxonomies. The first quarter in the top left was the most important, and

[92] By, for example, Ragen (1994).
[93] Boutell (1900, 34, 175–176).

Fig. 4.12 Top left illustrates how a shield is quartered and, in fields A, B, and D, compound quartered, in B into six fields (the term "quartering" is used also for more than four divisions). Top right illustrates the arms of the son and grandson of a baron (A), if both the baron and his son marry heraldic heiresses (B and C). Below the arms of Richard Neville (1428–1471). Unconventionally, Richard placed the arms of his father-in-law in the first quarter and his father's in quarters 2 and 3. This reflected that he considered his title as Earl of Warwick, acquired through marriage, more important than his inherited title as Earl of Salisbury. Drawn by author, shield of Richard Neville based on image by MostEpic with eagles from Ipankonin

usually illustrated the direct patrilineal ancestry.[94] This is what we could expect by combining the principles of spatial agency bias, which put more powerful actors to the left, and the primary metaphor "Control is Up."

However, heraldic shield-diagrams are arguably much more difficult to decode than tree diagrams. As Peirce noted, icons combine diagrammatic and symbolic aspects.[95] The genealogical trees are relatively easy to decipher since the diagrammatic element dominates in them, while it is obscured by symbolism in the heraldic shield. In fact, you might need a tree diagram in order to decipher the development of the quarterings in an old family.

Feudal Japan developed an indigenous tradition reminiscent of European heraldry. The Japanese *mon* was from the twelfth century a symbol of the family. It originated as embroidery on clothing, but moved to weapons and flags, the opposite of what was the case in Europe. With roots among the nobility, the *mon* spread to other classes in the seventeenth century. The *mon* was a general symbol for families, and did not follow strict rules for manipulating symbols in order to display detailed heritage equivalent to European quartering. However, similar tendencies existed, as junior branches of the family often adapted a modified version of the senior branch's *mon*. Japanese artists studied the constructions of *mon* in detail and wrote books about their composition.[96]

The *mon* are also often similar to diagrams of geometrical problems produced by Japanese mathematicians beginning in the early seventeenth century. A large share of these problems investigated how to fit geometrical shapes within or around each other. From the last third of the seventeenth century, such problems were often displayed on votive tables outside temples. Historians of geometry have referred to these as recreational problems, in contrast to the practical problems that otherwise occupied Japanese mathematicians.[97] However, for an artist constructing *mons*, how to fit shapes into each other was a highly practical problem. For example, the diagram described in the problem "a circle of radius r is surrounded by a loop of five equal circles of radius R. Find r in terms of R" is identical to the *mon* of Hochi Umebachi (Fig. 4.13).[98] That these dia-

[94] Boutell (1900, 171). Folk taxonomies, see section "The Shape of Life" in Chap. 5.

[95] Section "What Is a Diagram? Conceptual History and Semiotic Definition" in Chap. 2.

[96] Slater and Znamierowski (2007, 478–479).

[97] Scriba, Schreiber and Schreiber (2015, 143–150).

[98] The connection between the temple problems and *mon* construction has been explored in Tabing (2018).

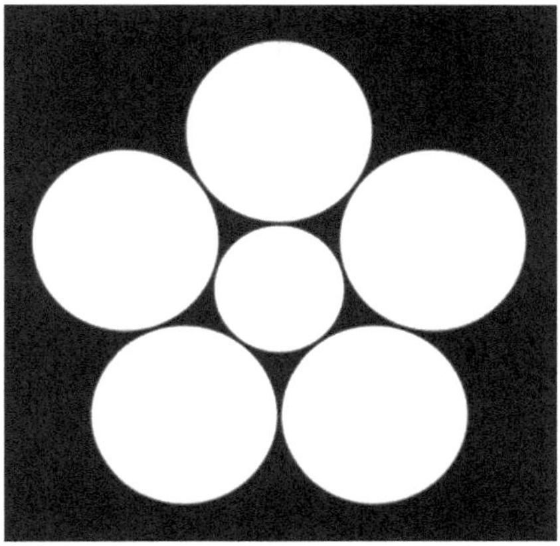

Fig. 4.13 The *mon* of Hoshi Umebachi, identical to a Japanese geometrical temple problem. Drawn by Magasjukur2

grams were hanging outside temples meant that their iconography reached broad circles of the population. Regardless of the direction of influence, the structural similarities of Japanese *mon* and geometry is an example of a shared habitus comparable to that of Gothic architecture and scholastic philosophy.[99]

In its visual expression, the *mon* is closer to the logo than to the coat of arms, and in the modern corporate world the Japanese tradition has triumphed over the Western. Companies that before would have used full achievements of arms now use symbols similar to those of a fifteenth-century Shogun.[100] It has been claimed that the use of tree diagrams in European science had its roots in the genealogical tradition of nobility.[101] The French Revolution in Europe and the Meiji Restoration in Japan might be seen as turning points where the visualization skills formerly deployed for feudal genealogy and heraldry were recruited for the needs of science, administration, industry, and commerce in bourgeoise society.

Another step was taken during the second Industrial Revolution, which was associated with more systematic and scientific business practices. This development was aided by the organization chart, which descent is made clear in this quote from the *Bulletin of the Efficiency Society* from 1912:

[99] Section "The Scholastic Diagram Habitus".
[100] Slater and Znamierowski (2007, 478–479).
[101] See Hellström (2019, 281–282).

> [T]he organization chart or diagram . . . resembles a genealogical tree in a very close way. Authority reaches down through the several branches of an organization like descent of blood, and, if properly planned, it will be as irregular for a factor in an organization to be in doubt as to the person in authority over him as for the child to deny the parentage of his father.[102]

The hierarchical structure of genealogical charts represented not only descent but also patriarchal authority, and the latter aspect was adopted by organization charts, which began to proliferate in the 1910s with the rise of the field of business administration.[103]

Llullism: Incomprehensible Originality

The square of opposition and different types of trees are examples of common diagram types that represent the main currents of medieval thought. However, lessons can also be learned by studying a unique outlier, since it can contribute to a discussion regarding why some diagram types become common while others remain marginal. The Catalan scholar Raymond Llull (1232–1315/16) has both been accused of heresy and beatified by the Catholic Church. Verdicts on his contributions to mathematics range from hopelessly behind his contemporaries (in geometry)[104] to 650 years ahead of his time (in voting theory).[105] Opinions have been divided also on Llull's diagrams, and his great art that they are related to. To a large extent, this is a consequence of his diagrams being so different from the common ones that their mere unfamiliarity has hindered comprehension. Llull's diagrams constitute a case that help us explore the limits of the innate, common to all spatial cognition, and the need for cultural learning of diagrammatic practices.

Llull's art can be described as a method for producing a wide range of statements on a subject by combining a number of axiomatic elements, or "necessary reasons," common to the three monotheistic religions of Judaism, Christianity, and Islam. These were signified by letters, which were sometimes inscribed on concentric disks, and combinations of them

[102] Efficiency Society (1912, 6).

[103] Early examples are the charts in Christensen (1912). Another diagram type associated with the emergence of industrial management in the second industrial revolution, the flow chart, is discussed in section "Diagrams in Economic Theory" in Chap. 6.

[104] M. Gardner (1982).

[105] Hägele and Pukelsheim (2001, 80); Tangian (2014, 80).

could be produced by rotating the disks. Llull hoped that religious truths could be revealed using this method, which would make it possible to convert Jews and Muslims to Christianity.[106] Llull grew up on Mallorca after it had recently been conquered from the Muslim Almohads by the Catalans, and Jews played a prominent role in the island's cultural and economic life.[107]

It is believed that Llull got inspiration for his rotating diagrams from the *zā'irjah* used by his Arab contacts. This device, judging from its name of Persian origin, could, with the help of rotating disks inscribed with letters, produce rhymed answers to astrological questions. As Llull's volvelles, the *zā'irjah* was dependent on a human operator for turning the wheels and interpreting the raw output of letter combinations. Allegedly, the answers the device gave were so impressive that it made the philosopher Ibn Khaldun dance on his terrace in joy.[108]

Llull's art and diagrams went through three stages of development, which Anthony Bonner called the quaternary, ternary, and final phases. Llull's highly original system was difficult for his contemporaries to comprehend, and later observers have struggled at least as hard to make sense of it. This applies particularly to Llull's rotary diagrams, the volvelles, which have been understood in several different ways. In the Renaissance, Agrippa von Nettesheim presented Llullism as an alternative to Petrus Ramus's rhetorical-mnemonic-logical system.[109]

The historian Frances Yates (1889–1981), who also interpreted Llull's system as mnemonic, pointed out that neither Llullism nor Ramism used the images that were so common in the artificial memory of the medieval and ancient rhetoric tradition, but aided memorization with the help of a logical structure.[110] Others, such as Gottfried Willhelm Leibniz, have focused on the role of Llullian wheels in combinatorics.[111] Michael Evans has claimed that while tree diagrams were used to illustrate the medieval logical practice of *distinctio*, the division of subjects into their most basic elements, Llull's circular rotational diagrams were used for the process of *interogatio*. This meant the systematic testing of an idea against evidence

[106] Bonner (2007).
[107] Bonner (2007, 1).
[108] Link (2010).
[109] Bonner (2007, 290).
[110] Yates (1966) 2014, 175–196.
[111] Doucet-Rosenstein (1981).

and the opinions of authorities, a practice mirrored by mentally or physically rotating concentric circles in a diagram.[112]

In his *Logic Machines and Diagrams*, originally published in 1958, Martin Gardner was not impressed with the Llullian diagrams' ability to perform logical operations but instead emphasized how rotary designs such as Llull's can be used to stimulate spontaneous invention and creativity.[113] However, Bonner claims that Gardner was misled by relying heavily on E. Allison Peer's outdated biography of Llull, ignoring newer research and Llull's original texts. According to Bonner, Gardner's text and the fact that it was reprinted basically unchanged in 1982 when even more new research could have been incorporated have given logicians and mathematicians an unfoundedly negative view of Llull's contribution.[114]

The following description of Llull's diagrams is based on Bonner, who has provided the most thorough analysis of Llull's art. However, this is only a brief overview, and to fully understand how Llull's art was used in practice, it is recommended to consult the examples in Bonner's book. According to Bonner, there was a close relationship between Aristotle's topical logic and Llull's art. Aristotle's topics discussed dialectic techniques for how to arrive at probable conclusions from generally accepted premises. *Inventio* was a method for finding arguments or counterarguments on almost any subject that could arise in a debate. This included the use of *loci* or places for guiding the user through the apparatus of arguments.[115] These loci can be described as themes that orators could reuse in different debates. Llull's diagrams were constructed to display and produce possible combinations of such themes.

Bonner identifies five categories of figures used in Llull's art:[116]

(a) Circular, non-rotating figures with letters, often with interior lines connecting concepts written on the circumference (Fig. 4.14).
(b) Triangular figures, tables with binary combinations of the terms of the circular figures.
(c) The revolving figures or volvelles (Fig. 4.14, bottom right), of which there are never more than one per work, used to generate

[112] M. Evans (1980).
[113] Gardner (1982, 20, 27).
[114] Bonner (2007).
[115] Bonner (2007, 16).
[116] Bonner (2007, 22–23).

third- or higher-order combinations. They were constructed with the help of concentric disks and a pin, and could produce all the combinations of the other figures. This was possible since the letters on the inner disks (A, Z, Y, X, V, T, S) each represent a different diagram.

(d) The tabula of the ternary phase, which is a tabular display of all the third-order combinations (without repetitions) of the revolving figure of that phase.

(e) The elemental figure, which shows the possible combinations of the four elements—earth, air, fire, and water (Fig. 4.14, middle right).

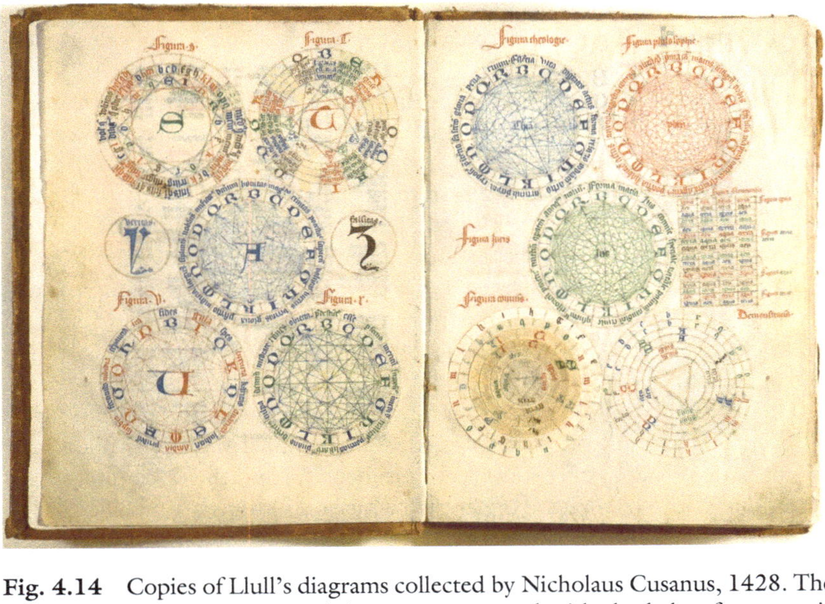

Fig. 4.14 Copies of Llull's diagrams collected by Nicholaus Cusanus, 1428. The revolving volvelles (bottom right) were constructed with the help of concentric disks and a pin in the middle and could be used to produce all the combinations of the other diagrams in Llull's art. For example, the letter S in the inner disks could be used to produce the combinations in the diagram S (top left). Codex Cusanus 83: Raymundi Lulli Opera. St. Nikolaus-Hospital-Cusanusstift Bibliothek, Bernkas. Photo: Erich Gutberlet; St. Nikolaus-Hospital/Cusanusstift, Bernkastel-Kues

The figures had four purposes. First, they were visual aids. Llull wrote that "it is only natural for the intellect to understand better by means of a visual and auditory demonstration than by one that is purely auditory." The second was mnemonic, which made the figures attractive to Renaissance thinkers. By showing the structure of the world, the figures made it more intelligible and easier to remember. The third function was to display the different sets into which the fundamental concepts of the Art were divided. Thus, under Fig. A (which symbolized God) we find the divine attributes, and under Fig. V the virtues and vices. The fourth function is to show the possible relationships between the components of each set. Initially, Llull did this graphically, by interconnecting lines. In his later ternary phase, the mere circularity, for instance of Fig. A (from which the lines had disappeared), showed the connections between the terms. There was also a layer of color-coding, when Llull mentioned, for instance, "the green triangle of Figure T" (referring to the triad of difference, concordance, and contrariety), or "red V" (referring to the vices).[117]

What united Llull's reasoning during all three periods was its foundation in the axiomatic "necessary reasons."[118] However, the road from them to conclusions was different in each period. During the first period, also called the quarternary, the system of diagrams reached its maximum complexity. In this version of the art, Llull's standard method for demonstrating something was to first suppose it was true, and then suppose its negation, and investigating the consequences from these suppositions within the art's system of demonstration. For example, "if A (God) exists" led to concordances among positive terms, while "if A does not exist" led to contradictions, demonstrating that A does indeed exist.[119]

The case of Llull illustrates that although diagrams can be useful tools for communicating complex ideas, novel diagrams might through their unfamiliarity became obstacles to comprehension and the dissemination of ideas. The criticism that the quarternary phase of Llull's art received at the University of Paris contributed to the simplified ternary form of the art, which contained fewer and more simple diagrams. Twelve of the 16 figures were removed.[120] In the ternary period, Llull also abandoned the

[117] Bonner (2007, 24).
[118] Bonner (2007, 276).
[119] Bonner (2007, 272).
[120] Bonner (2007, 5).

method of looking for concordances and contradictions, and instead worked his way from definitions to conclusions.[121]

Lull's new logic, which succeeded the ternary art, was yet another step in the direction of achieving comprehensibility, as he rid himself of all diagrams and presented his ideas in the form of syllogisms, a common form of scientific discourse among his contemporaries. While in the quarternary phase a reader of Llull's works needed to learn a complex system of novel interconnected diagrams, this learning process was shortened considerably in the ternary phase. In the last period of the new logic, readers who had normal schooling at the faculty of philosophy could jump straight into Llull's syllogisms without any special training.

Llull's example illustrates that it is not necessarily the most advanced and expressive diagrammatic systems that become dominating but those that are most comprehensible, which means diagrams that are both simple, in their spatial structure and their adherence to cognitive principles, and familiar through their likeness with previously known diagrams.

Circles of Inclusion

From the mid-1960s to the early 1980s, when the "New Mathematics" was in vogue, many students came in contact with Venn diagrams. These were used to illustrate set theory, which became a unifying theme in the new education.[122] The diagrams consisted of circles or ovals whose overlap or lack thereof signified inclusion, intersection, or exclusion of sets, such as "Some A are B." The New Mathematics was criticized for formulating simple propositions in convoluted and complicated ways that were difficult to understand.[123] However, the diagrams associated with the New Mathematics were not very complicated and definitely not new. In fact, they have a history in logic and philosophy that spans several thousand years.

John Venn (1834–1923) is far from the only one to have developed similar diagrams. In fact, the simplified versions used in schools were often more similar to Euler diagrams than Venn diagrams. During the same time period as Venn, Charles Sanders Peirce also experimented with similar figures, which he called "existential diagrams." Venn and Peirce diagrams use

[121] Bonner (2007, 274).
[122] Kilpatrick (2012); Brand and Sherlock (1972).
[123] See, for example, Feynman (1965).

shadings or markings of surfaces in order to increase their formalism, which, however, makes them less intuitively understandable.[124] It can be argued that Euler's diagrams made better use of diagrammatic spatial thinking. In Venn's and Peirce's diagrams, the placement of the circles is fixed, and intersections, inclusions, and exclusions are conveyed with symbols. Thus, they are a kind of intermediary between symbolic and diagrammatic logic. Euler's diagrams are closer to the purely diagrammatic, since more information is conveyed through the spatial location and extension of the circles (Fig. 4.15).

It is less known that Arthur Schopenhauer also used diagrams of intersecting circles, since he did not use them in his popular books, but only in lectures for an academic audience. While logical diagrams such as those of Venn, Peirce, and Euler almost never used more than three circles, Schopenhauer's diagrams were filled with a multitude of them. However,

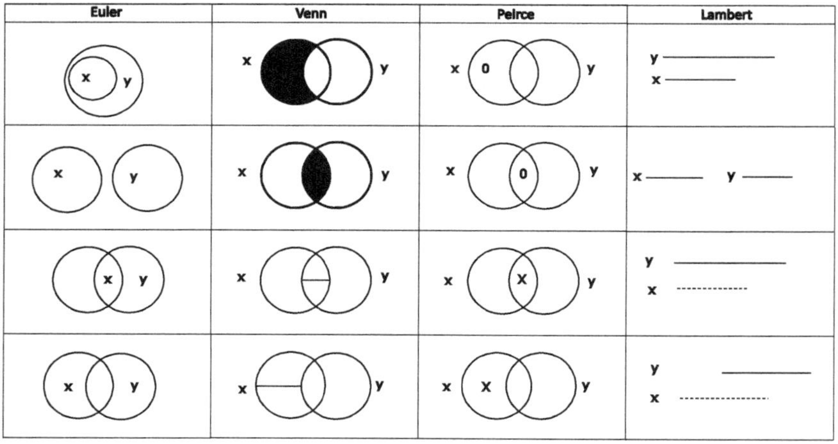

Fig. 4.15 The logic diagrams of Euler, Venn, Peirce, and Lambert. From top to bottom, the diagrams display that all x are y, no x are y, some x are y, and some x are not y. In the Euler and Lambert diagrams, the diagrammatic element dominates, since the spatial location of the circles or lines conveys information, while Venn and Peirce diagrams rely on symbolic notation. Drawn by author, based on Moktefi and Shin (2012) and Englebretsen (2020, 31)

[124] For the difference between Euler-type and Venn-type diagrams, see Moktefi and Shin (2012, 623); Dau (2009).

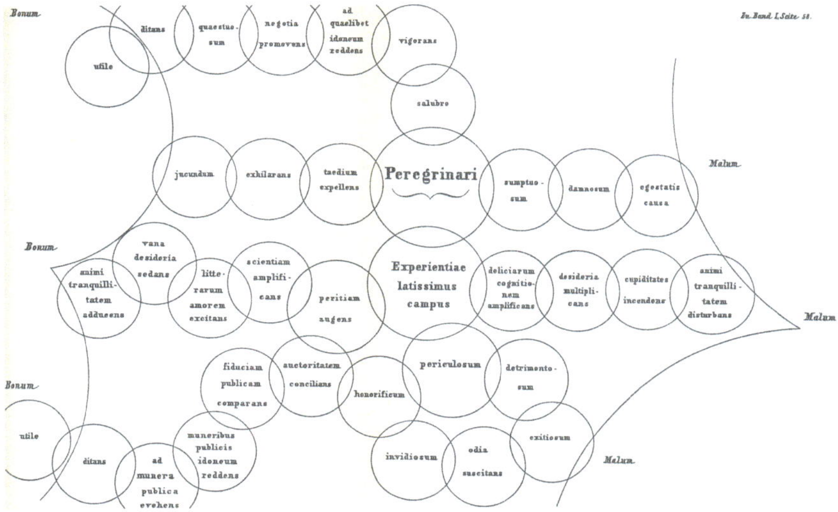

Fig. 4.16 Schopenhauer's diagram of how to use a series of overlapping arguments to convince someone that traveling has either good or bad consequences. Schopenhauer (1961)

they were not primarily embedded in logic, but were attempts to illustrate the rhetorical art of persuasion. For example, in the diagram of the good and evil of traveling, found among Schopenhauer's Berlin lectures from the 1820s, he wanted to point out that with the help of a series of overlapping concepts, it is possible to arrive at any conclusion in an argument (Fig. 4.16).[125]

Schopenhauer claimed that all possible relationships between concepts can be represented by diagrams, and there was an "extremely fortunate" isomorphic relation between circle diagrams and conceptual spheres. He listed the popularizers of this idea as Gottfried Ploucquet (in square form), Johann Heinrich Lambert (in line form), and Leonhard Euler (in circle form).[126]

Although Schopenhauer was aware of some of his predecessors, Jens Lemanski has identified three periods in the history of Euler-type diagrams that seem to have been isolated from each other. The first period, from

[125] Lemanski and Moktefi (2018).

[126] Dobrzański and Lemanski (2020).

1531 to 1614, began with Juan Luis Vives *De censura veri et falci*, where he published some diagrams of V-shapes included in each other. Line diagrams apparently inspired by Vives were produced by Bartholomäus Keckermann in 1601 and 1603 and Johann Heinrich Alsted in 1614. However, the circle diagrams of Nicolaus Reimers from 1589 seem to have been a separate invention. What all these four diagrams had in common was that they were used to illustrate Aristotle's *dictum de omni* and *dictum de nullo*, the principle that whatever applies to a whole kind applies to any of its subkinds.[127]

The second period from 1660 to 1712 consisted of two schools: the first, to which Leibniz belonged, was centered around Erhard Weigel in Jena, and the second was formed around Christian Weise in Zittau. Leibniz was familiar with diagrams from both the Weigel and Weise schools. Weise's disciple Lange wrote a history of logic diagrams that included everything from his own period but none from the first period, which illustrates that the two periods were isolated from each other. Weigel, who was probably the pioneer in this period, did not use abstract geometric forms but letters. For example, an "a" could be included in a "B."[128] Leibniz used both circle and line diagrams. He was inspired by the system of Raymond Llull,[129] which included some diagrams of intersecting circles. Leibniz's line diagrams were, however, in some ways superior, as they were simpler and facilitated comparisons of multiple sets. It is easier to draw and visually compare a large number of parallel lines than the same number of intersecting circles. Since Leibniz's work remained unpublished for long, line diagrams became known through the work of Johann H. Lambert, whose dotted line diagrams were even simpler (Fig. 4.15).[130]

Lambert was part of the third period that began in 1760. Euler, Lambert, and Ploucquet all used Euler-type diagrams (Euler did so in his notebooks already in the 1730s), but they made no references to the earlier periods in their initial publications. However, after publication and having become aware of each other, Lambert and Ploucquet began searching for predecessors. Ploucquet found Lange from the second period but also referred to Raymond Llull from the Middle Ages. Kant used similar diagrams in the late 1700s, and sometimes even referred to them as Euler

[127] Lemanski (2017, 50–55).
[128] Lemanski (2017, 55–58).
[129] See section "Llullism: Incomprehensible Originality".
[130] Baron (1969); Moktefi and Shin (2012, 629).

diagrams, although he did not limit himself to Euler's type. The field of diagrammatic logic soon became engulfed in a dispute between Kantians and Leibnizians. According to Lemanski, that we today refer to such diagrams as Euler diagrams is probably related to the Kantian victory in this struggle.[131]

Lemanski's study is limited to the period from 1530 to 1800, but similar diagrams can be found earlier. The practice of illustrating shared properties by drawing intersecting circles was common in early medieval European diagrams. For example, one type of Carolingian diagram displayed the seasons, cardinal directions, and properties (coldness, hotness, humidity, dryness) as semicircles. Since several seasons are warm, their semicircles were intersected by the semicircle of heat. The earliest example of this diagram is found in *De Natura Rerum*, originally written in 612–615. Since the author, Isidorus of Seville, refers to the diagram in the text, it was likely there in the original manuscript.[132]

Diagrams of intersecting semicircles or arcs were also used to illustrate musical ratios in Boethiuss' *De Institutione Musica*, written in the early 500s CE. Such arcs became so common that they were also used in books on other topics such as arithmetic or other parts of music theory where they made less sense.[133] As in the case of the square of opposition, some versions of the arch diagrams became so complicated that they obscured rather than aided understanding.[134]

Diagrams representing intersections and exclusions can be found also outside Western Europe. The Byzantine monk Hierotheos (d. ca. 1300) devised a variety of circle diagrams that outlined the relationship between father, son, and the Holy Spirit in Orthodoxy, and contrasted it with the Latin dogma.[135] The Baghdad scholar Abu al Barakat (ca. 1080–ca. 1165) drew line diagrams similar to those of Leibniz and Lambert.[136]

There are indications of similar examples from the Indian subcontinent. Dominik Wujastyk mentions a simile from the old Buddhist canon, which talks about a large elephant footprint containing many smaller footprints:[137]

[131] Lemanski (2017, 58–63).

[132] Kühnel (2005); Murdoch (1984, 52–53); Embach (1969, 12–13).

[133] See for example Murdoch (1984, 106–112) and section "The Musical Prelude to Diagrams of Scientific Systems" in Chap. 5, Fig. 5.1.

[134] Murdoch (1984, 105).

[135] Safran (2022).

[136] Hodges (2020).

[137] Wujastyk (2018).

> Brothers, all the footprints of moving animals are contained inside the footprint of an elephant, and the elephant's footprint is considered to be the first amongst them because of its size. In just the same way, Brothers, all good things are included within the Four Noble Truths.[138]

Wujastyk believes this simile predates the death of the Buddha in 400 BCE. He points out that although no preserved Sanskrit manuscripts on logic contain drawn diagrams, it is known that a now lost seventh-century Chinese treatise on Indian logic, *Explanations and Diagrams on Logical Demonstration and Refutation,* did contain diagrams.[139] The author of the *Explanations* was not a Buddhist monk but Lü Ts'ai, a physician, astronomer, and musician at the Emperor's court. He also directed the 120 dancers of the national ballet with the help of diagrams. The layman Lü Ts'ai stirred up resistance among the Buddhist establishment and was forced to apologize for having written the treatise.[140] It is possible that the unconventional background of Lü Ts'ai, for example, his experience of embodied cognition through the ballet diagrams, primed him for applying diagrams to a new field.

The repeated inventions and rediscoveries of Euler-type diagrams across time and space have oftentimes been entirely independent of each other. This reflects that the use of overlapping figures in space as an analogy for overlapping concepts resonates with human spatial cognition, and particularly with the primary metaphor "Categories are Containers."[141]

References

Anacker, Stefan. 2005. *Seven works of Vasubandhu: The Buddhist psychological doctor.* Delhi, India: Motilal Banarsidass.

Archibald, J. David. 2014. *Aristotle's ladder, Darwin's tree: The evolution of visual metaphors for biological order.* New York: Columbia University Press.

Arethas of Caesarea and Michael Share. 1994. *Aretha Kaisareias Scholia Eis Tēn Porphyriou Eisagōgēn Kai Tas Aristotelous Katēgorias: Arethas of Caesarea's Scholia on Porphyry's isagoge and Aristotle's categories: (Codex Vaticanus Urbinas Graecus 35): A Critical Edition, vol. 1.* Athens: Akad. Athēnōn.

[138] Wujastyk (2018, 57).

[139] Wujastyk (2018, 59).

[140] Waley (1952) 2005, 107–111.

[141] The link between Venn diagrams and the primary metaphor "Categories are Containers" is discussed in Lakoff and Núñez (2000, 32).

Baron, Margaret E. 1969. A note on the historical development of logic diagrams: Leibniz, Euler and Venn. *Mathematical Gazette* 53 (384): 113–125.
Bonner, Anthony. 2007. *The art and logic of Ramon Llull: A User's Guide.* Leiden: Brill.
Bourdieu, Pierre. 2002. Habitus. In *Gotik. Arkitektur. Skolastik. Habitus*, ed. Vagn Lyhne, 143–172. Århus: Klim.
Boutell, Charles. 1900. *English Heraldry.* 6th ed. Boston: Page.
Brand, T. E., and A. J. Sherlock. 1972. VK'+ V'K. Venn diagrams or Karnaugh maps, but not both? *Mathematics in School* 1 (7): 4–7.
Bray, Francesca, Vera Dorrofeeva-Lichtmann, and Georges Métailié, eds. 2007. *Graphics and text in the production of technical knowledge in China: The Warp and the Weft.* Leiden: Brill.
Byrne, Patrick H. 1997. *Analysis and science in Aristotle.* New York: SUNY.
Carruthers, Mary J. 2008. *The book of memory: A study of memory in medieval culture.* 2nd ed. Cambridge: Cambridge University Press.
Celaya, Juan de. 1525. *Expositio … in primum tractatum Summularum magistri Petri Hispani.* Brunello.
Christensen, J. Cornelius. 1912. *University business administration.* Topeka: Kansas state agricultural college.
Cicero, Marcus Tullius, H. Rackham, and E. W. Sutton. 2014. *De Oratore.* Cambridge, MA: Harvard University Press.
Couzin, Robert. 2021. *Right and left in early Christian and medieval art.* Leiden: Brill.
Creese, David E. 2010. *The monochord in ancient Greek harmonic science.* Cambridge: Cambridge University Press.
Crisp, Ryan Patrick. 2003. Consanguinity and the Saint-Aubin genealogies. *Haskins Society Journal: Studies in Medieval History* 14:105–115.
Dau, Frithjof. 2009. The advent of formal diagrammatic reasoning systems. In *International Conference on Formal Concept Analysis*, 38–56. Berlin: Springer.
Dobrzański, Michal, and Jens Lemanski. 2020. Schopenhauer diagrams for conceptual analysis. In *International conference on theory and application of diagrams*, ed. Ahti-Veikko Pietarinen, Peter Chapman, Leonie Bosveld-de Smet, Valeria Giardino, James Corter, and Sven Linker, 281–288. Cham: Springer.
Dodwell, Charles Reginald. 1993. *The pictorial arts of the West, 800–1200.* New Haven, CT: Yale University Press.
Doucet-Rosenstein, Diane. 1981. *Die Kombinatorik als Wissenschaftmethode Bei Raimund Lull und G.W. Leibnitz.* Dissertation. Münich: University of Münich.
Dunér, David. 2019. The axiomatic deductive ideal in early modern thinking. In *Cognitive history: Mind, space, and time*, ed. David Dunér and Christer Ahlberger, 3–34. Berlin: De Gruyter Oldenbourg.
Eco, Umberto. 2014. *From the tree to the labyrinth: Historical studies on the sign and interpretation.* Boston: Harvard University Press.

Efficiency Society. 1912. The need of organization. *Bulletin of the Efficiency Society* *1* (3): 5–6.

Embach, Carolyn. 1969. *On the nature of things* (De Natura Rerum), by Isidore of Seville (ca. 560–636 CE) Trans. C. Embach. https://www.researchgate.net/publication/315663902_On_the_Nature_of_Things_De_Natura_Rerum_by_Isidore_of_Seville_ca_560-636_AD_Translated_by_Carolyn_Embach_1969.

Emmer, Michele. 1982. Art and mathematics: The Platonic solids. *Leonardo* 15 (4): 277–282.

Englebretsen, George. 2020. *Figuring it out: Logic diagrams*. Berlin: De Gruyter.

Evans, Michael W. 1980. The geometry of the mind. *Architectural Association Quarterly* 12 (4): 32–55.

Evans, Steven N. 2008. *Probability and Real Trees: École d'Été de Probabilités de Saint-Flour XXXV-2005*. Berlin: Springer.

Even-Ezra, Ayelet. 2021. *Lines of thought: Branching diagrams and the medieval mind*. Chicago: The University of Chicago Press.

Feynman, Richard. 1965. New textbooks for the "new" mathematics. *Engineering and Science* 28 (6): 9–15.

Ganeri, Jonardon. 2004. Indian logic. In *Handbook of the history of logic, vol. 1: Greek, Indian and Arabic Logic*, ed. Dov M. Gabbay and John Woods, 309–396. Amsterdam: Elsevier.

Gardner, Martin. 1982. *Logic machines and diagrams*. 2nd ed. Brighton: The Harvester Press.

Gobet, Fernand, and Gary Clarkson. 2004. Chunks in expert memory: Evidence for the magical number four … or is it two? *Memory* 12 (6): 732–747.

Gontier, Nathalie. 2011. Depicting the tree of life: The philosophical and historical roots of evolutionary tree diagrams. *Evolution Education & Outreach* 4 (3): 515–538.

Granger, Herbert. 1980. Aristotle and the genus-species relation. *The Southern Journal of Philosophy* 18 (1): 37–50.

Hacking, Ian. 2007. Trees of logic, trees of Porphyry. In *Advancements of learning: Essays in honour of Paolo Rossi*, ed. John L. Heilbron, 219–261. Firenze: Leo S. Olschki.

Hägele, Günter, and Friedrich Pukelsheim. 2001. Llull's writings on electoral systems. *Studia Lulliana* 41:3–38.

Hayes, Richard P. 1988. *Dignaga on the interpretation of signs*. Dordrecht: Kluwer.

He, Xi. 2020. *Lineage and community in China, 1100–1500: Genealogical innovation in Jiangxi*. Abingdon: Routledge.

He, Huanhuan, and Leonard W. J. van der Kuijp. 2016. Once again on the "Hetucakraḍamaru": Rotating the wheels. *Journal of Indian Philosophy* 44 (2): 267–302.

Hellström, Petter. 2019. *Trees of knowledge: Science and the shape of genealogy*. Uppsala: Uppsala universitet.

Hodges, Wilfrid. 2020. Medieval Arabic notions of algorithm: Some further raw evidence. In *Fields of logic and computation III: Essays dedicated to Yuri Gurevich on the occasion of his 80th birthday*, ed. Andreas Blass, Patrick Cégielski, Nachum Dershowitz, Manfred Droste, and Bernd Finkbeiner, 133–146. Cham: Springer.

Holladay, Joan H. 2019. *Genealogy and the politics of representation in the High and late Middle Ages*. Cambridge: Cambridge University Press.

Hoskin, Michael. 1997. Astronomy in antiquity. In *The Cambridge illustrated history of astronomy*, ed. Michel Hoskin, 22–47. Cambridge: Cambridge University Press.

Ibn Fariun. 1985. *Compendium of sciences*. Frankfurt am Main: Institute for the History of Arabic-Islamic Science at the Johann Wolfgang Goethe University.

Jha, Aditya Kumar. 2020. A mathematical model of Dignāga's hetu-cakra (within the framework of Simon's bounded rationality). *Journal of Indian Council of Philosophical Research* 37 (3): 471–479.

Katz, Victor J. 1998. *A history of mathematics: An introduction*. 2nd ed. Reading: Addison Wesley Longman.

Kelly, Lynne. 2012. *When knowledge was power*. Dissertation, LaTrobe University. https://opal.latrobe.edu.au/articles/thesis/When_knowledge_was_power/21858909.

Kepler, Johannes. 1597. *Prodromus dissertationvm cosmographicarvm continens Mysterivm cosmographicvm*. Frankfurt: Godefrid Tampach.

Kilpatrick, Jeremy. 2012. The new math as an international phenomenon. *Zdm: The International Journal on Mathematics Education* 44(4): 563–571.

King, Daniel. 2013. Why were the Syrians interested in Greek philosophy? In *History and identity in the late antique Near East*, ed. Philip Wood, 61–82. New York: Oxford University Press.

Kinsella, Karl. 2016. *Edifice and education: Structuring thought in twelfth-century Europe, vol. 1: Text*. Oxford: Keble College.

Klapisch-Zuber, Christiane. 2000. *L'ombre des ancêtres: essai sur l'imaginaire médiéval de la parenté*. Paris: Fayard.

Krämer, Sybille. 2016. Is there a diagrammatic impulse with Plato? "Quasi-diagrammatic-scenes" in Plato's philosophy? In *Thinking with diagrams: The semiotic basis of human cognition*, ed. Sybille Krämer and Christina Ljungberg, 163–177. Boston: De Gruyter.

Kühnel, Bianca. 2005. Carolingian Diagrams. In *Seeing the invisible in late antiquity and the early Middle Ages*, ed. Giselle de Nie, Karl F. Morrison, and Marco Mostert, 359–389. Turnhout, Belgium: Brepols.

Lackner, Michael. 2007. Diagrams as an architecture by means of words: The *Yanji Tu*. In *Graphics and text in the production of technical knowledge in China: The warp and the weft*, ed. Francesca Bray, Vera Dorrofeeva-Lichtmann, and Georges Métailié, 341–377. Leiden: Brill.

Lakoff, George, and Rafael E. Núñez. 2000. *Where mathematics comes from: How the embodied mind brings mathematics into being*. New York: Basic Books.

Laponce, Jean A. 1981. *Left and right: The topography of political perceptions*. Toronto: University of Toronto Press.

Lay Yong, Lam. 1994. Nine chapters on the mathematical art: An overview. *Archive for History of Exact Sciences* 47 (1): 1–51.

Lemanski, Jens. 2017. Periods in the use of Euler-type diagrams. *Acta Baltica Historiae Et Philosophiae Scientiarum* 5 (1): 50–69.

Lemanski, Jens, and Amirouche Moktefi. 2018. Making sense of Schopenhauer's diagram of good and evil. In *Diagrammatic representation and inference: Diagrams 2018*, ed. Peter Chapman, Gem Stapleton, Amirouche Moktefi, Sarah Perez-Kriz, and Francesco Bellucci, 721–724. Cham: Springer Nature.

Link, David. 2010. Scrambling TRUTH: Rotating letters as a material form of thought. *Variantology* 4:215–266.

Lloyd, Geoffrey. 2014. *The ideals of inquiry: An ancient history*. Oxford: Oxford Academic Press.

Londey, David, and Carmen Johanson. 1984. Apuleius and the square of opposition. *Phronesis* 29 (2): 165–173.

Lusthaus, Dan. 2003. *Buddhist phenomenology: A philosophical investigation of Yogacara Buddhism and the Ch'eng Wei-Shih Lun*. London: Routledge.

Lyhne, Vagn. 2002. Pierre Bourdieu. In *Gotik. Arkitektur. Skolastik. Habitus*, ed. Vagn Lyhne, 131–141. Århus: Klim.

Miller, George A. 1956. The magical number seven, plus or minus two: Some limits on our capacity for processing information. *Psychological Review* 63 (2): 81–97.

Moktefi, Amirouche, and Sun-Joo Shin. 2012. In *A history of logic diagrams. In Logic: A history of its central concepts*, ed. Dov M. Gabbay, John Woods, and Francis Jeffry Pelletier, 611–682. Elsevier.: North-Holland.

Murat, Zuleika. 2015. The compendium Historiae in Genealogia Christi and its iconographical tradition: Legacy of classical antiquity in a mediaeval biblical anthology. In *Actual problems of theory and history of art 5*, 404–411. St. Petersburg: St. Petersburg University.

Murdoch, John E. 1984. *Album of science: Antiquity and the Middle Ages*. New York: Charles Scribbner's Sons.

Natali, Carlo, and Douglas S. Hutchinson. 2013. *Aristotle: His life and school*. Princeton, NJ: Princeton University Press.

Netz, Reviel. 1999. *The shaping of deduction in Greek mathematics: A study in cognitive history*. Cambridge: Cambridge University Press.

O'Meadhra, Uaininn. 2012. Medieval logic diagrams: In Bro church, Gotland, Sweden. *Acta Archaeologica* 83(1): 287–316.

Ong, Walter J. 1958. *Ramus. Method, and the decay of dialogue: From the art of discourse to the art of reason*. Cambridge, MA: Harvard University Press.

Panofsky, Erwin. 1951. *Gothic architecture and scholasticism*. Latrobe: Archabbey Press.
Ragen, Brian Abel. 1994. Semiotics and heraldry. *Semiotica* 100:5–34.
Reisch, Gregor. 1503. *Margarita Philosophica*. Strassbourg: Johann Schott.
Rosenberg, Daniel, and Anthony Grafton. 2010. *Cartographies of time: A history of the timeline*. New York: Princeton Architectural Press.
Safran, Linda. 2020. A prolegomenon to Byzantine diagrams. In *The visualization of knowledge in medieval and early Europe*, ed. Marcia Kupfer, Adam S. Cohen, and J. H. Chajes, 361–382. Turnhout: Brepolis.
Safran, Linda. 2022. Diagramming Byzantine orthodoxy. In *The diagram paradigm: Cross-cultural approaches*, ed. Jefferey F. Hamburger, David J. Roxburgh, and Linda Safran, 489–518. Washington, DC: Dumbarton Oaks.
Savage-Smith, Emilie. 2002. Galen's lost ophthalmology and the Summaria Alexandrinorum. *Bulletin of the Institute of Classical Studies* 77:121–138.
Saxl, Franz. 1942. A spiritual encyclopaedia of the later Middle Ages. *Journal of the Warburg and Courtauld Institutes* 5:82–142.
Schopenhauer, Arthur. 1961. *Die Welt als Wille und Vorstellung. Erster Band* [The World as Will and Representation, vol. 1]. Wiesbaden: Brockhaus.
Scriba, Christoph J., Peter Schreiber, and Jana Schreiber. 2015. *5000 years of geometry: Mathematics in history and culture*. 15th ed. Basel: Springer.
Siu, Man-Keung. 1993. Proof and pedagogy in ancient China. *Educational Studies in Mathematics* 24 (4): 345–357.
Slater, Steven, and Alfred Znamierowski. 2007. *The world encyclopedia of flags and heraldry*. London: Lorenz Books.
Stcherbatsky, Feodor Ippolitovic. 1984. *Buddhist logic, vol. 1*. New Delhi: Oriental Books.
Tabing, Felicia. 2018. Mathematics in drafting Japanese crest designs. In *Proceedings of bridges 2018: Mathematics, art, music, architecture, education, culture*, ed. Eve Torrence, Bruce Torrence, Carlo Séquin, and Kristóf Fenyvesi, 463–466. Phoenix: Tesselations Publishing.
Tangian, Andranik. 2014. *Mathematical theory of democracy*. Berlin: Springer.
Tiles, Mary, and Yuan Jinmei. 2004. Could the Aristotelian square of opposition be translated into Chinese? *Dao: A Journal of Comparative Philosophy* 4(1): 137–149.
Verboon, Annemieke R. 2010. *Lines of thought: Diagrammatic representation and the scientific texts of the Arts Faculty, 1200–1500*. Dissertation. Lieden: Leiden University Scholarly Publications.
Verboon, Annemieke R. 2014. The medieval tree of Porphyry: An organic structure of logic. In *The tree: Symbol, allegory, and mnemonic device in medieval art and thought*, ed. Pippa Salonius and Andrea Worm. Turnhout: Brepols.
Waley, Arthur. (1952) 2005. *The real Tripitaka: And other pieces*. Oxfordshire: Routledge

Watson, Arthur. 1934. *The early iconography of the Tree of Jesse*. Oxford: Oxford University Press.

Wilkins, John S. 2009. *Species: A history of the idea*. Berkeley: University of California Press.

Williams, Jean Anne Hayes. 2000. The earliest dated Tree of Jesse image: thematically reconsidered. *Athanor* 18:17–23.

Willman, Marshall. 2021. Logic and language in early Chinese philosophy. In *The Stanford encyclopedia of philosophy*, ed. Edward N. Zalta. Stanford: Stanford University. https://plato.stanford.edu/archives/spr2021/entries/chinese-logic-language/. Accessed 30 July 2024.

Wujastyk, Dominik. 2018. The elephant's footprint: An ancient Indian logic diagram. *Studia Orientalia Electronica* 6:56–61.

Yates, Frances A. (1966) 2014. *The art of memory*. London: The Bodley Head.

Yi, Hwang. 1988. *To become a sage: The ten diagrams on sage learning by Yi T'oegye*. Trans. M.C. Kalton. New York: Columbia University Press.

CHAPTER 5

Printing Diagrams of Systems

Printing has been invented several times in different cultural areas. However, the consequences for the reproduction of diagrams have depended on printing technology, market conditions, and even length of alphabet. Important differences between Europe and China in these respects affected the ability for scholars and printers to include and update diagrams in their works.

Marshall McLuhan and Walter Ong saw the invention of printing as an important watershed in the development from an aural to a visual phase in human culture, and in doing so they overstated the impact of printing on spatial cognition (section "Printing and Spatial Cognition"). For example, McLuhan claimed that before print, manuscript culture "was less tempted to seek the means of reducing non-visual processes of mind to diagrams."[1] Examples from earlier chapters in this book show that human cognition was visual and spatial long before the invention of print, and that diagrams flourished even earlier. However, printing brought real changes to the spatial layout of book pages, such as a structured presentation. This facilitated the emergence and spread of systems, a new genre of scientific literature that was also inspired by music theory (section "The Musical Prelude to Diagrams of Scientific Systems"). These systems were often illustrated

[1] McLuhan (1962) 1967, 159.

J. Holmén, *A History of Diagrams*, Palgrave Studies in the History of Science and Technology,
https://doi.org/10.1007/978-3-031-99691-7_5

with tree diagrams printed with bracket types. Section "Petrus Ramus and the Editable Print Diagrams" discusses how such diagrams were used in the pedagogical philosophy of Petrus Ramus. It also explores why bracketed diagrams were so dominating in scientific and scholarly literature from the 1500s to the late 1700s. The field most intimately associated with bracketed tree diagrams is biological taxonomy, which is discussed in section "The Shape of Life". Chapter 5 ends with a discussion of what is arguably the most successful and resilient diagram of a scientific system ever—the periodic system of the elements (section "The Periodic System of the Elements").

Printing and Spatial Cognition

Still today, some researchers of diagrams refer to Marshall McLuhan's claim that "the media is the message," when they want to emphasize that message is transformed when mediated through a new information technology, such as the printing press.[2] However, McLuhan's claims about the importance of printing were actually more fundamental than that. In the mid-twentieth century, Walter Ong and McLuhan, Ong's master's thesis supervisor, developed the theory that human thought had undergone an evolution from aural to the visual and spatial, and that the Renaissance with the invention of printing was an important milestone in this development.[3] According to Ong, "[t]he history of philosophy itself has largely been the history of a search after more and more visualist or spatialist analogies by which to represent and deal with the real universe and the universe of the mind."[4]

The direct causality between print and "the visual quantities imposed on our psychology" was even more strongly emphasized by McLuhan.[5] In *The Gutenberg Galaxy: The Making of Typographic Man* from 1962, McLuhan leaned heavily on Ong's doctoral dissertation about Petrus Ramus from 1957. However, he mixed in some wild ideas, primarily on the psychology of illiterate Africans, from the ethnopsychiatrist J.C. Carothers. The research of Carothers, as well as the whole field of

[2] See, for example, A.M. Smith (2020) and Kupfer (2020).
[3] Ong (1957).
[4] Ong (1957, 135).
[5] McLuhan (1962) 1967, 246.

ethnopsychiatry, was discredited by the late 1960s.[6] However, the theories that McLuhan constructed on this shaky foundation have survived longer.

McLuhan argued that intensive use of the visual sense had created a spatial and visual mode of thought in civilized man. This stress had increased when humans become sedentary, and again with the introduction of writing. However, it took off rapidly with the invention of print. McLuhan specified that only phonetic writing printed with replaceable types completely separated form from content and isolated the visual sense from the others. Thus, he saw Chinese and Japanese culture as tribal because of their ideogrammatic writing. Russia was described as a primarily oral culture, because of the high levels of illiteracy until recent times. For unspecified reasons, Germans were also said to have "retained the core of auditory tribal unity." Although the development toward visual and spatial thought had taken millennia in human history, McLuhan claimed that it could take place within one generation for the individual. He referred to Carothers idea that learning to read had changed the psychology of educated Africans, while illiterate Africans could still not understand a film or think in three dimensions. However, McLuhan believed that the discovery of electromagnetism had reverted the process and was now leading humanity into a new tribal phase less dominated by the visual mode of thought—the global village.[7]

McLuhan seems to have twisted and turned his arguments in order to ascribe unique cognitive qualities to the Western victors of World War II. Today, especially his insistence that Chinese and Japanese writing does not produce the same visuospatial transformation as phonetic alphabets seems strange. When Asia excels in international student assessments and comparisons of IQ tests, including questions related to spatial thinking, some aspects of the ideogrammatic writing systems have been used to explain this. For example, that Chinese signs sometimes involve rotated versions of other signs have been claimed to help in the mental rotation exercises of IQ tests.[8]

Later research has also emphasized the spatial structure of the medieval manuscripts that preceded printing. According to Even-Ezra, the fact that scholasticism was so preoccupied with subdividing texts with the help of tree diagrams suggests that texts were seen as nonlinear spaces, not as

[6] Carothers (1959); McCulloch (1995, 1).
[7] McLuhan (1962) 1967, 18–44, quote on 29.
[8] Li and O'Boyle (2011).

sequences of events.[9] She even suggests that "some highly structured theological works, such as the *Summa Halensis* and Thomas Aquinas's *Summa theologiae* and their fourteenth-century successors, could not have been written or read without visualizing their structure externally."[10]

For the author of a book basing its premise on the invention of printing, McLuhan was surprisingly uninterested in its actual origins. He seemed to believe that movable type was unique to phonetic alphabets, and claimed that Korea but not China invented movable type, since Korea has a phonetic alphabet. However, movable metal types were used in both countries by the twelfth century at the latest, and the Korean phonetic alphabet was not introduced until 1443. McLuhan also supposed, in line with his theory that the Chinese are "tribal," auditory, and tactile, that they became fascinated by printing for mystical reasons, as the repetitive nature of printing functioned as an equivalent to prayer wheels. In fact, Chinese printing was a large commercial industry, while in Korea it remained a state monopoly.[11]

Ong, who in general made a more careful, nuanced, and well-researched argument than McLuhan,[12] pointed out that it is a mystery why European printing was developed so late, in spite of the fact that all of the necessary elements—punches, molds, lead castings, presses, paper, and ink—had been known in Western civilization since antiquity. His answer was that the invention had to wait for a reorientation of the human spirit that made it possible to think of knowledge in spatial terms. Ong explained this reorientation by claiming that the Middle Ages, since it was oriented toward past authors, was a manuscript culture to a greater extent than antiquity. This created an association of knowledge with space and visual apprehension. Combined with the development of highly quantified logic, for example, that of Peter of Spain, and of organized university studies, this prepared the Western world for the Gutenberg era.[13] Thus, increased

[9] Even-Ezra (2021, 182).

[10] Even-Ezra (2021, 195).

[11] McLuhan (1967, 34–35). His main source on Chinese writing was Carothers (1959), although the idea about prayer wheels is not found there. Information about Chinese and Korean printing is found in T. Christensen (2014); Kim-Renaud (1997); and Tsien (1985, 211). Sohn (1959) would have been available to McLuhan in 1962.

[12] For example, Ong had better knowledge of Asian printing and did not as McLuhan proposes that printing caused fundamental changes in spatial cognition, although he connected it to many other important developments. See, for example, Ong (1982).

[13] Ong (1958, 308–309).

focus on the visuospatial was not only an effect of the invention of print, but also one of its causes.

Ong's explanation disregards that printing was invented much earlier in China. Describing the history of Chinese printing, Tsuen-Hsuin Tsien also discusses why Europe was so late in developing print, despite not requiring any new complicated technology. However, he arrives at an entirely different answer than Ong. Tsien argues that the demand for texts was much lower in Europe than in China, and that the European guilds held back the development of new technology. There were European guilds of block printers, who printed, for example, playing cards, but they did not admit typographers. Reproduction by other than manual means was in many instances forbidden outright.[14]

There was indeed an increased industrialization of book production in the centuries preceding the introduction of printing in Europe, for example, through a putting-out system.[15] The orientation toward reproducing old manuscripts and the growth of the university system that Ong mentions probably paved the way for print, but not by radically altering human spatial thought. Instead, it gradually increased the demand for books until a tipping point was reached where it made economic sense to develop printing, just as it had earlier in China.

McLuhan's perspective on the human relationship to space is the exact opposite to the one adopted in this book. McLuhan saw full visuospatial ability as unique to modern Western Europeans, who had acquired it by exposure to phonetic print. Rather than that the homogeneous grid pattern of print books impressed a new visual and spatial order on readers, as McLuhan claimed, my view is that the rigid grid structures of print books were created in resonance with humanity's innate system of spatial cognition. The same grid pattern found in printed books has existed in the writing systems of most cultures for millennia and is also common in other cultural artifacts throughout history.

Many of the patterns in rock paintings and other early art have a grid structure. They are found among the 32 ice age signs from the period 40,000 to 10,000 BCE identified by Genevieve von Petzinger.[16] These grids are unlikely to imitate nature, since such patterns are rare in the natural world. Grids can be found in tesselations such as crystals, honey-

[14] Tsien (1985, 9).
[15] Eisenstein (1983, 9–10).
[16] Von Petzinger (2017).

combs, and fish scales and in the cracks of, for example, drying clay, but seldom at the right angels preferred by humans. They are rare in comparison to fractals, meanderings, and wave patterns, which cover most of the physical universe in the form of clouds, oceans, sand dunes, mountains, and vegetation. Spirals, spots, and stripes are also far more common patterns in the plant and animal kingdoms than grids. Still, grids are ubiquitous in early human paintings. Such grid patterns could be found also in some of the first civilizations. The city plans of ancient cultures often followed grid patterns, and the Egyptian hieroglyph for district, *hesp*, contained a grid symbol.[17]

It is likely that the grid patterns found in human signs and artifacts from all ages reflect a bias in how humans perceive the world and orientate in it. Several mechanisms might contribute to this. First, neurological experiments with grid cells indicate that mental maps are indexed according to a grid system. Second, experiments have revealed that we perceive horizontal and vertical lines better than slightly tilted ones, and that we tend to perceive tilted lines as more horizontal and vertical than they actually are.[18] If lines in the preferred vertical and horizontal orientations are combined, the result is a grid pattern. Thus, inherent properties of the human cognitive apparatus seem to impose a grid structure on many of our cultural artifacts.

McLuhan considered it wrong to suppose that "Euclidean space, linear, flat, straight, uniform, is rooted in our minds at all. Such space is a product of literacy and is unknown to pre-literate or archaic man." Likewise, Western notions of space and time were absent from archaic man and from Chinese culture.[19] In McLuhan's view, Euclidean space was a mere parenthesis in the history of human consciousness, since it was dissolved with Einstein's theory of relativity in 1905, which allowed for curved space.[20] McLuhan's idea that space was perceived differently in the Middle Ages might not be entirely without merit. For example, his claim that to the minds of the Middle Ages, space did not contain objects, but objects contained space, corresponds well with the findings of later research. However, this change is not explained with the introduction of print, but by the

[17] D.E. Smith (1958, 316); Wallis Budge (1920, 511).

[18] Section "The Neurology of Spatial and Abstract Thinking" in Chap. 2; Tversky (2001, 99–102).

[19] McLuhan (1962) 1967, 177–178.

[20] McLuhan (1962) 1967, 253.

difference between traditional Euclidean geometry and the advances in analytical geometry in the 1700s, including the introduction of the coordinate system.[21]

McLuhan's main mistake seems to be that as a communication theorist and professor of literature, he overemphasizes the impact of media and communication on the human mind. It is not impossible that oral and print culture affected human thinking in the way that he suggests, with oral presentation promoting a more aural, narrative form of thinking and print emphasizing a spatial one. However, human minds have not been formed only or even primarily through communication, as the research on embodied cognition illustrates. In oral cultures, where people instead of reading and writing talked to each other and listened to narrated stories, people still had to live in, interact with, and survive in a complex world in three-dimensional space. Illiterate people had to orientate on hunting and foraging trips according to inner mental maps and perform complex three-dimensional tasks such as weaving carpets and baskets or making fish nets. Before Europeans developed printing, they were already constructing complex machinery such as windmills, spinning wheels, locks, and clocks, activities that involve complex spatial cognition. Cultural phenomena performed in three dimensions, such as games and dances, also provided a broader stimulus of spatial cognition than what was offered through one-dimensional and unidirectional oral narratives. On top of that, oral cultures often linked their stories and songs to spatial locations, making them truly multidimensional. Even in the case of straightforward oral narratives, they were still accompanied by hand gestures, and the stories themselves were given a spatial structure to facilitate memorization.[22]

The theory of the typographic man and his special relationship to space is even more thoroughly undermined by research that indicates that basic principles of geometry are gradually developed during childhood regardless of the cultural setting. For example, members of the Munduruku tribe in the Amazon, who lack formal schooling, seem to be far from "tribal" in McLuhan's sense, but develop a spatial cognition similar to that of the US population.[23] Evolutionary cognitive archaeologists also see evidence that

[21] McLuhan (1964) 2005, 176. For the importance of analytical geometry, see section "Statistical Graphics" in Chap. 6.

[22] See sections "Language and Gestures" and "Terrain and Architecture: Structuring Knowledge in "Oral" Cultures and Early Civilizations" in Chap. 3.

[23] Spelke (2011, 306–309); Izard et al. (2011).

modern abilities for spatial cognition have been present in hominins for at least half a million years.[24]

Was then the true impact of print a mere question of quantity, that more people could now come in contact with books than before, since it was faster and cheaper to reproduce them? A. Mark Smith is of the opinion that the introduction of printed books also had a real influence of the spatial layout of particularly scientific books, and the role of diagrams in them. In medieval manuscripts, the text was dense since the medium of writing was expensive. Paper, introduced in the eleventh century in parts of Europe, was initially expensive but gradually fell in price. Medieval scribes made use of every inch of their expensive pages, and crowded them with tightly stacked words, often abbreviated. The medieval scholarly manuscript might be compared to a compressed file, which the reader had to extract in his mind by reading aloud. Manuscripts with a higher budget could be somewhat more generous with space, but this was more common for luxury bibles than for scientific texts. The logical structure of the books, with chapters and paragraphs, was indicated with markings such as colorized letters that interspersed the compact text. The content had a logical spatial structure, but it had to be reconstructed by the readers, either in their minds or by drawing tree diagrams in the margins.[25]

However, when the price of paper dropped and the printing presses made many more books available, the scarcest resource was no longer the medium of writing or the time of the scribe, but the time of the reader. Smith argues that in order to make the texts more accessible, easier to understand without specialized knowledge, and faster to read, printers introduced a standardized spatial layout, punctuation marks, indices, and more accurate and better placed diagrams. These advances made it easier to navigate the texts and focus on selected parts.[26] Others have pointed out that some of these features, such as indices, become common in twelfth-century manuscript culture.[27]

One aspect of this simplification was that new Latin typefaces were used. For example, in older handwriting, letters with short vertical staffs

[24] Wynn and Coolidge (2022, 9).

[25] A.M. Smith (2020, 211–228); for diagrams in the margins, see Even-Ezra (2021, 7). Smith bases his conclusions on Bloom (2001, 212), and although Smith seems to have misread how quickly the price of paper was dropping – it was a sixth that of parchment already in the 1280s, not at the time of Gutenberg – the general argument is still sound.

[26] A.M. Smith (2020).

[27] Even-Ezra (2021, 24).

such as *m*, *n*, *w*, and *i* often flowed into each other, which was efficient for the writer and economized space, but made it difficult to see where one letter ended and the next began. However, printers solved this by using typefaces where all letters were separated.[28] Thus, printers perfected the grid shape pattern of the book page, but they did it consciously in order to improve readability. This was not a necessary consequence of the technology, since until the twentieth century some books continued to be printed with fraktur typefaces, reminiscent of medieval handwriting. The grid pattern of separated letters did not create spatial cognition, as McLuhan believed, but was created in resonance with innate principles of human spatial cognition that had always been there. This revolution took place when readability could be promoted since writability and economics of paper were no longer restraining factors.

Thus, readers were no longer supposed to read the entire text aloud for themselves, as if listening to a lecture from the author. Knowledge became divided into manageable parts, which were related to each other but could be studied separately. Smith goes so far as to suggest that Kepler's discoveries in optics would not have been made if he had not had access to a new printed edition of Alhacen's optics (originally from ca. 1030) with better structure and diagrams. According to Smith, the new piecemeal presentation in the printed version helped Kepler disentangle light from sight, an approach that has dominated optics since.[29] These new spatially structured arrangements of texts paved the way for a new learned genre called "systems," which replaced older forms of learned textual presentations that had been based on oral lectures and dialogues.

The Musical Prelude to Diagrams of Scientific Systems

The theory that print gave rise to three-dimensional spatial thinking was refuted in the previous chapter. In this chapter, we will address a similar idea: that the astronomical discoveries of Galileo Galilei caused a shift toward thinking in terms of systems, which were regularly presented in diagrammatic form. However, in order to fully understand the co-development of the system concept and diagrammatic forms of

[28] A.M. Smith (2020).
[29] A.M. Smith (2020, 212, 226–227).

representation, we need to consider the work of another Galilei in another century, working in a different field of knowledge.

Clifford Siskin has plotted the increase in the use of the word system in the seventeenth and eighteenth centuries, when it proliferated to a range of fields. He claims that "system" was until then used exclusively to describe the entire Ptolemaic cosmos, with heavenly bodies circling the Earth. However, in the new heliocentric system, planets orbited the sun and moons orbited the planets, creating subsystems. With Galileo Galilei's discovery of moons around Jupiter in 1610, there were suddenly several systems of different sizes. According to Siskin, this made the concept scalable and adaptable. Scholars now began to name their works "systems" rather than "dialogues" or the like, which had been the case earlier. For example, Siskin points out that disciples of Petrus Ramus, such as Johann Heinrich Alsted, liked to present their systems of logic in the form of tree diagrams.[30]

However, it is not entirely correct that the word "system" pre-Galileo exclusively—or even primarily—referred to the entire Ptolemaic cosmos. In a footnote, Siskin recognizes that it could also describe systems of musical intervals.[31] In fact, surveying texts from the fifteenth and sixteenth centuries, it becomes clear that the dominating use of the concept was to describe musical systems. Even if we search for texts containing "Ptolemaeus" and "system," we are as likely to find treatises on Ptolemaeus's musical theory as on his cosmology.[32] Music was at this time still considered a mathematical science, investigating the same divine harmony of proportions as astronomy. This view was held by many of the first astronomers of the scientific revolution, and it was taken to its extreme in Johannes Kepler's synthesis *Harmonices mundi* from 1632, which connected the laws of music and astronomy to symmetrical geometric shapes. Kepler's third law was developed with the help of an analogy between the angular velocities of planets and musical harmonies. Thus, Kepler could draw a diagram that represented the planetary orbits as tunes in a musical score.[33]

[30] Siskin (2016, 29, 38). For Ramus and tree diagrams, see section "Petrus Ramus and the Editable Print Diagrams".

[31] Siskin (2016, 254).

[32] For example, Maffei (1511, 34).

[33] Kepler (1619); Dreyer (1953, 405ff); Leatherdale (1974, 17).

However, although the term "system" had been sometimes used to describe musical scales, it proliferated after the publication in 1581 of a dialogue on music by Vincenzo Galilei (1520–1591).[34] There, he presented and discussed the virtues of ancient Greek and modern musical systems, the latter represented by his former teacher Gioseffo Zarlino (1517–1590). Vincenzo came in contact with Greek musical systems through recent translations from Byzantine manuscripts, with the help of the Florentine classicist Girolamo Mei.[35] Galilei sided with the ancient monophonic music against the polyphonic music of Zarlino. In his first two books on music theory, Zarlino had not used the word system, but it occurs frequently in his 1588 response to Vincenzo's dialogue (Fig. 5.1).[36] It seems likely that he was influenced by Galilei in this respect. Both Vincenzo and Zarlino had great influence on Vincenzo's son Galileo Galilei.[37] Later, Galileo came to write a dialogue of his own where he discussed the Ptolemaic and Copernican cosmological systems. According to Paolo Gozza, Zarlino's *Dimostrationi harmoniche* (1571) was the most direct precedence for Galileo's scientific dialogues.[38]

Zarlino also seems to have inspired René Descartes first work, a short compendium on music that built on older theories but was innovative in using circular logarithmic diagrams of musical intervals. While the compendium was rooted in the tradition of classical music theory and humanism, it also introduced the deductive method that Descartes was to use in later works.[39] Descartes's compendium on music was also a transitional work, passing the torch from ancient musical theory to modern science. Similar works with logarithmic musical diagrams were also produced by the cartographer Gerardus Mercator and by Isaac Newton, of whom at least Newton was directly influenced by Descartes.[40]

Thus, the initial impetus for the proliferation of the concept "system" was not that the astronomical discoveries of the scientific revolution multiplied the cosmological systems, but that Renaissance archival discoveries and translations multiplied the musical systems. In order to explain how these systems were related to each other, music theoreticians such as

[34] Galilei (1581).
[35] See, for example, Palisca (2003, xxxvii–xli).
[36] Zarlino (1588).
[37] Greco (2018).
[38] Gozza (2000a, 95).
[39] Kent (1961); Muzzulini (2015).
[40] Wardhaugh (2008).

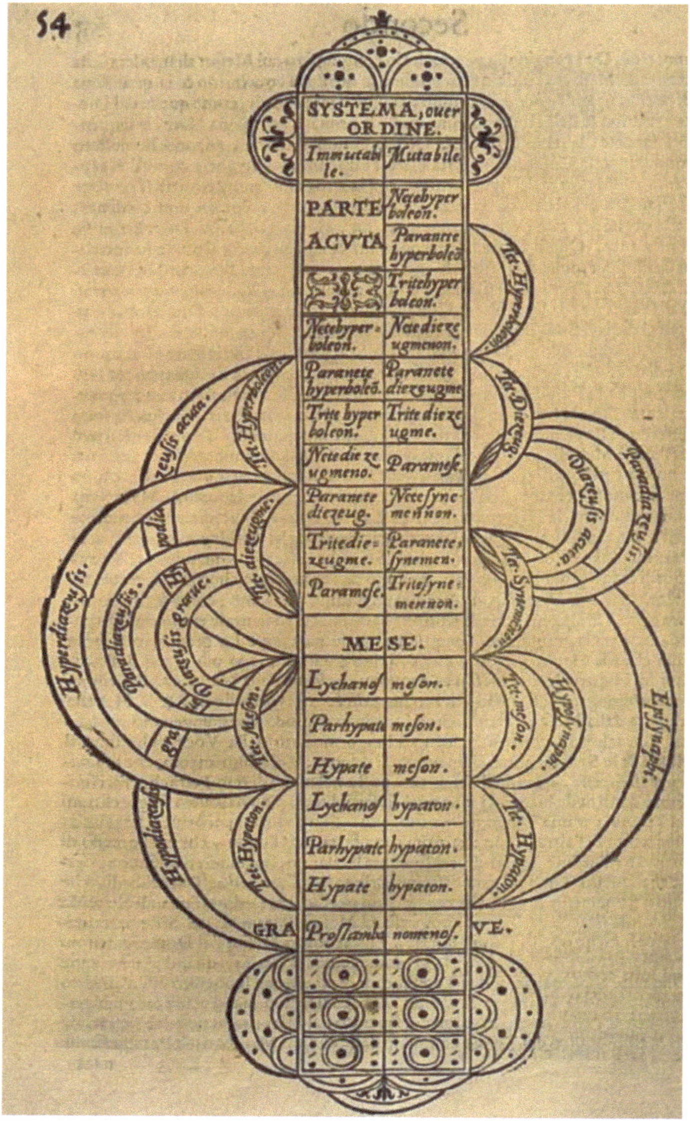

Fig. 5.1 Zarlino's musical system. The concept "system" was used in music theory before it became adopted by the sciences. The overlapping semicircles convey information about range and inclusion comparable to that of the circle and line diagrams discussed in section "Circles of Inclusion" in Chap. 4. Zarlino (1588, 54)

Galilei the elder and Zarlino made frequent use of diagrams. The Renaissance genre of musical treatises was constituted of two parts, called history and method. While the first part provided a background of music's origins, aims, and virtues, the method section presented theories of the musical scales.[41] This part bore many similarities to the later system genre investigated by Siskin.

During the Renaissance and the Enlightenment, the natural sciences developed from being qualitative and dialogical to quantitative and systematic. However, the sciences borrowed the concept "system" from music theory, which during the same period developed in the other direction. According to Paolo Gozza, from the Renaissance to the Enlightenment, music "abandons the Department of Exact Science and enters the Department of Fine Arts."[42] The late 1500s and early 1600s was a period when the arcs of these developments could meet, allowing for a transmission of concepts and styles, including diagrammatic forms.

However, "system" is not the only concept that modern science has adopted from music theory. Although the word "diagram" has Greek roots, it entered the European languages through the Latin *diagramma*. This rare word is missing from most Latin dictionaries since it was introduced to the language relatively late. The Greek *diágramma* was used in Aristotle's *Nicomachean ethics* and *Organon* and in Aristoxenes's musical theory, but these works were lost in Europe until their rediscovery in the High Middle Ages. Aristoxenes's *Elementa Harmonica* was not translated to Latin until 1564.[43] In contrast, Euclid's *Elementa* was preserved through the Dark Ages. Although it was full of diagrams, he mostly called them σχῆμα, schema, which was translated into *figura* in Latin.[44] In the rare Latin dictionaries where you can find *diagramma*, its primary meaning is "musical scale." In English, the word "diagram" was first used in the 1610s.[45] Thus, the concepts system and diagram arrived to the scientific revolution simultaneously, and from the same source—music theory.

[41] Gozza (2000b, 28–35).
[42] Gozza (2000c, xii).
[43] Livesey (1989, 25n20).
[44] For example, Euclid (1301).
[45] Barnhart and Steinmetz (1988); Georges and Georges (1913).

Petrus Ramus and the Editable Print Diagrams

Petrus Ramus (1515–1572) was professor at the College de France in Paris. He wanted to reform university education by replacing Aristotle's texts with his own textbooks, where knowledge was presented according to a method that often involved the construction of branched diagrams. The understanding of Ramus has been dominated by Walter Ong's (1958) biography of him. Ong did not consider Ramus particularly brilliant or important, but rather treated him and his followers as a symptom of what he considered to be the true driving force in cultural history during this period: the invention of print. Ong's quite negative views of Ramus are balanced by newer research such as that of Howard Hotson.[46] Regardless of Ramus's individual importance, the fact that he has been the object of so much research makes him a good lens for studying the relationship between print culture and diagramming practices.

Hotson has criticized Ong for neglecting the specific pedagogical, social, and geographical contexts of Ramus's method. According to Hotson, Ramism was associated with a seven-year liberal arts curriculum for boys ages 8 to 15 that was particularly common in cities of middling size and influence in northwestern Germany. The project was inspired by an egalitarian intention of making knowledge available to broader circles in society through an efficient and cost-effective education. Ramus was himself the grandson of a charcoal burner and wanted to enable such social mobility by lowering the cost of education. Ramism was first adopted in lower schools such as *gymnasia* and *gymnasia illustra* and only later worked its way into universities.[47] Thus, Ramism should first and foremost be understood as a pedagogical method that achieved social aims through simplification, for example, with the help of diagrams (Fig. 5.2).

According to Ramus, "'[t]he method of teaching . . . is the arrangement of various things brought down from universal and general principles to the underlying singular parts, by which arrangement the whole matter can be more easily taught and apprehended.'"[48] Ong perceived this as an oversimplification of philosophy for didactical purposes. He saw Ramus's tendency to turn dialectics into didactics with the help of spatial schemas as the culmination of a development that had begun in

[46] Ong (1958); Hotson (2007, 9).

[47] Hotson (2007, 35–41).

[48] Quoted from Ong (1958, 245).

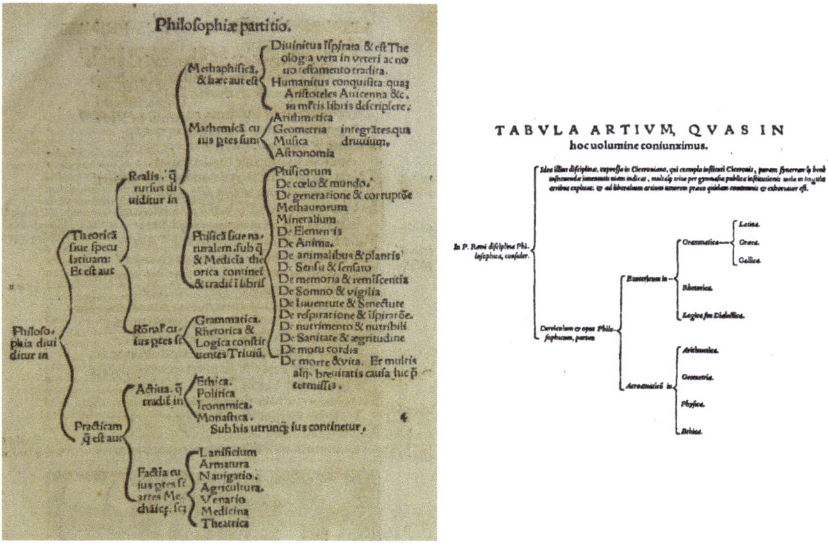

Fig. 5.2 Printing the branches of knowledge. Left: Diagram with printed text and woodcut brackets from Gregor Reisch's *Margarita Philosophica*, 1503. Right: In later prints, such as this outline of the liberal arts curriculum by Ramus from 1576, brackets were instead constructed with lead types. Ramus's pedagogical diagrams had a limited number of branches at each node, often simple dichotomies. Ramus (1576)

scholasticism. At medieval universities, the faculty of Arts was primarily concerned with how to teach logic and physics to adolescent boys. In Ong's view, this created a mindset of simplification where it was considered that "a diagram or its equivalent is adequate to any situation."[49]

An important step in the direction of pedagogical simplification was taken by Rudolph Agricola (1443–1485), who made the topics or places in Aristotelian logic, which functioned as headings for stores of knowledge, the core of logic. According to Ong, Ramus and his followers believed that any area of learning could be analyzed by spatial arrangement through definitions and divisions. In Ramus's words, such a diagram "provides the collation and arrangement of many and various arguments cohering to one another and linked by an unbroken chain as to lead to one

[49] Ong (1958, 147).

determined end." We would label it a tree diagram, and, alluding to his own name, Ramus used the same metaphor, as he described how golden apples dangled from the boughs (*rami*) of the tree of knowledge.[50] Ayelet Even-Ezra has pointed out that Ong was mistaken in claiming that Ramus's tree diagrams constituted a radical break spurred on by print culture. In fact, horizontal tree diagrams were often drawn in the margins of manuscript by medieval scholars trying to understand and structure the texts they were reading, and sometimes they were placed there by the original scribes.[51]

Ramus's system developed over time, as he moved in the direction of splitting up his categories into a web of dichotomies. According to Ong, these dichotomies had no clear philosophical foundation and were not conducted according to any obvious principle, but were dictated by their pedagogical appeal.[52] However, Hotson claims that Ong adopted the negative view of Ramus's method held by his contemporary adversaries. In contrast, others such as Francis Bacon, Comenius, and Leibniz lauded the three laws of truth, homogeneity, and reciprocity that Ramus had developed for organizing the treatment of any discipline.[53] Ramus was also more careful to respect cognitive limitations than, for example, Gregor Reisch, who used a physical/medical bracket with 17 subdivisions (Fig. 5.2). Thus, this diagrammatic method had old roots, both in manuscript and print culture, but Ramus perfected it.

Ramus criticized old systems of logic, primarily Aristotle's system, and promoted his own on the grounds that it was a better aid for memorization. Ancient and medieval philosophers had also suggested mnemonic techniques, but Ramus claimed that they were artificial, while his system of divisions was a natural order reflecting the organization of concepts in the mind. Ong described Ramus's system as a map of the mind that culminated in diagrams. He claimed that many earlier memorization techniques had been based on sound recall, and that was why Ramus perceived his system based on analogy between the mind and objects in space more real.[54] Ong's emphasis on sound in earlier mnemonic techniques is influenced by his and McLuhan's hypothesis that humanity transitioned from

[50] Ong (1958, 104–105, 174, 187ff, quote on 187).
[51] Even-Ezra (2021, 7).
[52] Ong (1958, 199–202).
[53] Hotson (2007, 45).
[54] Ong (1958, 194–195).

an auditory to a visual period with the introduction of print culture. As we have seen, many earlier memorization techniques were also spatial, such as the method of loci.[55] However, they had built on constructing artificial spaces for memory storage through graphic imagery rather than hierarchical structure.

Although we can establish that the bracketed tree diagrams used by Ramus and his followers were not a novelty of print culture, it is still debated why this diagram type came to dominate in printed works from the 1500s to the 1700s. Hellström has argued that the shift from figurative to abstract diagrams in the sixteenth and seventeenth centuries was associated with advances in printing technology. It was easy to produce abstract tree diagrams through printing using letter and bracket types, while other forms of diagrams required engravers.[56]

However, it is also possible that the technical advances in printing reinforced ideological preferences for the abstract over the figurative. According to Elizabeth Eisenstein, "[p]rinting made it possible to dispense with the use of images for mnemonic purposes and thus reinforces iconoclastic tendencies already present among many Christians."[57] Ramus converted to Protestantism in 1561, and was killed during the massacre of St. Bartholomew in 1572. Frances Yates claimed that "[t]he extraordinary success of Ramism, in itself rather a superficial pedagogic method, in Protestant countries like England may perhaps be partly accounted for by the fact that it provided a kind of inner iconoclasm, corresponding to the outer iconoclasm."[58] Just as images on church walls, which had been used to tell stories to and aid the memories of largely illiterate congregations, were replaced by the written word, the memory-evoking inner images recommended by ancient rhetoricians were replaced by Ramus's simple diagrams. Yates points out that in his only theological work, Ramus criticized Catholic images and Greek idolatrous worship, and suggests that this was relevant for his attitude to images in the art of memory.[59]

In order to explain the connection between early printing and diagramming, we need to look further into the technology and market conditions of printing. The curly braces or brackets, { }, that we find on today's

[55] See section "Terrain and Architecture: Structuring Knowledge in "Oral" Cultures and Early Civilizations" in Chap. 3.

[56] Hellström (2019, 237f).

[57] Eisenstein (1979, 64).

[58] Yates (1966) 2014, 231.

[59] Yates (1966) 2014, 23–234.

keyboards were invented by medieval scholars engaged in producing horizontal tree diagrams.[60] In the age of print, these diagrams became easy to reproduce as adjustable brace types were invented. Bracketed diagrams in the first printed books were combinations of printed text and woodcut brackets (Fig. 5.2 left).[61]

Soon thereafter, such brackets or braces began to be constructed with the help of lead types. According to Joseph Moxon's printing manual from 1683, published when this practice had been common for one and a half centuries, printers kept bracket types of different lengths, spanning up to four lines. Short of a century later, Philip Luckombe's manual for printers from 1771 recommended that printing houses had braces of up to six lines. Longer ones could be constructed with the help of brass rules, according to Moxon, 16 inches long, which could be severed to print lines of desired length. By combining them with the middles and corners of brackets, any length of bracket could be constructed. However, Luckombe claimed that in his day brass rules were becoming unfashionable, since they did not always print well and some printers thought it "not pertinent to their profession to work in Brass." Instead, braces were more often constructed by combining middle, corner, and rule types. These were a real test of a founder's skill, since it was difficult to combine them in such a way that the brace seemed as if cast in one piece.[62]

Luckombe claimed that

> Braces Are used chiefly in Tables of Accounts, and other such - like Matter that consists of a variety of Articles, which would require much circumlocution, were it not for the curious method of Tabular Writing, which is practised in England to greater perfection than in any other Nation. Braces stand before, and keep together, such Articles as are of the same import, and are Subdivisions of preceding Articles.

Braces standing after lines could be used to summate, for example, pecuniary or mercantile items.[63] Ong claimed that Ramism was particularly popular in the merchant classes, and that Ramus's publishers had great part in the success and spread of his book. For example, the "merchant-printer-humanists" of the Flemish Wéchel printing firm had an

[60] Even-Ezra (2021, 9, 194–199).
[61] Reisch (1503).
[62] Luckombe (1771, 284), Moxon (1683) 1958, 26–27, 35, 335.
[63] Luckombe (1771, 283).

active role in the spread of Ramism. Ong suggests that artisan and bourgeois culture desired a simplified philosophy in which reality could be explained through analogy with mechanical constructs.[64] It can also be argued that Ramism was a pedagogical system that mirrored the merchant classes' diagrammatic language of accounting.

A factor that contributed to the success of brackets was that they could be used for two different forms of the diagramming practice of linking. They can function as connectors, the lines connecting nodes in tree diagrams, or as boundaries, summating or keeping like items together, as the circles in Venn diagrams or semicircles in music theory.[65]

In response to claims that Ramus and Ramism influenced the scientific revolution by encouraging experimentation, Ong pointed out that Ramus himself seldom mentioned observation and experimentation and that his basic positive view on inductive reasoning was shared by most medieval thinkers, such as Thomas Aquinas. According to Ong, Ramus's influence on the scientific revolution was real but indirect and unintended, as his visual approach to the field of knowledge was favorable to the emergence of modern science.[66]

It can be argued that illustrations in medieval manuscripts and woodblock printed diagrams had also appealed to the visual sense. Several scholars have questioned the visual novelty of printed books compared to what preceded them: most typographic and diagrammatic techniques used in printing had direct precedents in medieval manuscripts.[67] However, as mentioned above, the central property of Ramus's method was that it was expressed in the form of editable, mass-reproducible diagrams. That capacity was central for the experimental and observational sciences that were to emerge in the following centuries, since they resulted in a cumulative knowledge production that necessitated constant revision.

In addition, diagrams constructed with standard types also made it easy and economical to produce good quality diagrams in small edition prints, since no engravers needed to be hired. In both Europe and China, scientific texts were mostly small editions. In Europe, most printed texts were for centuries religious, and the best-selling scholarly works were ancient or

[64] Ong (1958, 97, 311).

[65] For more on linking and other diagramming practices, see Engelhardt and Richards (2021). For Venn diagrams and semicircles in music theory, see sections "Circles of Inclusion" in Chap. 4 and "The Musical Prelude to Diagrams of Scientific Systems".

[66] Ong (1958, 268–269).

[67] See, for example, Eisenstein (1979, 92).

medieval texts, not new scientific treatises. In China, type printing, which allowed for easy editing, was only economical for large editions, due to the high investment costs associated with the large number of characters. This meant that in Europe, new scientific ideas were disseminated by movable type printing, where both text and diagrams were easily edited and updated. In China, scientific communication was confined to block printing, where neither text nor diagrams were easily changed. It has been claimed that the invention of woodblock printing in China, after spurring an initial burst of creativity in the production of diagrams, led to a stagnation as printers continued to use old woodblocks for diagrams in new books, even when they were less appropriate.[68] Initially, European printing suffered from similar disadvantages. Andrew Pettegree has claimed that the management of texts and illustrations was in many ways easier in manuscripts than in print. He illustrated this with a page of diagrams in an early Venetian printed book, displaying "an uneasy combination of densely massed text with woodcuts of mixed quality and relevance."[69] Diagrams constructed with the help of bracket types solved this problem by being editable in the typesetting process.

As the availability of old woodblocks in China induced printers to reproduce the same diagrams over and over, the comparative ease of constructing branching diagrams with bracket types in Europe enticed writers and printers to present their ideas in that spatial form. This might also help explain why Ramus attempted to squeeze every conceivable topic into the same diagrammatic structure, and why his method reached such a wide dissemination: it was perfectly adapted to the medium of the day.

Ong claimed that although Ramus's method was supposed to be "natural," his diagrams were altered between every edition of his books.[70] However, this rapid pace of change was a feature and not a bug of bracket diagramming. It was a technology well suited for fast and cost-efficient updating of diagrams in low-volume scholarly print publications, something that would be invaluable during the rapid knowledge accumulation of the scientific revolution.

Ramus's program was continued by successors such as Keckermann and Alsted. A culmination in the attempt to summarize all human knowledge in bracketed tree diagrams was reached with Alsted's *Encyclopedia* from

[68] Bray (2007, 68–69).
[69] Pettegree (2010, 59).
[70] Ong (1958 199–202).

1630, where the first 26 pages were covered with 38 trees of different aspects of knowledge.[71]

Their advantages made brackets the dominating forms of diagrams for several centuries. Hellström has observed that tree diagrams became more abstract with the introduction of bracket-type print in the sixteenth and seventeenth centuries, but then again more naturalistic when copperplate printing became more common in the late eighteenth century.[72] Although printing with engraved copperplates had been used from the fifteenth century, it had the disadvantage that, unlike woodcut illustrations, it could not be mixed with types on the same page and printed in an ordinary printing press. In woodcuts, you removed all the white parts of the image and put the ink on the elevations, which made them compatible with type. In copper engravings, you removed the black parts of the image, and filled the crevasses with ink. Squeezing the ink out of the crevasses required harder pressure during printing, making it incompatible with type printing. This meant that you either had to print illustrated pages twice in two separate presses, or have the illustrations on separate pages from the text. On the other hand, copperplates made it much easier to engrave, for example, a diagram, as you only needed to remove the lines, not everything but the lines. In addition, you could easily change the diagram on a copperplate to the next edition by engraving new lines, while changes in a woodcut required that you inserted new plugs of wood. Woodcuts allowed for very long series of prints, while copperplates deteriorated with use. However, copperplates were easier to store since woodcuts could dry, crack, and get eaten by worms. In the late 1500s, woodcuts began to be replaced by copper engravings in books, as readers demanded better quality engravings.[73]

Copper prints did, however, demand more technical skill, and only in the late 1700s did capacity to produce high-quality prints spread widely across Europe. In France, the revolution in 1789 collapsed publisher's copyright and deregulated the printing trade, which led to an increase in book illustrations.[74] In the late 1700s and early 1800s, machines were developed that simplified the engraving of textures and sped up the copper

[71] Alsted (1630).
[72] Hellström (2019, 238).
[73] Griffiths (2016, 21, 182–183).
[74] Griffiths (2016, 24, 192).

print process.[75] These advancements in printing around the year 1800 facilitated the development of new forms of diagrams that did not use a bracket format, such as the first data visualizations.

In summary, the long period of dominance of bracketed diagrams had roots among medieval university scholars who wanted to simplify complicated philosophy into a form palatable to the growing numbers of adolescent boys in their classrooms. With the invention of print, bracketed diagrams became easy to produce compared to other diagrams. Printers could use them at a low cost also in small editions, since no engravers needed to be hired. The scientific revolution created a need for systematization of new knowledge, which the bracketed tree structures were well adapted for. They could easily be changed from one edition to another, for example, when new lands, species, or astronomical objects were discovered. In the preceding period, bracketed diagrams had been used for logic and textual analysis, and, as Walter Ong has pointed out, for structuring knowledge based on everyday experiences common to all.[76] That the same diagramming technique could be used for so long and for so many different purposes reflects that tree structures resonate with the limitations of the human consciousness: we need to subdivide the complex world into a hierarchy of manageable chunks of information. Thus, Ramus's system was "natural," but not in the sense that it reflected an order in the external natural world. Instead, it was adapted to our natural way of perceiving and thinking about the world. Using the terminology of cognitive anthropology, we can say that Ramus was visualizing folk taxonomies.[77] Being an educator rather than a researcher, Ramus did not himself engage in structuring new knowledge from new discoveries. However, the emerging sciences adopted the same approach. The research field most intimately associated with tree diagrams is natural taxonomy.

The Shape of Life

Cognitive anthropologists have found that, in all living languages, names of plants and animals are taxonomically arranged in hierarchies with a small number of ranks.[78] Similar taxonomic structures are also found in

[75] Griffiths (2016, 39–40); Sarfatti (2022, 159).
[76] Ong (1958, 195).
[77] See section "The Shape of Life".
[78] Boster (2005, 110).

historical sources. That Aristotle arrived at a hierarchical system of plants and animals similar to that discovered by cognitive anthropologists is not surprising if we consider that he used an "anthropological" method. His collections of popular proverbs constituted an important source for his studies of nature. He justified this with the wisdom of the crowd, as he claimed that "nobody is able to reach the truth satisfactorily, nor completely misses it; but each one says something true about nature, and while individually they contribute nothing, or not much, to the truth, from the conjunction of all we get a considerable amount."[79] Aristotle's use of "citizen science" should be understood in the context of the mass participation that characterized all aspects of public life, political, administrative, and military, in Athens and many other Greek city-states.

Aristotle's ten books on animals, best known under their Latin title *Historia Animalium*, provided several attempts at classification. Using the theory of the four elements and their corresponding bodily fluids, he divided animals into, for example, bloodless and blooded, which roughly corresponds to vertebrates and invertebrates. He also used a division into species and genera. However, these terms, which we also know from Aristotle's logic, had a relative meaning since species in turn formed genera on the next level in the hierarchy. In addition, Aristotle pictured life as a continuous hierarchical scale or ladder of increasing complexity from lifeless matter via plants up to animals. Different plants formed a continuous scale from those similar to the lifeless world to the animal-like plants. No ancient illustrations of such ladders have survived, if they ever existed, but they are clearly described in texts by Aristotle and later authors such as Pliny the Elder.[80]

During the Middle Ages and the Renaissance, the Aristotelian ladder was expanded to a great chain of being that went from the lowest animals to the angels and God. From this period, we also have surviving diagrams of ladders or chains.[81] Still in the 1700s, naturalists such as the Swiss philosopher Charles Bonnet viewed the order of nature as a ladder.[82] However, by then the ladder coexisted with other forms of visual representations.

Holger Funk has shown how from the sixteenth century, botanists adopted the dichotomous horizontal tree diagrams printed with bracket

[79] Natali and Hutchinson (2013, 25–26).
[80] Archibald (2014, 1–2).
[81] Lovejoy (1936); Archibald (2014, 3–5).
[82] Archibald (2014, 7–8).

types that Agricola and Ramus, for example, popularized. Such dichotomies were first used in medical texts, and most early botanists were medical professionals.[83] Arab medical writings had included tree diagrams, possibly based on examples in earlier Greek Alexandrian treatises.[84] However, such diagrams were now increasingly used to structure the collection of data. In the words of Funk:

> The efforts of humanists like Agricola and Ramus were directed against what they believed were abstract, logical quibbles by mediaeval scholars, of no avail in the education of students. Against this tradition they developed their own "dialectica"—concrete, practice-oriented, and immediately applicable to mundane matters of all kinds.[85]

The botanical dichotomous diagrams should help students learn to identify, name, and memorize plants. In addition, the brackets were used to create a typology of different plants based on their properties. However, this was not yet a classificatory system in the modern sense, since scholars in the sixteenth century due to the heterogeneity of plants consciously avoided definitions that were too precise in favor of typical descriptions. The Aristotelian concepts of species and genera were used, but their meanings were still fluid and relative.[86]

However, exact classification was the aim of Linnaeus (1707–1778) in his *Systema Naturae*. He was inspired by Aristotle's nomenclature and logic structure, but even more so by medieval scholastic logic. In scholasticism, classifications were often performed according to a five-level schema: genus summum, intermedium, proximum, species, and individum.[87] The cognitive anthropologist Scott Atran claims that the five-level hierarchy of scholasticism appealed to Linnaeus because it resonates with the principles of folk taxonomies. Atran attributes this folk taxonomy to what he calls "common sense," domain-specific cognition given to all humans.[88] However, the principles of categorization found in folk taxonomies do not seem to be domain specific, but are applied to all kinds of domains. For example, Linnaeus also likened his five-level categorization

[83] Funk (2014, 100–112).
[84] Savage-Smith (2002).
[85] Funk (2014, 106).
[86] Funk (2014, 106–107).
[87] See Larson (1971, 150).
[88] Atran (1990, 253).

to the five levels of ranks in the military and the five levels of administration in a kingdom.[89] Another cognitive anthropologist, Roy D'Andrade, argues that the limitation of biological folk taxonomies to five levels is related to limitations in short-term memory rather than to any structure of plants.[90]

Linnaeus used bracketed tree diagrams as tools for memorization and elaboration of his natural system. His early publications especially were endowed with dichotomous diagrams, such as the diagram of his sexual system in the first edition of *Systema Naturae* (1735). In later editions, the amount of text increased—the first edition had 15 pages, the tenth 830 pages, and the thirteenth posthumous edition 3600 pages—and the brackets disappeared. However, the bracket heads became heads of tables. According to Matthew D. Eddy, this suggests that Linnaeus's perception of the system remained fundamentally spatial, and the omission of diagrams in later editions was due to typographical concerns and space restraints.[91] Staffan Müller-Wille and Isabelle Charmantier have interpreted Linnaeus's diagramming practices as a method for coping with severe information overload, which was inflicted upon him by his vast network of collaborators and disciples.[92]

After Darwin, we interpret trees of life as literal family relationships. However, the pre-Darwinian trees of life filled other purposes. Some "artificial" systems were merely intended as tools for identifying species, such as the one of Jean-Baptiste de Lamarck (1744–1829) from 1778. Such keys (*claves*) can still be found today, for example, in Stuessy's plant taxonomy from 2009. Other tree diagrams were supposed to represent a natural order, such as that of Antoine Laurent de Jussieau (1748–1836) from 1789.[93] However, the naturalness of such systems was not evolutionary, but tied to the plan of creation. As Petter Hellström has phrased it, "[e]ighteenth-century naturalists were not trying to reveal the history of nature, they were trying to reveal its *shape*."[94]

When displayed in diagrammatic form, Jussieau's natural method was very similar to Lamarck's artificial one.[95] The classificatory tree diagram

[89] Larson (1971, 150).
[90] D'Andrade (1995, 43, 93, 240).
[91] Charmantier (2011, 404); Linné (1735) 1960; Eddy (2010, 241–242).
[92] Müller-Wille and Charmantier (2012).
[93] Funk (2014); Lamarck (1778); Stuessy (2009); Jussieau (1789, lxxi).
[94] Hellström (2019, 66).
[95] Hellström (2019, 72–73).

was firmly established as the dominating form of scientific diagram across disciplines, and it should therefore be no surprise that it could fill several functions within biology. Although Linnaeus considered his own sexual system artificial, in *Philosophia Botanica* he stated his conviction that there was a natural order. The view of it was still fragmented, but Linnaeus believed it would become perfected through the accumulation of knowledge. However, it was not best described by the sharp distinctions of a tree diagram, but by another spatial representation: "All plants exhibit their contiguities on either side, like territories on a geographical map." Some of the earliest attempts to visualize Linnaeus's system resembled ordinary maps, with species appearing as free-flowing, bounded islands in an archipelago.[96]

Giulio Barsanti has shown that the map was a very common analogy for the natural world in the late eighteenth and early nineteenth centuries. He claims that it formed an intermediary stage between the one-dimensional ladder of the Middle Ages and the tree that dominated from the second half of the nineteenth century.[97] However, since some trees are as old as the oldest maps,[98] it is difficult to see maps as an intermediate form preceding the trees. They were rather a parallel alternative that become less common with time.

Some of the diagrams that Barsanti classifies as maps are more reminiscent of network or tree diagrams with thin lines connecting the names of species and genera. However, in others, such as Giseke's map of Linnaeus's system from 1789, genera occupy delimited areas, as nations or islands on a geographical map. Relations and similarity between groups of organisms are expressed by spatial proximity on these maps.[99] The interpretation of such diagrams is quite intuitive to most observers since it is founded on the primary metaphors "Similarity is Closeness" and "Categories are Containers."

However, in 1801, the priest and schoolteacher Augustin Augier (1758–1825) published a natural order that was based on another

[96] English translation from Charmantier (2011, 399); Broberg (1990, 55); Jönsson (2014, 111f); Hellström (2019, 67).

[97] Barsanti (1992, 43–73).

[98] See, for example, the trees discussed in Funk (2014).

[99] Barsanti (1992, 43–73).

common diagrammatic analogy, that of the genealogical tree. In contrast to earlier botanists, Augier did not arrive at his model from actual plants, but from comparing existing classificatory systems with each other. He codified the systems into synoptic diagrams and arranged them so that he could view them all at the same time.[100]

While Augier compared his botanical tree to genealogical trees, nothing suggests that he considered groups of plants to be descendants of each other; the natural order he sought was still that of creation. The analogy with the genealogical tree was limited to its structure. Although the central part of Augier's publication was a two-dimensional fold-out tree diagram with naturalistic elements, the accompanying text explained that it should be understood in three dimensions. As Linnaeus, he believed that the biological world was continuous, and interpreted the gaps between the branches of the tree as species and genera not yet discovered.[101] The complete "tree" would thus look like a compact sphere, reflecting the perfect symmetrical order of creation.

From the Middle Ages to the early modern period, genealogical trees had clear religious and feudal connotations, as they were used to explain family relations of royals, nobles, or biblical figures. However, Hellström has claimed that with the French Revolution genealogical trees became secularized, and were taken up by the sciences, although the foundation for this development was laid earlier by the French encyclopedists.[102] Augier's genealogical tree is an example of this. Hellström emphasizes Augier's background as a teacher and member of a religious order dedicated to education. This sets him apart from professional naturalists and might have made him open to communicate through images and metaphors.[103]

Although Augier did not intend to illustrate descent, his choice of diagrammatic form might have swayed others who came in contact with his tree to associate natural order with biological descent. According to Hellström, it is not impossible that Lamarck and Georges Cuvier became influenced to use arboreal metaphors after Augier's work was presented at the academy of sciences in Paris in 1801.[104] A similar influence might have taken place regarding the genealogical aspect.

[100] Hellström (2019, 77).
[101] Hellström (2019, 78–79, 82–83).
[102] Hellström (2019, 275–276).
[103] Hellström (2019, 109).
[104] Hellström (2019, 115–121).

In the mid-1700s, however, some genealogical diagrams had been used as breeding charts for domesticated species. For example, Georges-Louis Leclerc Buffon (1707–1788) had described his diagram of dog breeds as a kind of genealogical tree. In addition to descent, it displayed the geographical origin of different breeds. By the early 1800s, Lamarck had adopted the view that not only races and varieties could change and evolve, but also species.[105]

In the first half of the nineteenth century, the advance of geology and paleontology made it possible to construct diagrams of the relative prevalence of different groups of animals at various geological periods. In 1837–1838, Heinrich Georg Bronn constructed a diagram of parallel lines[106] similar to the timelines of Joseph Priestley.[107] One needed only connect the beginnings of the shorter lines to the longer ones in order to construct a tree diagram. Bronn did so in 1858, but before then the "Paleontological chart" in Edward Hitchcock's geology from 1840 and Louis Agassiz's genealogy of fishes from 1844 had also been published. A modern reader would interpret these diagrams as evolutionary, but, although they realized that the world was old, neither Hitchcock nor Agassiz believed in the transmutation of species. This is revealed by the fact that Agassiz did not attach the side branches to the trunks of his tree since they, according to him, had not transformed from one another, but were affiliated through the creative intelligence of God. In editions of Hitchcock's geology after the publication of Charles Darwin's *The Origin of Species* in 1859, the paleontological chart was removed. The diagram could no longer be used in the old way since it was now associated with evolution.[108]

A similar tree diagram of the animal kingdom was published in 1857 by Anna Maria Redfield (1800–1888), a Canadian with a master's degree from Ingham, the first university in the United States to admit women. Although later editions of Redfield's work were published after Darwin had made his theories public, Redfield maintained her creationist views.[109] Redfield, one of the first women to receive a university education, was also among the first to publish a tree diagram. This underscores how closely

[105] Hellström (2019, 84–85, 124).
[106] Archibald (2014, 77–79).
[107] See section "Timelines: Spatial Representations of Time" in Chap. 7.
[108] Archibald (2009, 2014, 70–74, 76–77); Hitchcock (1840, 99f).
[109] Archibald (2014, 74–76).

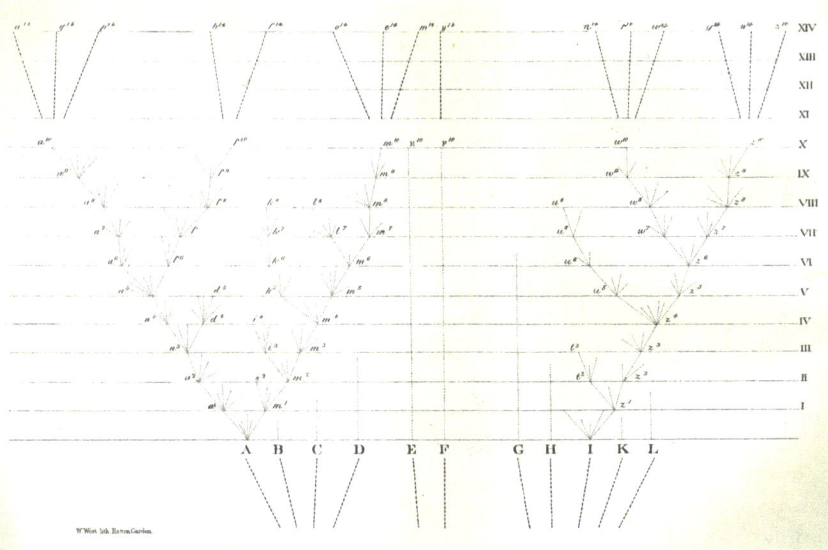

Fig. 5.3 Darwin's tree from the *Origin of Species.* The shorter branches at each node represent variation, and the longer ones continuing to the next node, the survivors of natural selection. Darwin (1866)

related tree diagramming is to the university tradition with roots in medieval scholasticism.

These trees all attempted to display the particular relationships between species and genera. Darwin was probably the first to adopt the medieval concept of the tree of life to describe a diagram displaying the relationship between living species.[110] However, his only published diagram of life (Fig. 5.3), found in *The Origin of Species*, is in fact purely abstract: it illustrates the principle of species formation and differentiation. All branches are unlabeled. Several other sketches of tree diagrams are known from Darwin's notebooks and unpublished works, the earliest from 1837. Most were used as tools in Darwin's process of coming to terms with the principles behind evolution and, in some cases, hybridization. Darwin also experimented with finding the most suitable biological metaphor for his diagrammatic tree structure. Sometimes he likened it to an ordinary tree,

[110] Hellström (2012, 234–252).

but at other times he preferred seaweed or corals. On a coral reef, the lower layers are dead, just as the ancestors of the species that are alive today.[111] Darwin was not the first to propose evolution; his original contribution was to identify natural selection as the mechanism behind it. In his diagrams, natural selection is displayed by the bushiness at each ramification, where several varieties coexist for a short time, of which a smaller number survive longer and form the ancestors of later species.

Although Darwin himself did not attempt to display the relationship between existing species in diagram form, several of his early acolytes did. The perhaps most influential and prolific drawer of trees was the German Ernst Haeckel (1834–1919), who also coined the term "phylogeny" to describe the investigation of evolutionary relationships between species. Today, trees of life are therefore often called "phylogenetic trees." Archibald describes the late 1800s as the gilded age of evolutionary trees.[112] This period coincided with what has been called the golden age of statistical graphics.[113] The traditions of evolutionary trees and statistical graphics were merged by George Gaylord Simpson in 1937. Using biometric data and statistical methods, he discussed how species differentiation could be interpreted from fossils of different ages and morphologies.[114]

For a long time, the structure of evolutionary trees was based on the knowledge and authority of the leading biologists that constructed them. However, in the mid-twentieth century, the German entomologist Willi Hennig and his followers developed cladistics, a systematic way of constructing phylogenetic trees based on an analysis of shared similarities and characters.[115] Cladistic diagrams, which attempted to display the evolutionary history of species, faced competition in the 1960s and 1970s from the phenetic diagrams of numerical taxonomy. In these diagrams, classification was based on observable phenotypical qualities. In the extreme empiricist approach of the pheneticists, the evolutionary history behind the observable qualities of living species was considered too theoretical and difficult to investigate. However, the rise of molecular genetics

[111] Voss (2010, 60–64, 90–91, 108–119); Archibald (2014, 80–112).

[112] Archibald (2014, 119–129).

[113] See section "Statistical Graphics" in Chap. 6.

[114] Archibald (2014, 152–154).

[115] Archibald (2014, 159–161).

provided a new tool for evolutionary research, and the phenetic approach was abandoned.[116]

Taxonomy based on similarities in molecular structures began with proteins in the blood; there were some forerunners in the early 1900s, but it took off in the 1960s. However, molecular methods turned more useful when it became possible to study DNA directly. The earliest method, DNA hybridization, was introduced in the 1950s but produced its best results from the mid-1970s to the mid-1980s. From then on, it was overtaken by PCR technology, which in combination with improved computing power automated the tedious laboratory work.[117]

These new methods have allowed studies of a multitude of genes among a vast range of related species. The resulting diagrams of evolution have often been in conflict with the ones constructed with the help of traditional paleontological and taxonomic methods. For example, DNA studies have suggested that the main orders of mammals were already formed in the late cretaceous period, before the extinction of the dinosaurs.[118]

Although different tree diagrams are still put forward in support of conflicting theories of evolution, the diagrammatic form of the tree as an analogy for the development of life has also come under question. The process of hybridization, although known to biologists for a long time, has through modern methods proven to be far more influential than previously thought. This makes the diagrams of evolution more similar to webs than to trees. For example, Doolittle's diagrams from 1999 illustrate the difference between the traditional view of evolution and what molecular biology has revealed about horizontal gene transfer among simpler organisms (Fig. 5.4).[119]

Regarding the role of analogies in science, William Leatherdale has claimed that "[t]he successful analogical act may be accomplished in a moment of insight, but its full implementation, elaboration and qualification may take one or several lifetimes, and is not restricted to the first discoverer."[120]

In the case of the tree of life, there have been multiple analogical acts spanning over millennia. In the Middle Ages, the structure of a diagram of

[116] Archibald (2014, 165–167).
[117] Archibald (2014, 168–179).
[118] Archibald (2014, 179–189).
[119] Archibald (2014, 191–197).
[120] Leatherdale (1974, 17).

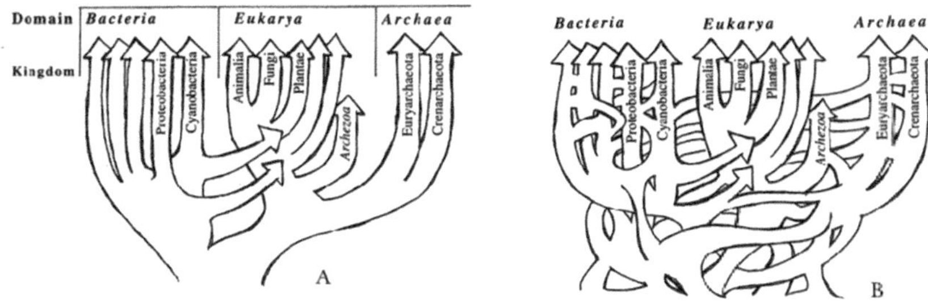

Fig. 5.4 Gene transfer among simpler organisms. The older view (A) had a tree structure, while the new one (B) is a web of gene transfer between species and across domains. Doolittle (1999)

family relations was compared to a tree, creating genealogical trees. However, around the same time, tree diagrams also became used for logical classification. For early naturalists, tree diagrams were a way of exploring the natural order of the living world. They referred to a natural classification rather than a natural genealogy, and they were firmly rooted in the classificatory tradition of tree diagramming. That taxonomical diagrams have their roots in medieval logical diagrams is revealed by the name of our species, *Homo sapiens* (The thinking man). Linnaeus was inspired to use this nomenclature by the tree of Porphyry, where rationality differentiated humans from other animals (Fig. 4.5).[121]

It took centuries before all dimensions of the tree analogy had been explored in the new domains into which it ventured. Lamarck's eighteenth-century diagrams still lacked the genealogical dimension and were mere classification tools, but his diagrams from the early nineteenth century acquired a dimension of time and descent.[122] With the theory of evolution, trees of life transitioned from one old tradition of tree diagramming to another, as they were transformed from analogies of logic classification to analogies of genealogy. Although this transfer should have been simplified by the fact that tree diagrams had a long history of being used for both purposes, the shift still took generations. The merger of the genealogical tree, the classifying tree, and the (geological) timeline into the evolutionary tree is an example of conceptual blending, a way of reaching new

[121] Broberg (1983).
[122] Hellström (2019, 73, 119, 120).

inferences by coactivating different conceptual domains.[123] The classificatory tree is based on the primary metaphor "Organization is Physical Structure," the timeline on "Linear Scales are Paths," and the genealogical tree has elements of both. The new evolutionary tree blended all of them in what amounted to one of the most fundamental paradigm shifts in the history of science.

It is worth nothing that although a branching classificatory structure is today strongly associated with biological taxonomy, eighteenth- and early nineteenth-century naturalists tried to apply this logical structure to all of nature, dead or alive. For example, Linnaeus and Buffon also attempted classifications of chemicals and minerals. Their efforts were hampered by the rudimentary state of chemical knowledge at the time, and they were, for example, mixing compounds and elements.[124] However, their attempts were steps in a development that by the mid-eighteenth century would result in one of the most successful spatial representations of nature that have ever been constructed: the periodic table of the elements.

The Periodic System of the Elements

Few diagrammatic representations have assumed such a central role in their discipline as the periodic system of the elements has within chemistry. The fundamental principle behind the system is the periodic law, which states that chemical properties recur periodically among the elements. This periodicity has been represented in a great variety of diagrammatic forms, over a thousand according to an estimate from 2006.[125] However, the periodic table is the most common, and today new students of chemistry usually encounter its medium-long form. In the table, the elements are arranged by atomic number (i.e., the number of protons in their nuclei), from the lightest, hydrogen, at the top left to the heaviest at the bottom right. Each column in the table is called a group, and the elements within each group have the same valency, that is the number of chemical bonds they can form. The valency is explained by the number of available places for electrons in the atom's outer electron shell. As a consequence of this, elements within a group share similar chemical properties. Each row

[123] Lakoff and Johnson (1999, 47).
[124] Spronsen (1969, 27).
[125] Scerri (2006, 277).

is called a period, and is composed of elements with the same number of electron shells.

However, this textbook description of the periodic system cannot explain how it initially evolved. The periodic system has not been constructed by scientists who, after discovering the number of protons in atomic nuclei and the structure of the surrounding electron clouds, wanted to summarize their findings in a neat table. Terms such as "electrons," "electron shells," "atomic nuclei," and "protons" did not appear until the early 1900s, and the periodic system had its breakthrough in the 1860s. The development of the initial system was a difficult puzzle based on clues such as similarities in chemical properties, valency, and atomic weights. The periodic system became a tool to that enabled the discovery of the structure of atoms half a century later.

The periodic system of the elements is one of those discoveries that was made more or less simultaneously by a number of researchers because the accumulation of empirical knowledge had made the breakthrough possible. In this case, improved measurements of atomic weights, which had been widely disseminated at the Congress of Karlsruhe in 1860, made the following decade a period of breakthroughs for different periodic systems. In 1862, the French geologist Alexandre Emile Béguyer de Chancourtois presented a system to the academy of sciences in Paris. It had the shape of a three-dimensional spiral, which he described as a "telluric screw." The spiral shape was periodical, but all groups did not align as columns in Chancourtois representation. Another factor which made his system less appealing to chemists was that he included some compounds among the elements. In addition, probably because of problems with reproducing the complicated diagram, the academy did not include any illustrations when they published his text in their series *Comptes Rendu*. This reduced the visual impact of Chancourtois's idea, and he chose to republish it privately with a diagram the year after. However, this private publication received even less notice than the previous one, and Chancourtois's contribution remained largely forgotten for decades.[126]

Soon thereafter followed attempts by John Alexander Reina Newland, William Odling, Gustavus Detlef Hinrichs, Julius Lothar Meyer, and Dmitri Ivanovitch Mendeleev. Most of them presented the periodic system in different forms of tables. However, Hinrichs, who had an interest in astronomy, presented a model that resembled the solar system. He also

[126] Spronsen (1969, 97–102); Scerri (2006, 70).

theorized that atoms were shaped like prisms and that properties of the elements were determined by the area of their base, which was the same for all atoms in one group, and by the length of their sides.[127] This theory was similar to Plato's solids and modern string theory in making diagrammatic geometrical shapes the fundamental constituents of the universe.

Presented last of the six in 1869, Mendeleev's periodic system came to dominate, to the extent that some have claimed he was its sole discoverer. Mendeleev needed the system for use in his lectures and chemistry textbooks. According to some of his biographers, Mendeleev arrived at the system by puzzling with little cards marked with the names of the elements, although this has been disputed by others. The groups (vertical columns) and periods (horizontal rows) of Mendeleev's system were based on the chemical properties of the elements. The later explanations for these similarities, such as electron configuration, were not known to Mendeleev. In contrast to most other pioneers of the periodic system, Mendeleev did not even think that he needed the atomic theory. Therefore, he did not use the term "atomic weight" as the foundation for the ordering of the elements, but "elemental weight."[128]

Mendeleev's table was thus not primarily meant as a theoretical system, but as a practical tool. The development of the system was closely intertwined with his work on a Russian chemistry textbook. Mendeleev claimed that the system could be used to, for example, determine the elemental weights of insufficiently analyzed elements, correct erroneously determined weights of elements, and examine the properties of unknown compounds. The system could deliver this, which made it superior to its predecessors.[129] Toward the end of his life, Mendeleev attempted to complement his periodic system, this time with a more theoretical agenda. For example, he included ether as the lightest element, preceding hydrogen in the system. Contrary to the practical endeavors of his youth, Mendeleev's late life theory building left no lasting impression on the history of science.[130]

In 1870, the year after Mendeleev's tables, Heinrich Baumhauer published a spiral periodical system.[131] In the following decades, a large

[127] Spronsen (1969, 103–144).
[128] Spronsen (1969, 132–134). For disputes on the puzzle cards, see Kaji (2018, 228).
[129] Spronsen (1969, 138–139).
[130] Spronsen (1969, 326).
[131] Spronsen (1969, 140–141).

number of other spiral- and screw-shaped systems were devised. According to van Spronsen, "[t]hese forms were intended to show the periodic system more clearly, but they were seldom successful and if at all, only partly. Some details were occasionally expressed more effectively, but often at the cost of over-all clarity." Systems shaped as cubes, wedges, arenas, pretzels, and all conceivable forms were constructed. H. Heribert's systems included moving parts, like Llull's medieval diagrams. One of Heribert's models was reminiscent of a slide ruler, while another was constructed from two strips of paper, one with lists of elements and their properties, and the other with openings. If the opening to the left was placed over one element, other elements in the same group were revealed in other holes.[132]

Philip J. Stewart claims that amateurs have favored spirals, while professional chemists have held fast to their table. He also points out that while tables and spirals were given equal importance in twentieth-century histories about the periodic system, the table has gained pre-eminence in the twenty-first, which is evident already from book titles that now refer to the "periodic table" rather than to the "periodic system."[133] This is an example of a general tendency in the history of diagrams: that the most familiar and simple versions prevail with time.

In 1974, Edward G. Mazurs did a comprehensive survey and classification of periodic systems. Since many tables were structurally identical or similar, he reduced their number to 146 basic forms. He also introduced a system of classification, where a Roman numeral annotated if it was a (I) short, (II) medium, or (III) long form of the table. Thereafter a capitalized letter described whether the system was presented as a figure in three dimensions (A), in two dimensions (B), or as a table (C). Finally, Arabic numerals and minuscules specified the class and type of table. Mazurs pointed out that since some variants were functionally identical, but presented in different forms (as three- or two-dimensional diagrams or as tables), it was possible to reduce the number of basic periodic systems to 33. For example, Chancourtois's helix from 1863 (IA1–1), Wallin's spiral from 1926 (1B1–1), and Newland's table from 1865 (1C1–1) all belong to class 1 (with eight groups and no subgroups) and type 1 (with the representative and transition elements in the same groups without separation

[132] Spronsen (1969, 161, 193). For Llull see section "Llullism: Incomprehensible Originality" in Chap. 4.

[133] Stewart (2018, 75).

into subgroups).[134] Thus, in Mazurs's classification, disparate diagrammatic forms are treated together as expressions of the same basic system. Mazurs classified different periodic systems with the same rigor as Mendeleev and others have displayed in their systematization of the elements. His hierarchical divisions also follow the simple principles of folk taxonomies.

Some attempts have also been made to unify the periodic system with biological taxonomy. Thierry William Preyer was a Darwinist physiologist, who tried to explain similarities between elements through ancestry. He believed the elements originated from each other, and did in the 1890s draw up a periodic system resembling a family tree. With the help of his system, he predicted the properties of 15 undiscovered elements, for the most part successfully. Robert M. Müller went even further in constructing a genealogy of elements, as he in 1944 tried to demonstrate the analogy between chemical elements and living organisms. Man was the crown of organic evolution, a position held by hydrogen among the elements, followed by carbon. By forming the foundation of life, hydrogen and carbon bridged the inorganic and organic worlds, just as humans bridged the organic and spiritual worlds. The different groups of elements included diverging elements, "tame" and "wild," just as the animal world. To illustrate his point, Müller constructed two analogous tree diagrams, one of elements and one of organisms.[135]

Analogies were also drawn between the periodic system and other realms. The physicist and astronomer Henry Wilde did in 1878 point out a correspondence between atomic weights and the distances between the planets. In Pythagorean fashion, several attempts have also been made to find analogies between music and properties of the elements. This line of thought was pioneered by Newland, one of the first discoverers of the periodic system. His system was called the Law of Octaves, and he drew an analogy between musical octaves and the distance in atomic number between the periods of his system. The pharmacologist Alfred Partheil in 1903 claimed a relationship between the spectral lines of elements and the wave numbers of tones on the musical scale. However, van Spronsen has pointed out that Partheil tried to prove his thesis by selecting elements as well as tones arbitrarily. Nonetheless, Partheil's ideas inspired E. Schmiz, who had also found relationships between atomic weight of homologous

[134] Mazurs (1974, 14–21).
[135] Spronsen (1969, 194–196, 201–202, 225); Müller (1944).

elements and the geometric shape of harmonic triangles. Schmiz claimed that nature was filled with harmonies, which was also reflected in architecture and art.[136]

Attempts to find a cosmic order behind similarities between musical octaves and physical phenomena, such as planetary orbits or the periodic system, all stumble on the fact that the division of the tonal scale into seven notes (the name octave, eight, comes from the second appearance of the first note in a higher pitch) is a cultural convention, with a foundation in Miller's magical number seven, a restriction imposed by the upper limit of working memory rather than by any fundamental principle in the universe.[137]

In the twentieth century, the chemical properties displayed in the periodic system, which originally had been derived empirically, could increasingly be deduced from quantum theory and quantum mechanics. However, Eric Scerri has emphasized that the process was less deductive than is often claimed. Quantum physicists had to use some ad hoc rules in order to fit their theories with the empirically derived structure of the nucleus and electron orbitals of the elements. According to Scerri, there is still no quantum mechanical explanation derived from first principles that can explain in which order the orbitals are filled with electrons as one moves up the periodic system.[138] In an opinion piece in *Nature*, Scerri claimed that "[e]ven 150 years on, theoretical chemists, physicists and philosophers still need to step in to comprehend the [G]estalt of the periodic table and its underlying physical explanation."[139] In addition to the similarities in properties in the vertical groups in the system, chemists have also found other spatial relationships that reveal similarities in properties, such as diagonal, zigzag, and even L-shaped knights-move patterns.[140]

It would be wrong to describe the periodic system as a theory or a paradigm. Rather, it is a diagrammatic representation that manages to compile and systematize empirical observations made by chemists over centuries. Thereby, it has become a yardstick against which all new theories and

[136] Spronsen (1969, 108, 193–194, 200–204).

[137] See section "Musical Notation as the First Time Series Graphs" in Chap. 6.

[138] Scerri (2006), sections "The Evolution of Spatial Differentiation and Mythical Space", "Brain, Culture, and Diagram" in Chap. 2, "Language and Gestures", "Terrain and Architecture: Structuring Knowledge in "Oral" Cultures and Early Civilizations", and "Quipu: Inca Infographics" in Chap. 3.

[139] Scerri (2019, 557–559).

[140] Scerri (2006, 265–274).

paradigms are measured. The periodic system has not only survived several shifts of paradigm regarding the structure of atoms, but has in fact remained a vital tool in theory development.

The periodic system is a mix of what Umberto Eco described as the deductive dictionary and the inductive encyclopedia.[141] Starting out as an encyclopedic way of systematizing observations by chemists, it is now mostly defined based on deduction from the laws of physics. However, the common medium-long form of the table still owes its shape to similarities in chemical properties; the best "dictionary" version deduced from laws of quantum mechanics would be the left-step table.[142]

References

Alsted, Johann Heinrich. 1630. *Johannis-Henrici Alstedii Encyclopaedia Septem Tomis Distincta*. Herborn: Friedrich Hulsius.

Archibald, J. David. 2009. Edward Hitchcock's pre-darwinian (1840) 'tree of life'. *Journal of the History of Biology* 42 (3): 561–592.

Archibald, J. David. 2014. *Aristotle's ladder, Darwin's tree: The evolution of visual metaphors for biological order*. New York: Columbia University Press.

Atran, Scott. 1990. *Cognitive foundations of natural history: Towards an anthropology of science*. Cambridge: Cambridge University Press.

Barnhart, Robert K., and Sol Steinmetz. 1988. *The Barnhart dictionary of etymology*. New York: H.W. Wilson Co.

Barsanti, Giulio. 1992. *La scala, la mappa, l'albero. Immagini e classificazioni della natura fra sei e ottocento*. Firenze: Sansoni.

Bloom, Jonathan M. 2001. *Paper before print: The history and impact of paper in the Islamic world*. New Haven: Yale University Press.

Boster, James S. 2005. Categories and cognitive anthropology. In *Handbook of categorization in cognitive science*, ed. Henri Cohen and Claire Lefebvre, 92–116. Amsterdam: Elsevier.

Bray, Francesca, Vera Dorrofeeva-Lichtmann, and Georges Métailié, eds. 2007. *Graphics and text in the production of technical knowledge in China: The Warp and the Weft*. Leiden: Brill.

Broberg, Gunnar. 1983. Homo sapiens: Linnaeus's classification of man. In *Linnaeus, the man and his work*, ed. Tore Frängsmyr, Gunnar Broberg, and Sten Lindroth, 156–194. Berkeley: University of California Press.

[141] Although Eco (2014) referred to tree diagrams, the principle can also be applied to other classificatory diagrams such as the periodic table.

[142] Scerri (2022).

Broberg, Gunnar. 1990. The broken circle. In *The quantifying spirit in the 18th century, vol. 7*, ed. Tore Frängsmyr, J. L. Heilbron, and Robin E. Rider, 46–71. Berkeley: University of California Press.

Carothers, John Colin. 1959. Culture, psychiatry, and the written word. *Psychiatry* 22 (4): 307–320.

Charmantier, Isabelle. 2011. Carl Linnaeus and the visual representation of nature. *Historical Studies in the Natural Sciences* 41 (4): 365–404.

Christensen, Thomas. 2014. *River of ink: Literature, history and art*. Berkeley: Counterpoint.

D'Andrade, Roy G. 1995. *The development of cognitive anthropology*. Cambridge: Cambridge University Press.

Darwin, Charles. 1866. *On the origin of species by means of natural selection, or, the preservation of favoured races in the struggle for life*. 4th ed. London: John Murray.

Doolittle, W. Ford. 1999. Phylogenetic classification and the universal tree. *Science* 284 (5423): 2124–2128.

Dreyer, John Louis, and Emil. 1953. *A history of astronomy from Thales to Kepler*. 2nd ed. New York: Dover.

Eco, Umberto. 2014. *From the tree to the labyrinth: Historical studies on the sign and interpretation*. Boston: Harvard University Press.

Eddy, Matthew D. 2010. Tools for reordering: Commonplacing and the space of words in Linnaeus's philosophia botanica. *Intellectual History Review* 20 (2): 227–252.

Eisenstein, Elizabeth L. 1979. *The printing press as an agent of change: Communications and cultural transformations in early-modern Europe, vol. 1*. Cambridge: Cambridge University Press.

Eisenstein, Elizabeth L. 1983. *The printing revolution in early modern Europe*. Cambridge: Cambridge University Press.

Engelhardt, Yuri, and Clive Richards. 2021. A universal grammar for specifying visualization types. In *Diagrammatic representation and inference: 12th international conference, diagrams 2021*, ed. Amrita Basu, Gem Stapleton, Sven Linker, Catherine Legg, Emmanuel Manalo, and Petrucio Viana, 395–403. Cham: Springer.

Euclid. 1301. *Euclidis elementa: interprete Adelardo, Bathoniens; accedit Campani commentarius*. Bibliothéque Nationale de France, MS Latin 7214. https://www.europeana.eu/sv/item/9200519/ark__12148_btv1b100352862. Accessed 29 July 2024.

Even-Ezra, Ayelet. 2021. *Lines of thought: Branching diagrams and the medieval mind*. Chicago: The University of Chicago Press.

Funk, Holger. 2014. Describing plants in a new mode: The introduction of dichotomies into sixteenth-century botanical literature. *Archives of Natural History* 41 (1): 100–112.

Galilei, Vincentio. 1581. *Dialogo di Vincentio Galilei ... della musica antica, et della moderna*. Florence: Giorgio Marescotti.

Georges, Karl Ernst, and Heinrich Georges. 1913. *Ausführliches lateinisch-deutsches Handwörterbuch: aus den Quellen zusammengetragen und mit besonderer bezugnahme auf Synonymik und Antiquitäten unter berücksichtigung der besten Hilfsmittel. 8., verb. und verm. Aufl. ed.* [Comprehensive Latin-German dictionary: compiled from the sources and with special reference to synonyms and antiquities, taking into account the best resources. 8th, revised and enlarged edition]. Hannover: Hahnsche Buchhandlung.

Gozza, Paolo. 2000a. 'Desiderio da Pavia' and renaissance musical theory. In *Number to sound: The musical way to the Scientific revolution*, ed. Paolo Gozza, 79–96. Dordrecht: Springer.

Gozza, Paolo. 2000b. Introduction. In *Number to sound: The musical way to the scientific revolution*, ed. Paolo Gozza, 1–63. Dordrecht: Springer.

Gozza, Paolo. 2000c. Preface. In *Number to sound: The musical way to the scientific revolution,* ed. Paolo Gozza, xi–xiii. Dordrecht: Springer.

Greco, Pietro. 2018. *Galileo Galilei: The Tuscan artist*. Cham: Springer.

Griffiths, Antony. 2016. *The print before photography: An introduction to European printmaking 1550–1820*. London: The British Museum.

Hellström, Petter. 2012. Darwin and the tree of life: The roots of the evolutionary tree. *Archives of Natural History* 39 (2): 234–252.

Hellström, Petter. 2019. *Trees of knowledge: Science and the shape of genealogy*. Uppsala: Uppsala universitet.

Hitchcock, Edward. 1840. *Elementary geology*. Amherst, MA: Adams.

Hotson, Howard. 2007. *Commonplace learning: Ramism and its German ramifications 1543–1630*. Oxford: Oxford University Press.

Izard, Véronique, Pierre Pica, Stanislas Dehaene, Danielle Hinchey, and Elizabeth Spelke. 2011. Geometry as a universal mental construction. In *Space, time and number in the brain: Searching for the foundations of mathematical thought*, ed. Stanislas Dehaene and Elizabeth Brannon, 319–332. London: Academic Press.

Jönsson, Ann-Mari. 2014. Linnés dikt till Giske. In *Svenska linnésällskapets årsskrift*, 109–122. Uppsala: Svenska linnésällskapet.

Jussieau, Antonii–Laurentii de. 1789. *Genera plantarum secundum ordines naturales disposita*. Paris: Herissant et Theophilum Barrois.

Kaji, Masanori. 2018. The origin of Mendeleev's discovery of the periodic system. In *Mendeleev to Oganesson: A multidisciplinary perspective on the periodic table*, ed. Eric Scerri and Guillermo Restrepo, 219–244. Oxford: Oxford University Press.

Kent, Charles. 1961. Introduction. In *Compendium of music (Compendium Musicae),* René Descartes. Trans. W. Robert. Rome: American Institute of Musicology.

Kepler, Johannes. 1619. *Harmonices Mundi Libri V*. Linz: Johann Planck.

Kim-Renaud, Young-Key, ed. 1997. *The Korean alphabet: Its history and structure.* Honolulu: University of Hawai'i Press.

Kupfer, Marcia. 2020. Introduction. In *The visualization of knowledge in medieval and early Europe*, ed. Marcia Kupfer, Adam S. Cohen, and J. H. Chajes, 361–382. Turnhout: Brepolis.

Lakoff, George, and Mark Johnson. 1999. *Philosophy in the flesh: The embodied mind and its challenge to Western thought.* New York: Basic Books.

Lamarck, Jean-Baptiste. 1778. *Flore françoise, ou, Description succincte de toutes les plantes qui croissent naturellement en France, vol. 1–3.* Paris: Imprimerie Royale.

Larson, James L. 1971. *Reason and experience: The representation of natural order in the work of Carl von Linné.* Berkeley: University of California Press.

Leatherdale, William Hilton. 1974. *The role of analogy, model and metaphor in science.* Amsterdam: North Holland.

Li, Yingli, and Michael W. O'Boyle. 2011. Differences in mental rotation strategies for native speakers of Chinese and English and how they vary as a function of sex and college major. *The Psychological Record* 61:2–19.

Linné, Carl von. (1735) 1960. *Caroli Linnæi Systema Naturæ: Sive Regna Tria Naturæ Systematice Proposita Per Classes, Ordines, Genera, & Species.* Uppsala: Bokgillet.

Livesey, Steven J. 1989. *Theology and science in the fourteenth century: Three questions on the unity and subalternation of the sciences from john of reading's commentary on the sentences.* Leiden: Brill.

Lovejoy, Arthur O. 1936. *The great chain of being: A study of the history of an idea.* Boston: Harvard University Press.

Luckombe, Philip. 1771. *The history and art of printing.* London: Adlard and Browne.

Maffei, Raffaele. 1511. *Commentariorum urbanorum octo et triginta libri cum duplici eorundem indice secundum tomos collecto.* Lyon: Sebastian Gryphium.

Mazurs, Edward G. 1974. *Graphic representation of the periodic system during one hundred years.* Tuscaloosa: University of Alabama Press.

McCulloch, Jock. 1995. *Colonial psychiatry and the "African mind."* Cambridge: Cambridge University Press.

McLuhan, Marshall. (1962) 1967. *The Gutenberg galaxy: The making of typographic man.* London: Routledge & Kegan Paul.

McLuhan, Marshall. (1964) 2005. *Understanding media: The extensions of man.* Abingdon: Routledge.

Moxon, Joseph. (1683) 1958. *Mechanick exercises on the whole art of printing*, ed. Herbert Davis and Harry Carter. London: Oxford University Press.

Müller, Robert R. 1944. *Über die Enfaltungsordnung und den Stambaum der chemischen Grundstoffe: Eine morphologische Betrachtung. Gestalt 15.* [On the order of development and the family tree of the basic chemical substances: A morphological consideration. Figure 15.] Halle: Niemeyer.

Müller-Wille, Staffan, and Isabelle Charmantier. 2012. Natural history and information overload: The case of Linnaeus. *Studies in History and Philosophy of Science. Part C, Studies in History and Philosophy of Biological and Biomedical Sciences* 43(1): 4–15.

Muzzulini, Daniel. 2015. The geometry of musical logarithms. *Acta Musicologica* 87 (2): 193–216.

Natali, Carlo, and Douglas S. Hutchinson. 2013. *Aristotle: His life and school.* Princeton, NJ: Princeton University Press.

Ong, Walter J. 1957. System, space, and intellect in renaissance symbolism. *Cross Currents* 7 (2): 123–136.

Ong, Walter J. 1958. *Ramus. Method, and the decay of dialogue: From the art of discourse to the art of reason.* Cambridge, MA: Harvard University Press.

Ong, Walter J. 1982. *Orality and literacy: The technologizing of the word.* London: Methuen.

Palisca, Claude V. 2003. Introduction. In Vincenzo Galileo, *Dialogue on Ancient and Modern Music.* Trans. C.V. Palisca, xv–lxix. New Haven, CT: Yale University Press.

Pettegree, Andrew. 2010. *The book in the renaissance.* New Haven, CT: Yale University Press.

Ramus, Petrus. 1576. *Professio Regia: Hoc est septem artes liberales.* Basel: Sebastian Henric Petri.

Reisch, Gregor. 1503. *Margarita Philosophica.* Strassbourg: Johann Schott.

Sarfatti, Tamar. 2022. *The description of Egypt from Napoleon to Champollion.* Cham: Palgrave Macmillan.

Savage-Smith, Emilie. 2002. Galen's lost ophthalmology and the Summaria Alexandrinorum. *Bulletin of the Institute of Classical Studies* 77:121–138.

Scerri, Eric R. 2006. *The periodic table: Its story and its significance.* New York: Oxford University Press.

Scerri, Eric. 2019. Can quantum ideas explain chemistry's greatest icon? *Nature* 565 (7741): 557–559.

Scerri, Eric R. 2022. Various forms of the periodic table including the left-step table, the regularization of atomic number triads and first-member anomalies. *ChemTexts* 8:1–13.

Siskin, Clifford. 2016. *System: The shaping of modern knowledge.* Cambridge, MA: The MIT Press.

Smith, D. Eugene. 1958. *History of mathematics.* Volume II. New York: Dover.

Smith, A. Mark. 2020. More than meets the eye: What made the printing revolution revolutionary. In *The visualization of knowledge in medieval and early Europe*, ed. Marcia Kupfer, Adam S. Cohen, and J. H. Chajes, 211–228. Turnhout: Brepolis.

Sohn, Pow-key. 1959. Early Korean printing. *Journal of the American Oriental Society* 79 (2): 96–103.

Spelke, Elizabeth S. 2011. Natural number and natural geometry. In *Space, time and number in the brain: Searching for the foundations of mathematical though*, ed. Stanislas Dehaene and Elizabeth Brannon, 287–317. London: Academic Press.

Spronsen, Johannes Willem van. 1969. *The periodic system of chemical elements: A history of the first hundred years*. Amsterdam: Elsevier.

Stewart, Philip J. 2018. Amateurs and professionals in chemistry: The case of the periodic system. In *Mendeleev to Oganesson: A multidisciplinary perspective on the periodic table*, ed. Eric Scerri and Guillermo Restrepo, 66–79. Oxford: Oxford University Press.

Stuessy, Tod F. 2009. *Plant taxonomy: The systematic evaluation of comparative data*. 2nd ed. New York: Columbia University Press.

Tsien, Tsuen-Hsuin. 1985. *Paper and printing: Science and civilization in China, vol. 5 part 1,* Series editor Joseph Needham. Cambridge: Cambridge University Press.

Tversky, Barbara. 2001. Spatial schemas in depictions. In *Spatial schemas and abstract thought*, ed. Merideth Gattis, 79–112. Cambridge, MA: MIT Press.

Von Petzinger, Genevieve. 2017. *The first signs: Unlocking the mysteries of the world's oldest symbols*. New York: Atria.

Voss, Julia. 2010. *Darwin's pictures: Views of evolutionary theory, 1837–1874*. New Haven, CT: Yale University Press.

Wardhaugh, Benjamin. 2008. Musical logarithms in the seventeenth century: Descartes, Mercator, Newton. *Historia Mathematica* 35(1): 19–36.

Wynn, Thomas Grant, and Frederick L. Coolidge. 2022. *An introduction to evolutionary cognitive archaeology*. New York: Routledge.

Yates, Frances A. (1966) 2014. *The art of memory*. London: The Bodley Head.

Zarlino, Gioseffi. 1588. *Sopplimenti musicali*. Venice: Franceschi.

CHAPTER 6

Spatial Representation of Data

"We live on islands surrounded by seas of data," Michael Friendly and Howard Wainer claim in the introduction to their history of data visualization.[1] It is this data flood that has made diagrams for data visualization necessary. However, such a situation is rather new. Innovations in data visualization only took off in the late 1700s, following in the footsteps of new measurement technologies that began to appear in the 1600s.[2] The previous period was characterized by diagrams for logical and textual analysis, with the High Middle Ages as a period of intense innovation in new forms of abstract diagrams. However, the 1500s to the 1700s saw very little innovation in abstract diagrams. Scientific texts were either accompanied by figurative illustrations, such as images of observed naturalia or of scientific equipment, or by tables of measurement results. The abstract diagrams that existed were mainly bracketed tree structures that systematized knowledge.

Friendly and Wainer see the roots to the rise of data visualization in the last 400 years in two factors. In addition to an increase in the amount of data recorded, they also believe that this period was associated with "a cognitive revolution we call 'visual thinking,' the idea that some problems and their solutions can be much more clearly addressed and

[1] Friendly and Wainer (2021, 1).
[2] Friendly (2006, 3).

J. Holmén, *A History of Diagrams*, Palgrave Studies in the History of Science and Technology,
https://doi.org/10.1007/978-3-031-99691-7_6

communicated in visual displays, rather than just in words or tables or numbers."[3] However, as we have seen in previous chapters, medieval scholars were well aware of the advantages of framing problems in diagram form, and even Aristotle emphasized how the contemplation of images aided the thought process. What then was new about the cognitive revolution of visual thinking in the last 400 years that Friendly and Wainer claimed allowed for the emergence of data visualizations?

This chapter of the book argues that the novel impetus came from the field of analytical geometry, which created the conception of a quantified space that could be used to visualize measurements. Although one of the most common diagrams of a function in statistical graphics, the time series diagram, had older precursors in the field of music (section "Musical Notation as the First Time Series Graphs"), the quantitative turn in diagramming only occurred after the invention of the mathematical function. This is discussed in section "Statistical Graphics", which also follows the development of statistical graphics until the computer age. In order to illustrate the difference between functions and the earlier qualitative diagramming practices, section "Diagrams of Heartbeats" compares how the beating heart was analyzed in diagrams under these two paradigms. Thereafter follows section "Diagrams in Economic Theory", which examines diagrams in economics. In that field, the multitude of diagrams makes it possible to discuss their role in theory formulation. In section "Stock Charts: Icons of Capitalism", dedicated to stock charts, particular focus is put on Japanese candlestick charts, since the old roots ascribed to them seem to contradict the thesis that the invention of the function was an important prerequisite for the development of time series charts. This chapter concludes with section "Diagrams of Social Hierarchies", which investigates how visualizations of social hierarchies were affected by the emergence of data visualization.

Musical Notation as the First Time Series Graphs

Diagrams appeared in music theory as early as ancient Greece. For example, Aristotle's disciple Aristoxenes (ca. 375–after 335 BCE) criticized earlier music theoreticians for not having used enough diagrams. Therefore, it is highly likely that he included diagrams in his original works. In the fourth century CE, Bacchius defined a diagram as "[a] representation of a

[3] Friendly and Wainer (2021, 5, 44–45).

(sc. musical) system. And we use a diagram so that, for students of the subject, matters which are hard to grasp with the hearing may appear before their eyes."[4]

Thus, it seems as if diagrams were in frequent use in ancient music theory, probably more so than in most other fields. Besides musical diagrams, only a small number of diagrams are mentioned or described in ancient Greek manuscripts outside the field of geometry, such as Aristotle's parallelogram of economic exchange.[5] As an applied field that early on achieved a high degree of systematization with close ties to mathematics, music theory functioned as a mediator that transmitted diagramming practices between pure geometry and other applied arts and sciences.

One of the most common types of diagrams today is the time series chart, where variables on the vertical axis are plotted against time on the horizontal axis. Time series charts do not appear fully formed until the eighteenth century. Some exceptions have been suggested, such as medieval astronomical diagrams. However, it is questionable whether these can be considered pure time series diagrams.[6] As was the case with the concept of "system,"[7] music is a more likely candidate than astronomy to be the true predecessor of time series diagrams. In fact, standard musical notation is a form of time series diagram with medieval roots. Musical notation has been found in Mesopotamia in 1400 BCE. However, this notation was symbolic, with cuneiform signs designating different tones. Although ancient Greek music theoreticians such as Aristoxenes used diagrams as pedagogical tools, the Greek system for music notation was also purely symbolic.[8]

However, in the Middle Ages, the relative pitch of chants began to be recorded with neumes (Fig. 6.1), a system with diagrammatic aspects. From the ninth to the twelfth century, similar neumatic systems were developed in Western Europe, Byzantium and Eastern Europe, Japan, and probably Tibet.[9] The earliest Western neumes appeared in Carolingian France as contour lines above the chanted text. The oldest preserved fragment is from the second quarter of the ninth century.[10] It is possible that

[4] Franklin (2002, 682).
[5] See section "Diagrams in Economic Theory".
[6] See section "Statistical Graphics".
[7] See section "The Musical Prelude to Diagrams of Scientific Systems" in Chap. 5.
[8] Bent et al. (2014).
[9] Bent et al. (2014).
[10] Rankin (2022, 104).

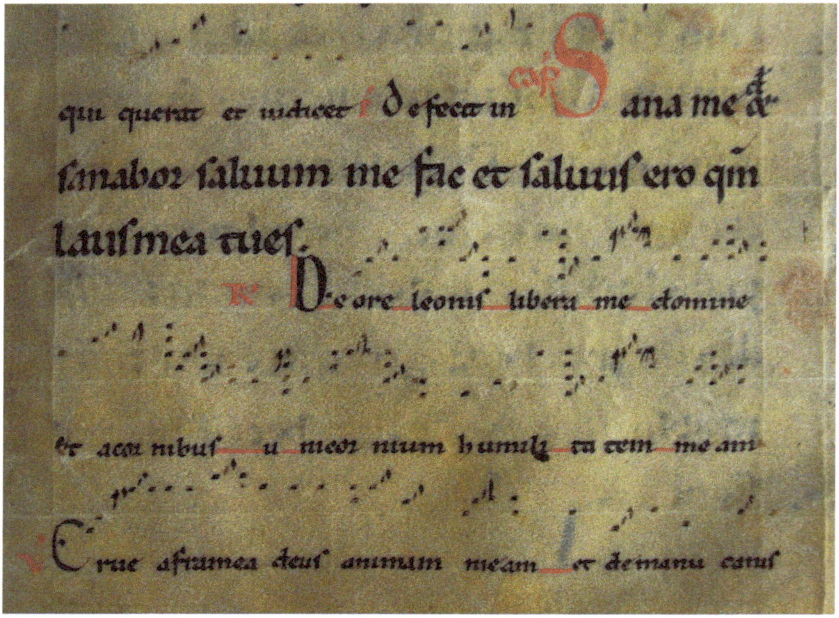

Fig. 6.1 Diastematic Gregorian neumes from Aquitania, twelfth century. Although there is no absolute key, the relative pitch is carefully represented by location in the vertical dimension. Manuscript from Arxiu Comarcal del Bages, Manresa. Photo: Carlets

the neumes were an entirely new invention, but it has also been suggested that they were inspired by accents, the hand gestures of the conductor, or a Frankish system of punctuation signs.[11] Some have claimed that the neumes were influenced by Tironian notes, a form of shorthand with roots in the first century BCE, which might have been used for musical notation on wax tablets in the early Middle Ages, of which none have been preserved.[12] Many of these suggested predecessors are sign systems with strong diagrammatic elements. In fact, accents can in themselves be considered tiny time series graphs. For example, an acute accent (´) illustrates how tone rises with time, reflecting how it was used in ancient Greece.[13]

[11] Taruskin (2005, 13–16).
[12] Haines (2020, 327–340).
[13] For ancient accents, see Taruskin (2005, 14).

Even if it is impossible to determine which theory of the neumes' origin is correct, they definitely did not evolve from earlier systems of entirely symbolic musical notation. The forgetting of ancient symbolic predecessors paved the way for a musical notation with a strong diagrammatic element.

The oldest adiastematic neumes gave only a vague idea of melodic contour and could probably only be used to reproduce familiar music. By the end of the tenth century, diastematic neumes could register relative pitch by careful vertical arrangement, and would allow you to sing an unknown melody, even though the absolute key was unknown.[14] These were similar to what Tufte calls "spark lines," dequantified data displays that lack a scale of measurement.[15]

Later, staff lines that notated absolute pitch were introduced. This invention is commonly attributed to the monk Guido of Arezzo, who described it in his *Micrologus* from 1028.[16] The historian of mathematics Peter Schreiber has observed that Western musical notation has the character of time functions. However, since Schreiber did not differentiate between diastematic and adiastematic neumes, he dated this invention to the introduction of staff lines.[17]

Thus, musical notation inherited its axis of time flow from left to right from the text, to which the musical notation was initially a mere appendix. With time, musical notation became an independent time series chart, where pitch was annotated as vertical position. The fact that "low" pitch (longer wavelengths of sound) is annotated below "higher" pitches (shorter wavelengths) could be arbitrary. However, it could also reflect that in singing, higher tones resonate higher in the head while lower tones resonate down in the chest. It is debatable whether this example of embodied cognition is a primary metaphor learned by everyone in childhood while learning how to use their vocal apparatus, or if it is a narrower experience shared by devoted singers across cultures but not obvious to the less musically inclined. In any case, the identification of lower tones with down predates the neumes, as it was found already in the accents used in ancient poetry.[18]

[14] Bent et al. (2014); Pryer (2011).
[15] Tufte (2006, 47).
[16] Taruskin (2005, 16).
[17] Schreiber (2003, 157–168).
[18] Taruskin (2005, 14).

The division of the tonal scale into seven steps (the name octave, eight, comes from the second appearance of the first note in a higher pitch) is a Western cultural convention. However, the number seven is in this case not entirely arbitrary. Comparative ethnological studies of music have revealed that almost all cultures have seven or less steps in their scales.[19] Most likely, it is Miller's "magical number seven" that dictates the upper limit of how many steps humans can handle.[20] Thus, the octave is not built into the universe but into our working memory's capacity to comprehend it.

In addition, it should be noted that the Western musical system actually has 12 steps, as there are half steps between some, but not all, full steps. Despite their names, the logarithmic differences in frequency between half and full steps are identical. At first, this system seems counterintuitive. However, when it is implemented into the layout of musical instruments, such as a piano, it makes perfect sense from a cognitive perspective. On the keyboard, the full steps are indicated by white and the half steps by black keys. Since there are not half steps between all full steps, the black half steps become spaced in groups of three or two. This makes it much easier for the pianist to navigate the keyboard than would be the case if it consisted of a long row of identical keys, which would be adequate from a pure physics standpoint. Thus, Western musical notation not only respects the magical number seven as the absolute upper limit our consciousness can handle, but also aligns with our preference for even smaller groups, ideally pairs or triads. The limitation of human working memory is also reflected in the keyboard's division into seven octaves.

With three extra keys before the first and one after the last octave, a standard piano has a total of 88 keys. This seemed a bit high to the pioneering cognitive anthropologist Anthony Wallace, who in 1961 introduced the theory that human discriminatory and taxonomic systems are limited to six binary combinations, that is, 2^6 or 64. His explanation for this exception was that contrary to ordinary folk taxonomies, the keyboard requires specialized training and is not expected to be mastered by everyone.[21] However, although Wallace was on to something in his theory that cognitive limitations affect human taxonomic systems, his two-to-the-power-of-six rule was crude and based on a limited empirical study of

[19] Savage et al. (2015).

[20] See section "Diagrams and the Limits of Consciousness" in Chap. 2.

[21] Wallace (1961).

kinship terms in a low number of ethnic groups. He did not reflect on how the keyboard was subdivided. In that respect, it follows the hierarchical principles of other folk taxonomies.

Although the development of spatial analogies and hierarchical categorization in the field of music offers many interesting inroads from a cognitive history perspective, the most important role of music theory in the history of diagrams is arguably that musical notation directly inspired the invention of time series charts. The first person known to have drawn a time series graph is Johann Heinrich Lambert, who, based on work done in the 1760s, published diagrams of the evaporation rate of water in 1771 (Fig. 6.2).[22] His thoughts on musical notation had been published in the section on semiotics in the second volume of *Neues Organon* (1764), only some years before he produced his evaporation diagrams. There, Lambert described the musical score as a two-dimensional figure of the tonal art, where one dimension described the linear succession of tones, and the other their height and depth.[23] Lambert saw musical notation as a model for a universal system of graphs that could be used to describe almost everything.[24]

Lambert's time series diagrams from the 1760s are in many ways prodigious anomalies. Although similar diagrams appeared in social and economic statistics in the early 1800s, it predates most similar efforts in the hard sciences by almost a century. Probably, Lambert's semiotic analysis of musical notation as not merely a symbolic system but a two-dimensional diagram primed him to take that intellectual leap, although it was only part of a perfect storm of conducive circumstances, which will be discussed further in the following chapter.

Statistical Graphics

The diagrams discussed in the earlier chapters of this book have been qualitative. The spatial thinking involved was not linked to numerical datasets as in modern data visualizations. This should not be surprising, since it took a long time for the concept of space to acquire the mathematical properties it has today. According to Vincenzo de Risi, it was Leibniz who

[22] Based on measurements from 1767 and 1768, published in 1771 in the 1769 yearbook of the Berlin academy. Lambert (1771).

[23] Lambert (1764, 32–33).

[24] Bender and Marrinan (2010, 84).

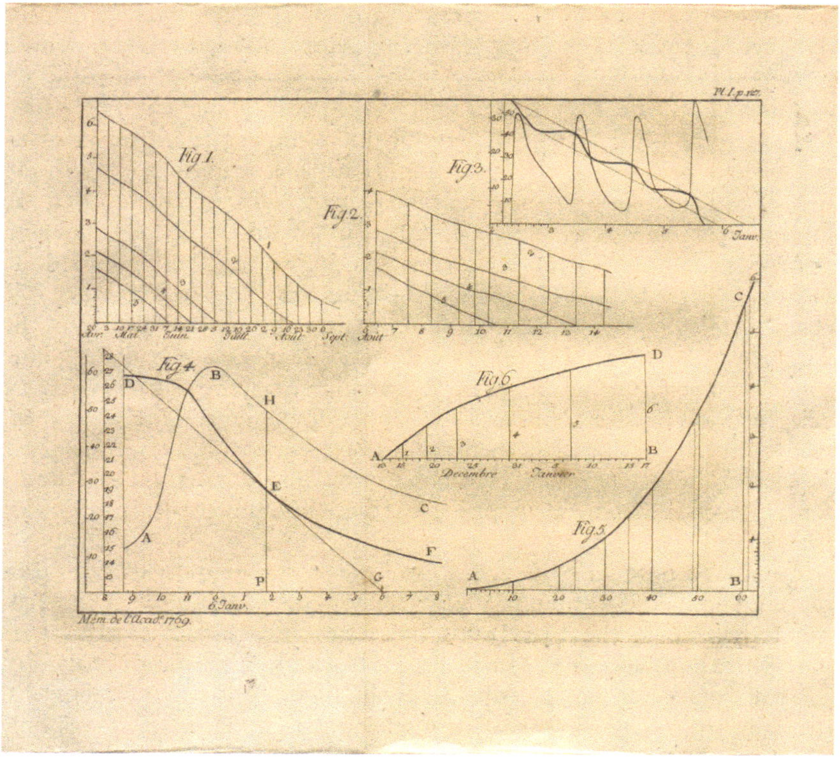

Fig. 6.2 Time series diagrams from Lambert's "Essai d'Hygrométrie ou sur la mesure de l'humidité." For example, the top right figure illustrates how the evaporation rate of water is affected by temperature. These diagrams were inspired by the development of the mathematical function. Lambert (1771)

first developed a concept of space as a structure with an extension produced by a certain order of distances between points.[25] Immanuel Kant considered Leibniz's work on space important but incomplete, and in 1768 he attempted to demonstrate that absolute space exists independently of the objects that occupy it and the relationship between them. Euler and his development of analytical geometry was another important

[25] Risi (2018, 253).

inspiration for Kant's views on space.[26] Descartes is commonly credited as having invented analytical geometry and the coordinate system in 1637. He is often mentioned together with Pierre de Fermat, who the same year circulated a manuscript on the subject. However, Descartes and Fermat merely initiated a development that, following the contributions of, for example, Newton, was perfected in the following century by Euler, who described analytical geometry in terms similar to those we use today.[27] Thus, by the mid-eighteenth century space had become established as a distinct entity which could be described mathematically. It is also around this time that space for the first time becomes used as a frame of reference for representing quantitative data.

Statistical diagrams such as pie charts, bar charts, and time series line graphs are today taken for granted. According to Tufte, 75% of graphics in newspapers and magazines are time series graphs.[28] Although these particular forms of diagrams were developed relatively late, they had notable historical forerunners. In fact, some basic preconditions for statistical graphics, such as identifying line length with number, are present already in infants, and thus seem to be innate.[29] Considering this, it is not surprising that lines were used in Euclid's *Elementa* to represent magnitudes (Book 5) or numbers (Book 7). It has also been suggested that Euclid's lost work *Porisms* included a functional approach to geometry, but that it was poorly understood by his followers.[30]

In the 1300s, Nicholas Oresme (1323–1382) used two-dimensional graphs, where one variable, "intensity," varied with the other, "extent." Although Oresme was probably preceded in the use of such two-dimensional graphs by his contemporary Giovanni di Casali, Oresme went much further in his use and interpretation of them, in some ways foreshadowing functions and analytical geometry.[31] There are different opinions regarding how close Oresme came to the modern concept of a function, where, for example, Peter Schreiber emphasizes the similarities and Jacob Klein the differences.[32]

[26] Kant (1768) 1992, 165–172; Boberg (2020, 121–135).

[27] Boberg (2020, 127–130); for the contributions of different mathematicians in this development, see Scriba et al. (2015, 337–348).

[28] Tufte (2001, 28). Based on a random sample of 4000 graphics in the years 1974–1980.

[29] Spelke (2011, 290).

[30] Scriba et al. (2015, 336).

[31] Clagett (1959, 332–333, 341–342).

[32] Schreiber (2003); Klein (1968) 1992.

One factor that sets Oresme's diagrams apart from modern functions is that they were not described by algebra. Still, Oresme's diagrams were similar to later diagrams of functions, as were his interpretations of them. For example, he concluded that in a diagram where the "latitude" (the horizontal axis) represented time and the "longitude" (the vertical axis) velocity, the area under the graph represented the distance traveled.[33] However, although this reminds us of a modern time function, to Oresme it was a purely philosophical exercise. He did not take data from measurements or experiments and plot them into graphs.[34] Schreiber ascribes this to a lack of interesting applications—there was not much "data" to visualize in the Middle Ages. Still, he considers the medieval attempts of Oresme and others superior to Descartes in some respects, since they also visualized abstract and nonmathematical matters.[35]

One explanation for why medieval diagrams visualized what we would describe as quantitative and qualitative entities with the same means is that no clear distinction between the two was yet developed. Diagrams of longitude and latitude similar to those of Oresme, to the modern eye clearly "quantitative" in their visual language, were also used to illustrate fundamentally qualitative ideas such as Galen's medical theories.[36] The differentiation of the modern concepts quantitative and qualitative did not begin until Descartes and Galileo in the early 1600s rediscovered and developed the distinction between primary and secondary qualities made by the Greek atomists.[37] To them, secondary qualities—such as sound and odor—were dependent on primary qualities, such as shape and number. For example, sound can depend on the length of a string.

Oresme had instead used the pair "qualities," which were permanent, and "motions," which were successive. He used lengths in order to represent both intensities in qualities and velocities in motions, but had no interest in measuring any of them or representing them quantitatively.[38] Galileo did similar work as Oresme, although based on actual measurements. However, they both worked in the setting of Euclidean geometry, aiming to prove the geometrical properties of the figures in their

[33] Clagett (1968, 47).
[34] Verboon (2010, 128–130).
[35] Schreiber (2003).
[36] Maclean (2006, 135–164).
[37] Campbell (2004).
[38] Clagett (1968, 14–15).

diagrams.[39] Today we would say that these diagrams display velocity as a function of time, but the concept of a function was not yet familiar to Oresme and Galileo. They still investigated the space inside figures, rather than figures in space.[40]

The new mathematical view of space was not fully formulated until *Introductio in analysin infinitorum* from 1748, where Euler synthetized and developed the advancements in analytical geometry. Important in this respect was the "function" concept, which had previously been used, for example, by Leibniz and by Euler's teacher Bernoulli, but which Euler made central to the field of analysis.[41] Still in 1748, Euler's definition of a function was heavily influenced by Bernoulli's and narrowly mathematical. However, in 1755 he produced a more general definition: "When certain quantities depend on others in such a way that they undergo a change when the latter change, then the first are called functions of the second."[42] This definition is highly relevant to the development of statistical graphics over the following century. We are used to see diagrams displaying one factor, usually measured on the vertical axis, that depends on another factor on the horizontal axis.

Statistical graphs are often functions, frequently with some other quantity displayed as a function of time. With the exception of musical notation, which we discussed in the previous chapter, such time series graphs did not come into use until after Euler's contribution to the field in the mid-1700s.[43] Several authors claim that only a single exceptional time series graph drawn before the 1700s has been found: a tenth- or eleventh-century diagram displaying the inclination of the sun, moon, and planets (Fig. 6.3).[44] This graph has been portrayed as a sole anomaly ever since attention was first drawn to it by Siegmund Günther in 1877.[45] However, John E. Murdoch presents several similar medieval graphs. Comparing them to each other, it is also evident that one can call into question whether these diagrams are true time series graphs. While some of them have the horizontal axis divided in 365 days, others apparently use 360 degrees. As the planets complete a revolution around the ecliptic in one

[39] Boberg (2020, 110).
[40] See, for example, the long note nr 323 in Klein (1968) 1992.
[41] Euler (1748); Eves (1990, 234–235); Youschkevitch (1976: 37–85).
[42] Euler (2000, vi).
[43] For Euler's introduction of functions, see Boberg (2020, 126–127).
[44] Tufte (2001, 28); Beniger and Robyn (1978, 1–11).
[45] S. Günther (1877, 19–25).

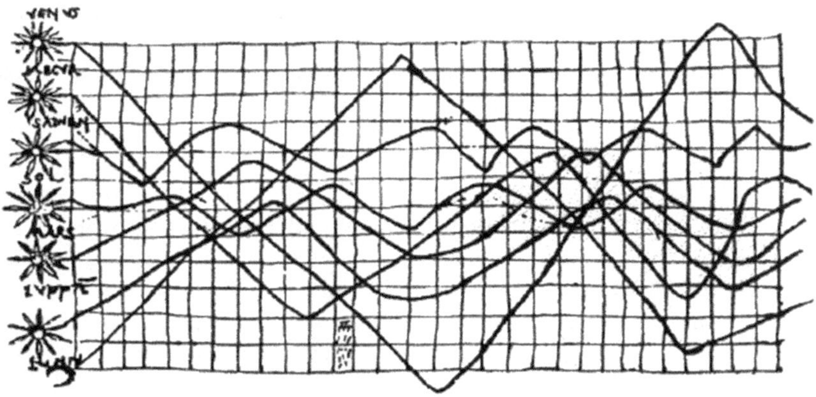

Fig. 6.3 Astronomical diagrams like this one, probably from the eleventh century, have been described as time series graphs, although it is debatable whether the horizontal axis displays time or degrees. In addition, they were not used to illustrate data but a verbal description from Pliny's *Natural History*. *De cursu zodiacum*, Bayerische Staatsbibiothek Clm. 14436

year, time and space are interchangeable on the horizontal dimension. In addition, Murdoch explains that these diagrams were used to visualize Pliny's description of the planets in his *Natural History*. For example, Pliny likened the suns trajectory to the snakelike course of a dragon.[46] Thus, these diagrams were not visualizing quantitative data measurements, but illustrated an ancient metaphor-laden text.

Friendly and Wainer believe that the development in the second half of the 1600s of graphic recording devices for weather measurements, with pens attached to revolving rolls of paper, was an important impetus for statistical graphics. They claim that Robert Plot was inspired by the visual format of such data recordings when he "plotted" the barometric pressure for the year 1684 in Oxford (Fig. 6.4).[47] Although Plot's diagram can be described as a form of time series graph, the difference between it and the graphs of Lambert illustrates some important points. In Plot's graph, the time axis is vertical from up to down. The measurements are marked on a grid pattern with 31 days for each month, including for February, which

[46] J. E. Murdoch (1984, 284–287).

[47] Friendly and Wainer (2021, 20–21). A closer description of the development of graphic recording devices can be found in Hoff and Geddes (1962).

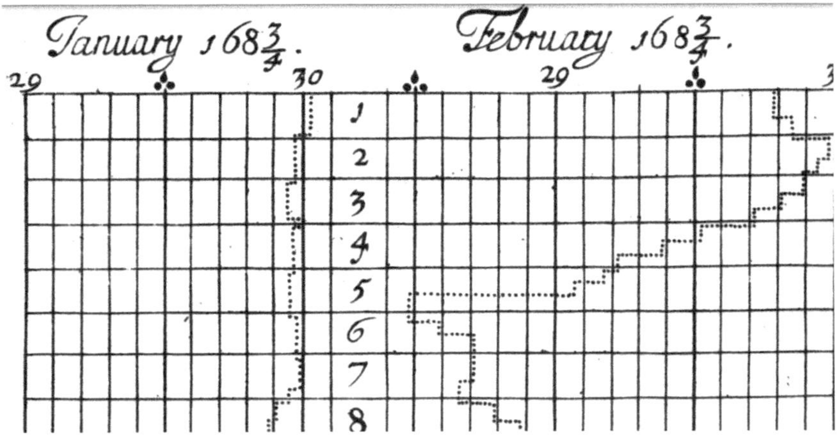

Fig. 6.4 Detail from Robert Plot's diagram of barometric pressure in Oxford 1684. The dotted line is clearly not the smoothed line of a function but a representation of discrete measurements. Plot (1685)

only had 28 days. Some days he made several measurements. However, no attempt is made to smooth the line through these measuring points, but the line is drawn with straight angles. If Plot was indeed inspired by graphic recording devices, he did not reproduce their smoothing of the curve as he chose to represent his measurements as a series of discrete numbers.

In the 1760s, the Swiss mathematician and philosopher Johann Heinrich Lambert drew graphs of the evaporation rate of water as a function of temperature, which he published in 1771 (Fig. 6.2). Here we can see the clear influence from analytical geometry and the concept of the function. Lambert is not merely plotting different measuring points, but is producing a curve that connects these points. His awareness of analytical geometry made Lambert consider individual measuring points examples of an underlying equation.[48] His graphs (Fig. 6.2) display the same smoothness as the illustrations of functions in Euler's work (Fig. 6.5).

Although scientists in the eighteenth century often seem to have used techniques of statistical graphics for their private analysis and record keeping, they did not include them in their books. Even graphs produced by mechanical recording devices were most often transformed into tables

[48] See, for example, quotes in Tilling (1975, 204–205).

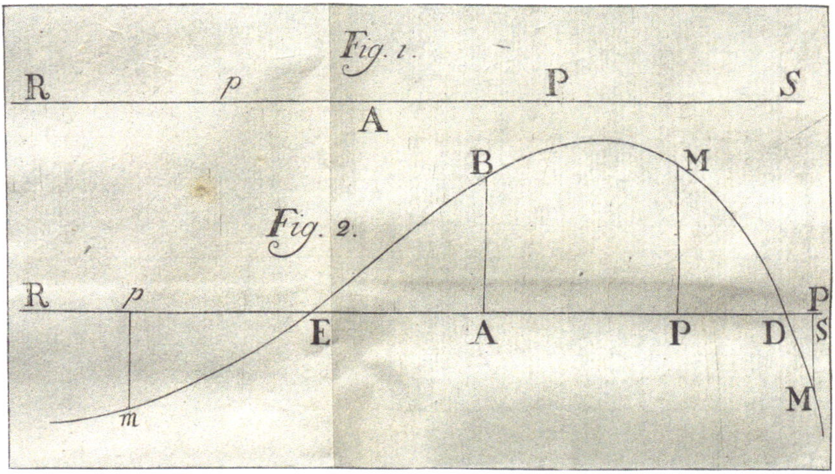

Fig. 6.5 Illustration of a function. Euler (1748, tab. 1)

before being put into print. According to Ian Spence and Howard Wainer, this aversion to graphs was in part founded in a philosophical distrust of sense perception, following the deductive tradition of Descartes. Other impediments were the technical problems with copper plate printing and, perhaps most importantly, concerns about accuracy.[49] Possibly, such concerns about accuracy also explain Plot's attempts to make his graph a series of discrete numbers.

In contrast, Lambert's use of diagrams was connected to his philosophy about the relationship between figures and symbols in scientific thinking. He convincingly argued that as long as the accuracy in the construction of the figure was greater than the accuracy of the data, interpretations from the figure could be as exact as calculations. His own work with the evaporation of water was, for example, limited by inaccuracy in temperature measurements, but by graphically smoothing a curve around the data he could still find a formula.[50] Lambert was active at the Berlin Academy in the mid-1700s, when it was headed by Euler.[51] It is not unreasonable to suggest that data visualization was born at the Berlin Academy through

[49] Spence and Wainer (2017, 46–47).
[50] Bullynck (2010).
[51] Truesdell (1972) 2007, 33.

the interchange between these two Swiss mathematicians, Euler bringing the theory of functions to completion and Lambert pioneering its use for analysis of scientific data.

That the Berlin environment was exceptionally conducive to the introduction of data visualization is underscored by the fact that Lambert's graphs were way ahead of their time. After him, there was a hiatus of almost a century before diagrams came in wider use for representing and analyzing scientific data. This time, the baton was picked up by British scientists such as James David Forbes, whose work on heat in the 1840s and 1850s was directly inspired by Lambert's. Unlike Lambert, Forbes attracted a wide following among natural scientists with his diagramming techniques. This was probably facilitated by an increased familiarity with visual data displays among the public. According to Laura Tilling, the 1830s was a turning point after which tabular data was often transformed into diagrams for easier perception, rather than as earlier when data from mechanically produced graphs had been tabulated before publication.[52]

However, scientists had been preceded in the use of data visualization by political economists and social statisticians. The pioneer in this field was the Scottish engineer and publicist William Playfair (1759–1823), the younger brother of mathematician John Playfair. When William was a child, John encouraged him to keep record of the thermometer by drawing lines of different lengths, giving him an early introduction to graphs.[53] William worked at James Watt's workshop as a draftsman in 1778–1781, and it is possible that he came in contact with charts or was even involved in drawing them. In 1764, Watt had drawn a graph of the relationship between steam pressure and the boiling point of water. This graph, which is unfortunately lost, was produced around the same time that Lambert drew his graphs of the evaporation rate of water and temperature. Watt claimed that "by laying down a curve from my data, I ascertained the particular [ratio] near enough for my purpose." The engineer Watt was satisfied with "near enough" for his practical purposes, while most scientists still preferred what they perceived as the higher accuracy of numbers and tables. In 1782, Watt included a curve in a patent application, and later he would invent his indicator, an automatic graphing device that measured steam pressure. Although Playfair's time with Watt was short and not very successful—Watt called him a blunderer in a letter to his

[52] Tilling (1975, 207–208).
[53] Funkhouser (1937, 289).

partner Boulton—he probably got inspiration for his statistical charts from the diagramming practices in the workshop.[54]

Influenced by the empiricist philosophy of the Scottish Enlightenment, Playfair published a *Commercial and Political Atlas* in 1786 that included 44 statistical charts. All but one of them plotted values on the vertical axis against time on the horizontal. The exception was a bar chart made necessary by a lack of time series data. This diagram had been inspired by Priestley's chronological chart from 1765.[55] Under advice from his friend and former employer James Watt, who was concerned about the accuracy of the charts, Playfair also included tables in the two first editions of his atlas. Playfair engraved the lines on the copper plates himself, and the concerns about accuracy were justified. However, Playfair's atlas established conventions that still today guide how data is displayed in time series line graphs.[56]

In his *Statistical Breviary* (1801), Playfair introduced circle charts, pie charts, and diagrams illustrating shared properties, similar to Venn diagrams. Possibly, he was inspired by the logic diagrams of Leibniz and Euler, although not necessarily, since we have seen similar diagrams being reinvented repeatedly throughout history.[57] Pie charts had been drawn by Raymond Llull in the Middle Ages and by Giordano Bruno during the Renaissance, but these had described qualitative, relative relations rather than absolute numerical data.[58]

Playfair's method was not an immediate success. In Britain, the concerns regarding the accuracy of his graphs seem to have been heightened by his careless and disreputable personality. It has been claimed that Playfair's diagrams were more positively received on the Continent. Indeed, the German naturalist Alexander von Humboldt, one of the foremost scientific celebrities of the time and inventor of isothermal charts, was positively inclined toward Playfair's diagrams. In his 1811 description of South America, he developed some of them further, such as the subdivided bar graphs. However, also on the Continent most statisticians remained skeptical. The wider adaptation of Playfair's methods had to wait

[54] Costigan-Eaves and Macdonald-Ross (1990, 324); quote from Muirhead (1854, lxx).

[55] Priestley's charts are discussed in section "Timelines: Spatial Representations of Time" in Chap. 7.

[56] Spence and Wainer (2017, 48–49).

[57] Spence and Wainer (2017, 54); Playfair (1801); section "Circles of Inclusion" in Chap. 4.

[58] Spence (2005).

until the second half of the nineteenth century, when Charles-Joseph Minard and Jacques Bertillon took up some of his inventions.[59] The science of statistics was during the first half of the century still entangled in a battle between two traditions regarding whether verbal or tabular numerical descriptions were preferable. The older school of Gottfried Achenwall, which had defined statistics as the description of states, was being replaced by a new science based on mathematics and probability, which was appropriating the name. Initially, none of these factions embraced graphs.[60]

In 1794, Dr. Buxton patented "coordinate paper," known today as squared paper, millimeter paper, or graph paper. This was before lined writing paper was patented in 1815. The first published graph using coordinate paper came in 1800, tracking barometric pressure.[61] Millimeter paper constituted a vital infrastructure for accurate data visualization. For example, the astronomer William Herschel, one of the first scientists to use diagrammatic data visualization, advocated its use.[62]

Another person with a background in astronomy, the Belgian mathematician Adolphe Quetelet, was also interested in social statistics. He developed several diagrams common today, such as the bell curve of normal distribution.[63] Florence Nightingale became inspired by Quetelet's application of statistics to public health and social issues. She made graphic presentations such as the polar area graph of soldier mortality in different months of the year during the Crimean war, where the area of each wedge represented deaths during one month (Fig. 6.6). The wedge diagrams replaced the earlier "bat wing" diagrams from her private report for the Secretary of War. There, deaths had been proportional to length of the lines radiating from the center, which exaggerated the difference in area. This is a common mistake in statistical graphics, which Nightingale realized and compensated for. Friendly and Andrews believe that she was inspired by William Farr's diagram of mortality in London from 1852, and that Farr was likely influenced by André-Michel Guerry's wedge diagrams of wind direction from 1829.[64] As discussed in section "Quipu: Inca

[59] Beniger and Robyn (1978, 3–4); Spence and Wainer (2017, 57). For details on the life of Playfair see Bellhouse (2023).

[60] Funkhouser (1937, 292).

[61] Friendly (2006, 24–25).

[62] Tilling (1975, 210).

[63] Quetelet (1846).

[64] Friendly (2008, 509); Friendly and Andrews (2021); Magnello (2012).

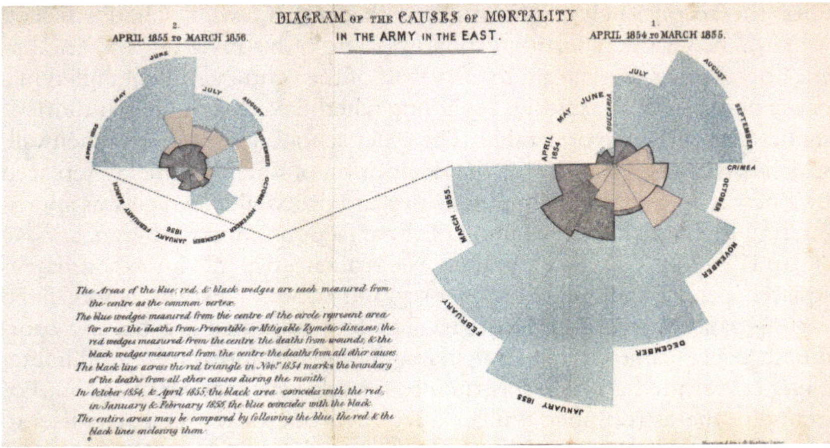

Fig. 6.6 Diagram by Florence Nightingale from 1859 of mortality in Crimea from April 1854 to March 1856. The areas of the wedges are proportional to deaths per month. Blue represents deaths from preventable disease, red deaths from wounds, and black other causes. Nightingale (1859, 320)

Infographics" in Chap. 3, Inca quipus might have been arranged as wedge diagrams.

Quetelet, Nightingale, Guerry, and Farr were active in social or moral statistics, which addressed problems such as illiteracy, disease, and crime. Although some actors in this field made use of graphs as early as in the first half of the 1800s, the period 1850 to 1900 experienced what Friendly describes as a "perfect storm" of data visualization. Development in the collection of data, statistical theory, technological advances, and inventions in techniques of visual representation combined to create a Golden Age of statistical graphics. Official statistical offices were established across Europe, and the growing amounts of data they collected could be analyzed with the help of new mathematical methods in statistics developed by, for example, Carl Friedrich Gauss and Quetelet. The period was characterized by many parallel developments and innovations, such as 3D graphics and logarithmic diagrams. These took place in an interchange with developments in the field of cartography. The new innovations were collected in statistical atlases, which could be printed in color in large editions thanks to advances in lithographic printing.[65]

[65] Friendly (2006, 13–14); (2008, 504–509).

According to Tufte, the work of Lambert and Playfair relieved graphical design from analogy to the physical world. In his early work, Playfair had likened his graphs of national income to piles of guineas, but he broke free of that in *The Statistical Breviary*. This meant that any variable could now be placed in relationship to any other variable quantity.[66] This paved the way for the scatterplot, which Friendly and Wainer have called the most scientifically important diagrammatic innovation of the late 1800s. The late century breakthrough for scatterplots followed a development where the basic principle was presented by Herschel in 1833. The idea of plotting two variables—of which neither was time—against each other was also used by Francis Galton in his investigations of heredity. From this work, important statistical concepts such as regression and correlation were born. Galton's scatterplots were semi-graphic tables, while later examples are usually plotted in coordinate systems.[67] Tufte claims that 40% of all graphics in modern scientific publications are relational, with two or more variables, none of which are space or time. With their help, it is possible to investigate the relationship between, for example, smoking and lung cancer.[68]

From 1853 to 1876, international statistical congresses were arranged with regular intervals. The seventh congress in 1869 recommended that publications be accompanied by graphs and maps. However, at the turn of the new century, the Golden Age came to an end. For example, the French *Albums de Statistique Graphique* was published for the last time in 1897, citing high production costs, and similar publications in other European countries and the United States were also discontinued or reduced in quality.[69]

Friendly considers the first half of the 1900s a dark age for data graphics. Statisticians came to rely more heavily on numbers and parameters, such as standard deviation and correlation. Although these examples of symbolic statistics were both born out of graphical analysis, the new concepts such as statistical significance became more or less equated with science, and graphics again were considered less scientific. However, Friendly also sees this period as one of consolidation, when graphs became more mainstream and spread in everyday publications such as school textbooks.

[66] Tufte (2001, 46).
[67] Friendly and Wainer (2021, 138–143).
[68] Tufte (2001, 46–47).
[69] Friendly (2006, 17, 20); Funkhouser (1937, 311).

Diagrams were also crucial for some scientific breakthroughs during this period, such as the discovery of atomic number by Henry Moseley in 1913 based on graphical analysis and E. W. Maunders's butterfly diagram, which was used to discover the systematic variability in sunspots.[70]

The presentation of data in charts, still frowned upon as inaccurate a century earlier, was now an accepted practice. Mary Eleanor Spear claimed in 1952 that the chart had long been recognized as the clearest and most effective way of presenting a subject visually. It could reveal hidden facts not obvious from the original data, and the graph-making process could also detect mistakes in statistical compilations that would otherwise have been overlooked.[71]

According to Friendly, a revival for statistical graphics began in the 1960s. One catalyst was the work of the US statistician John Tukey, who invented or developed several innovative graphical tools for data analysis, such as stem-leaf plots, boxplots, hanging rootograms, and two-way table displays. Another impetus came from Jacques Bertin, who in his *Semiologie Graphique* from 1967 organized and systematized the visual elements of data graphics. The art of manually producing data visualizations was, however, at this time disappearing, and the revival came in the form of computer-generated graphics. Computer processing of data began with the introduction of the programming language FORTRAN in 1957, and within a decade, the spread of university computers made it possible to construct a multitude of graphical forms, old and new.[72] Friendly believes that many older forms of graphic presentations that reappeared were invented anew, since their precedents from the Golden Age were largely forgotten.[73]

Advancements in statistics also paved the way for novel types of graphic presentations. Improvements in computation made it possible to develop methods that could project multidimensional data in lower dimensional displays. Karl Pearson had pioneered this in 1901 with principal component analysis,[74] but the methods became more useful once the calculations could be computerized. For example, in the 1960s, Jean-Paul Benzécri (1932–2019) developed correspondence analysis, which visualizes multivariate categorical

[70] Friendly (2006, 20–22); (2008, 529–531).
[71] Spear (1952, 3).
[72] Friendly (2006, 23).
[73] Friendly (2008, 31); Bertin (1983).
[74] Pearson (1901).

data—rows and columns in a table—in graphs of usually two dimensions (26.5). The method became popular in, for example, sociology, since it could be used to explore relationships between factors such as social background, income, family connections, and education[75] (Fig. 6.7).

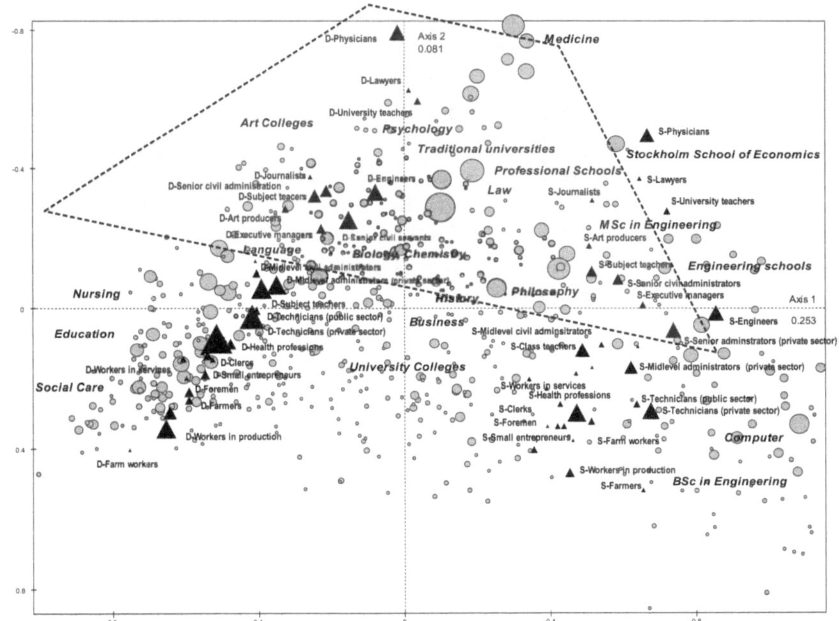

Fig. 14.1 The Swedish field of higher education and the subfield of elite education, autumn 2006. Correspondence Analysis, the plane of axes 1 and 2. 32 categories of social origin, sexes separated
Data source: Statistics Sweden
Active variables:
▲ Social origin. ***S*** **sons,** ***D*** **daughters**
○ Educational programmes/courses offered by higher education institution
The sizes of the triangles and circles are proportional to the number of students in a particular category

Fig. 6.7 Diagram illustrating the relationship between social origin and higher education, based on correspondence analysis. Börjesson, Broady and Palme (2020, 188)

[75] Friendly (2006, 23–25); Blasius and Schmitz (2015).

The technological development also facilitated the development of moving graphics. In 1952, Mary Eleanor Spear discussed animated graphics for television. She recommended the use of a flannel board to which charts of sandpaper-backed cut-outs could be built up one piece at a time.[76] In 1969, she described how gluing arrows to magnets could make them move on a chart during live presentations.[77] However, it was computerization that truly revolutionized moving graphics. For example, in the first decade of the twenty-first century, the professor of international health Hans Rosling used animated graphics for public health purposes similar to those of Quetelet, Farr, and Nightingale in the 1800s. While his nineteenth-century predecessors wanted to illuminate social problems and bring them under the public scrutiny, Rosling used moving graphics to illustrate the progress in global health since their days. For example, Rosling represented countries as circles with areas proportional to their population, and placed them in a two-dimensional graph with income per capita on the X-axis and life expectancy on the y-axis. When the film was playing, circles grew and moved from the bottom left to the top right corner of the graph, illustrating population increases as well as improved wealth and health.[78] This was in a way a double victory for data visualizations, since diagrams had, by raising awareness of problems and tracking the efficiency of countermeasures, contributed to the progress illustrated by the moving graphics.

The development of statistical graphics was enabled both by innate conceptions of space, such as the identification of line length with number, and by primary metaphors learned by all early in life, such as "More is Up." The close match between some forms of data visualizations and such primary metaphors helps explain why they have been invented several times independently. However, the habit of using space as a millimeter paper on which numerical data can be plotted, which we today take for granted, seems to be an invention situated in historical conditions that have only been present the last 300 years. The history of statistical graphics reveals how diagrams have evolved in interplay between enduring principles of human cognition, historical forerunners in niches such as musical notation, technological changes such as the invention of measuring devices, and paradigmatic shifts in the sciences, such as the introduction of

[76] Spear (1952, 240).

[77] Spear (1969, 272).

[78] See, for example, Friendly and Wainer (2021, 229–230).

analytical geometry and functions as well as the development of the distinction between quantity and quality.

Diagrams of Heartbeats

We can better understand the nature of the change in diagramming practices that the quantitative data revolution brought about by studying this transformation within one subfield of science. Already early on, the analysis of heartbeats had an important role in medicine, and diagrams were used for this purpose. For example, Galen (129–ca. 216) wrote several treatises on the pulse. In the Middle Ages, Arab, Byzantine Greek, and Latin scholars explained his treatises with the help of tree diagrams. In the Byzantine version (Fig. 6.8), pulse was divided into ten categories, of which one, heat, could not be found in Galen's original works. These ten categories were then subdivided into further categories. For example, the category "kind of motion" could be subdivided into swift, balanced, or slow.[79] Similar complex categorizations of pulse can be found in traditional Chinese medicine. Of the 24 to 30 categories, "the rapid" described the number of beats per stroke of breath, a quantitative but relative measurement. Others, such as "the string," which related to whether the pulse felt like the string of a musical instrument, were entirely qualitative.[80]

The Greek physician Herophilos (335–280 BCE) showed some interest in quantitative pulse measurements, and even constructed a water clock, Klepsydra, for the purpose. He also conducted dissections, which revealed the difference between veins and arteries. However, Herophilos's experimental approach left no permanent impression on ancient and medieval medicine.[81] Similar attempts with pulse measurements would not be conducted again until the Renaissance.

The Venetian friar Santorio Santorio (1561–1636) made an early attempt to redirect science, medicine, and pulse measurement in a more quantitative direction. In addition to making instruments for measuring winds, currents, weight, and humidity, he constructed a *pulsilogium,* a pendulum that could be synchronized with the heartrate during pulse measurements. Santorio's method was relative, since he was interested in changes in the rhythm of the pulse rather than beats per a fixed period of

[79] Bouras-Vallianatos (2022, 290–295).
[80] Tang (2012).
[81] Stefanou (2020).

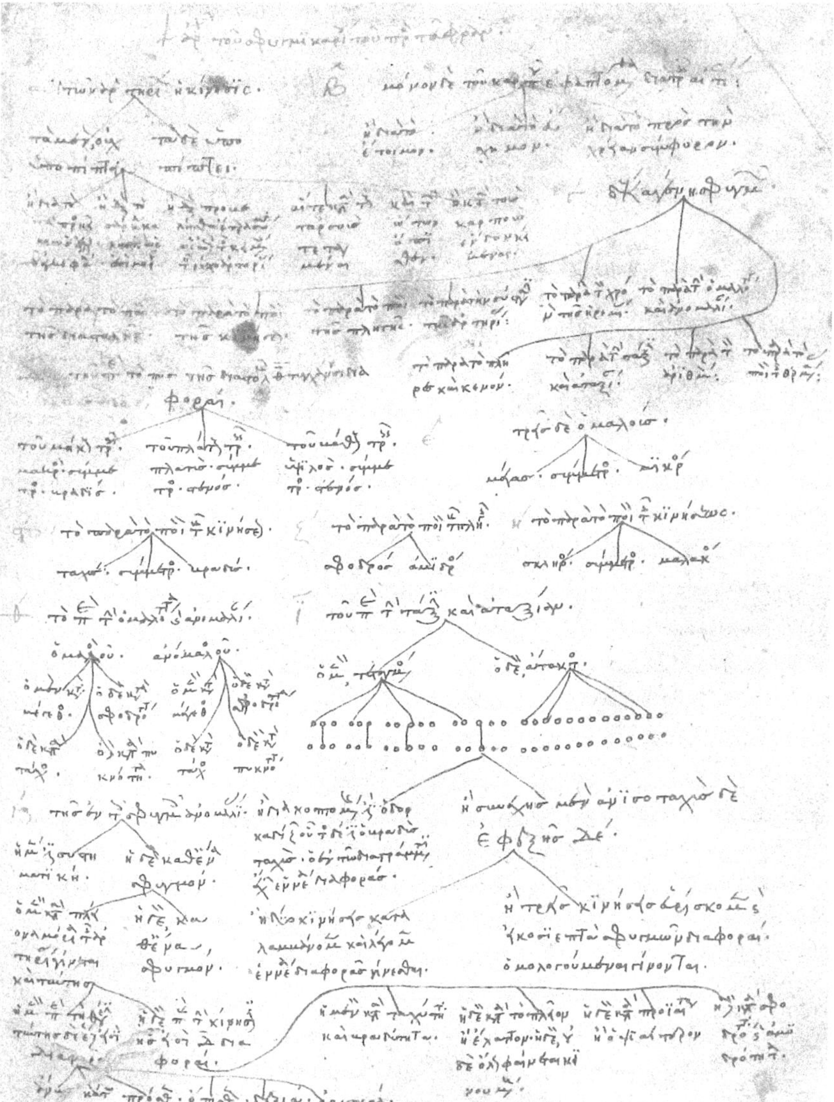

Fig. 6.8 Tree diagrams on pulse measurement from Byzantine manuscript of Galen's *On the Pulse for Beginners*, thirteenth century, that displays a hierarchical system of categories. From Bouras-Vallianatos (2022, 294)

time.[82] Santorio wanted to keep the details of his construction secret, and timed heartrate measurement remained rare for another century. Thus, there was yet no quantitative pulse measurement data that could form the basis of data visualizations.

In 1695, Samuel Watson, the watchmaker to the king of England, invented a stopwatch on behalf of John Floyer, a doctor who needed it for exact pulse measurement. Floyer published a book on heart rate measurement in 1707, and in the same year, Watson released his stopwatches to the wider market.[83] Watson's stopwatch was more portable and practical than ancient clepsydra or Santorio's *pulsilogium*, and became more widely used among physicians.

However, the theoretical interpretations that Floyer made of his measurements were still far from those of modern medicine. His theories on the pulse were influenced by both Galen and Chinese medicine, which at this time began to attract interest in the West. By comparing Chinese measurements with his own measurements in England, Floyer hypothesized that the healthy heart rate followed a gradient from 20 beats per minutes at the equator to 90 at the poles. He explained this with Galenic theories on heat and suggested that patients with a high pulse should be sent south and those with a low pulse north.[84] Thus, Foyer had a method for measuring heart rate, but no adequate theory for interpreting it or a suitable diagrammatic form for presenting it. Instead, he tabulated measured heart rates for individuals in a rather unstructured fashion, combined with columns for age and "habit of body," a category which contained verbal descriptions such as "A little fat," "Plump," or "A thin, spare man."[85] Just as ancient Galenic medicine did not reach its height of diagrammatic systematization until the Middle Ages, it took time for quantitative measurements to find their mature form of diagrammatic representation. In the nineteenth century, several mechanical devices were constructed for recording cardiac activity. Carl Ludwig's kymograph from 1847 was quite invasive, as its stylus was attached to a mercury manometer linked to a dog's artery. Etienne-Jules Marey's sphygmograph from 1860 was more practical, as it was portable and measured pulse at the wrist with the help

[82] Bigotti and Taylor (2017).
[83] Read (2018).
[84] Floyer (1707).
[85] Floyer (1707, 183).

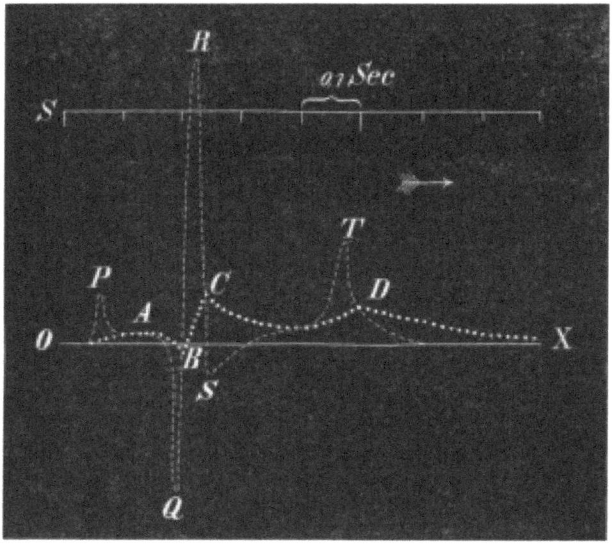

Fig. 6.9 This early electrocardiogram from a publication by Willem Einthoven illustrates the quantitative turn in heart rate measurement. Five years later Einthoven constructed the first practical and commercially available ECG machine, the string galvanometer. Einthoven (1896, 49)

of a lever attached to a stylus.[86] However, it was electrification that would make recordings of cardiac activity widespread.

In the late nineteenth century, several experiments were made with recording the electric signals from heartbeats (see, e.g., Fig. 6.9), but Willem Einthoven's string galvanometer from 1901 was the first practical ECG machine. It produced diagrams with the help of a drop-plate camera.[87] From 1932, direct-writing ECG machines were introduced, which drew graphs with pens on rolls of paper. Although they were initially viewed as less reliable, their practicality made them dominate after the war.[88]

By then, the tactile pulse measurements of the Middle Ages could be replaced by data recordings, but the problem of interpretation remained

[86] Braun (1992, 16–18).
[87] Burch and DePasquale (1990, 29–35).
[88] Burch and DePasquale (1990, 50–54).

strikingly similar. George Burch and Nicholas DePasque claimed in 1962 that there was a gap between empirical evidence of ECGs and the physical theories used to explain them. They even described some research fads in the field of ECG as "wrought in the most pseudoscientific environment and approach." Writing at the beginning of the computer age, Burch and DePasque also rekindled the old debate on the accuracy of diagrams. They considered it absurd that crude direct-writing recorders were used to feed information into high fidelity recorders for analysis in complex electronic computers with an accuracy far exceeding that of the direct-writing recorders.[89] These arguments about accuracy in the transfer between measurements and diagrams are similar to those expressed by Lambert in the 1700s.[90] Perhaps such discussions are characteristic of periods when the regimes of measurement, visualization, and analysis undergo rapid change, such as during the introduction of statistical graphics and the shift from analog to digital recording.

Since ECG diagrams do not remove the need for interpretation, they cannot be said to have replaced the classificatory trees in the Galenic tradition. Rather, they have replaced the fingers on the artery. Tree diagrams are still used in academic publications that describe, for example, how decision trees can be used in order to efficiently classify cardiac arrythmias.[91] The relationship between classificatory trees and ECG graphs in interpreting pulse data is illustrated by the "ECG Interpretation Flow Diagram" presented on a medical YouTube channel in order to teach viewers how to interpret different aspects of ECG graphs: it looks similar to a Galenic tree diagram, but with a small ECG graph at each node.[92] A more common form of ECG interpretation flow-sheet is illustrated in Fig. 6.10, which has a structure similar to Galenic diagrams such as Fig. 6.8.

The field of heart rate diagnostics illustrates that the quantitative revolution and the diagrams associated with it, such as functions, complement rather than replace older types of diagrammatic representations, since logical classification and systematization are still needed.

[89] Burch and DePasquale (1990, 50–54).

[90] Section "Statistical Graphics".

[91] Sahoo et al. (2020).

[92] Demos (2013).

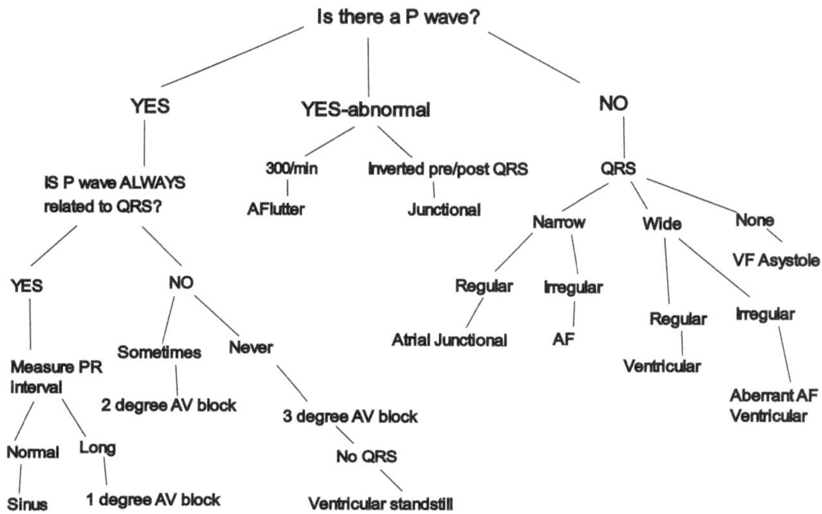

Fig. 6.10 ECG interpretation trees are common in medical education. Structurally, they are similar to the tree diagrams used in medieval medicine. Drawn by author based on several similar examples

Diagrams in Economic Theory

What is probably the oldest preserved reference to a diagram of economic theory is found in Aristotle's *Nichomachean Ethics*. On the subject of just exchange of goods, Aristotle wrote that "the acts of mutual giving in due proportion may be represented by the diameters of a parallelogram, at the four angles of which the parties and their wares are so placed that the side connecting the parties be opposite to that connecting the wares and each party be connected by one side with his own wares" (Fig. 6.11).[93]

According to Aristotle, an equilibrium in value between goods traded was necessary for the transaction to be just and for a deal to take place in the first place. According to Theocharis, Aristotle seems to have pictured what economists today would describe as equilibrium conditions of exchange. Theocharis also notes that after Aristotle, no one attempted to apply mathematics and diagrams to economic theory until the 1700s.[94]

[93] Requoted from Theocharis (1983, 4).
[94] Theocharis (1983, 3–4).

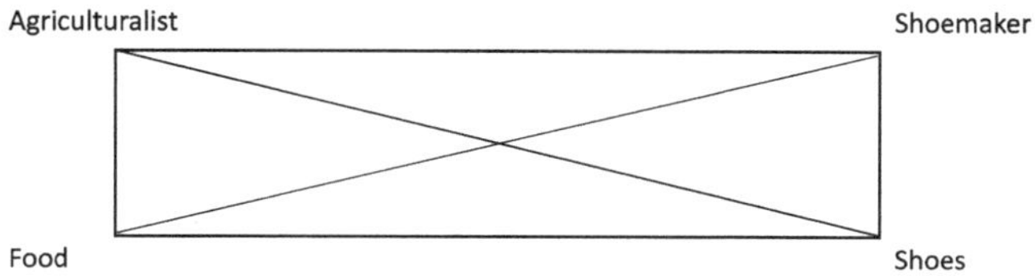

Fig. 6.11 Visualization of Aristotle's description of the "parallelogram of exchange." Drawn by author

For researchers of diagrams, there is an obvious similarity between this "parallelogram of exchange" and the square of opposition (Fig. 4.2) that was used in late antiquity to illustrate Aristotle's logic. It is in this context Aristotle's economics diagram has to be understood. We should not let the crossed lines in the middle of the diagram mislead us into considering it an ancient forerunner of the Marshallian scissors with its intersecting supply and demand curves. That kind of diagram had to wait until the invention of functions, which help explain why new advances in economic diagrams were not made until the 1700s. However, from the mid-1800s, the number of diagrams used to illustrate economic theory has exploded.

In the history of diagrams, we see some examples of diagrams that were invented as pedagogical illustrations of principles developed generations or even centuries earlier, such as the wheel of logical reasons, the square of opposition, and the tree of Porphyry.[95] However, we can also find examples of diagrams that were probably instrumental in the development of the theories that they were used to illustrate, such as James Watt's diagrams.[96] Compared to these examples, the development of economics is quite recent and relatively well documented, which makes it a useful case for studying the role of diagrams in theory development. In the introduction to the volume *Famous Figures and Diagrams in Economics*, Mark Blaug and Peter Lloyd discuss whether economic theories have been discovered from diagrams or from algebra, logic, or intuition. Although they acknowledge that it might sometimes be difficult to uncover the role of diagrams in theory discovery, they mention some cases where the role is clear. For example, the Engle curve and the Lorenz curve were both fitted to data, which would illustrate that the curves were there first, where after the theories were developed based on it. However, most economic

[95] Section "Tree Diagrams" in Chap. 4.

[96] See section "Statistical Graphics".

diagrams are not based on empirical data but on economic models, and in those cases, it is more difficult to know how the theory was conceived. There are some exceptions, such as the Scitovsky curve. Since the author did not include any mathematics in the paper beside geometry, Blaug and Lloyd find it likely that he arrived at his theory by the use of diagrams. They also believe that Marshallian stability analysis was discovered with the help of the diagram. In other cases, such as the Slutsky derivation of the income and substitution effect and the Nash equilibrium, it is evident that the theories were found using algebra, since the diagrams only appeared later.[97]

According to Blaug and Lloyd, in the last decades there has been a decline in the use of geometry and diagrams in the discovery of economic theories. The growing complexity of economic models has forced macroeconomists to formulate their models as systems of equations rather than as diagrams.[98] Diagrams using two-dimensional geometry can only handle two goods, and since higher dimensional theories have become common from the 1950s, diagrams are used more as low-dimensional illustrations than as tools of discovery.[99]

However, diagrams continue to excel as expository devices. Blaug and Lloyd claim that they are much more than illustrations: "In many areas of economic theory, the way in which economists understand economic concepts and propositions is through figures and diagrams." Diagrams also help economics students understand concepts that are incomprehensible to them in their algebraic forms. According to Blaug and Lloyd, the big advantage of diagrams is that they juxtapose information from two curves that cannot be easily comprehended in equations or words. Lately, the pedagogical power of economics diagrams has been strengthened by multicolored textbook figures and computer-based programs which allow overlays and other graphical techniques for conveying complex ideas.[100]

Perhaps the most common type of diagram in economics is that of intersecting supply and demand curves in a two-dimensional space of price and quantity (Fig. 6.12). Often, these diagrams are associated with Alfred Marshall, and they are sometimes even named after him: the Marshallian cross or the Marshallian scissors. However, similar diagrams were

[97] Blaug and Lloyd (2010, 5).
[98] See section "Causal Path Diagrams and Fuzzy Cognitive Maps" in Chap. 7.
[99] Blaug and Lloyd (2010, 7).
[100] Blaug and Lloyd (2010, 8–9).

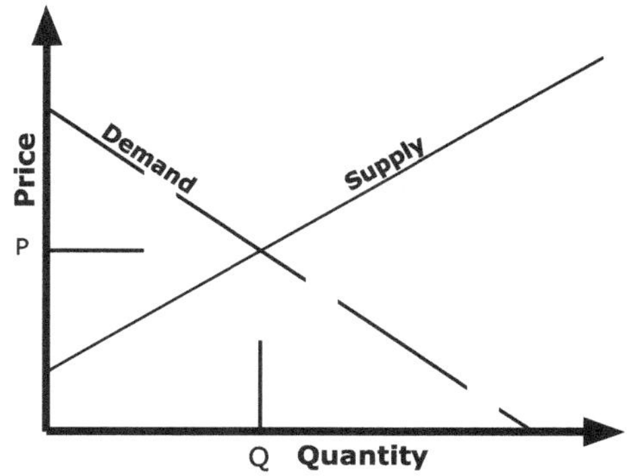

Fig. 6.12 Curves describing the relationship between supply and demand and the quantity and price of goods were developed independently by several economists in the nineteenth century. By Dallas.Epperson

produced earlier. The real pioneer was in all likelihood Antoine-Augustin Cornot, who used diagrams of this type already in 1838. Several others, such as Karl Rau, Jules Dupuit, Hans von Mangoldt, and Fleeming Jenkin, developed it further before Marshall presented his version in *Theory of Domestic Values* in 1879.[101]

Thomas M. Humphrey claims that these pioneers were all ignorant of each other's work, with the exception of Mangoldt who knew about Rau. Humphrey is also of the opinion that although modern economic textbooks are based on Marshall, his theories were more or less identical to those of his predecessors, and it would have made little difference for the discipline if they had been introduced in textbooks instead.[102] This situation is similar to the many parallel discoveries of the periodic system of the elements, which took place during the same time period, the mid-1800s. In both cases, it was the last discoverers, Mendeleev and Marshall, who became the public face of the theories and diagrams.

The field of economics is closely intertwined with politics. This is illustrated by the history of the Laffer curve, a diagram describing the relationship between tax rate and tax revenue. It was introduced in a 1978 book by Jude Wanniski. At the time, Wanniski pursued a career as a political consultant after having been forced to resign as columnist at the *Wall Street Journal* because he had distributed leaflets for a Republican senatorial candidate. The Laffer curve was intended to illustrate that government tax revenues might actually decrease if tax rates go up, since high tax rates

[101] Humphrey (1992, 16–21).
[102] Humphrey (1992).

remove the incentives for production. The curve was named after Arthur Laffer, an economist then working as an adviser for the Republican Party. He is supposed to have drawn it on a napkin at a meeting with two of Richard Nixon's political aides in a Washington restaurant in 1974. The curve, which was developed in the sphere of policymaking, was initially frowned upon by academic economists. With time, less simplified and more mathematically advanced versions of the curve were produced in academia. However, the original, simple, and, by most accounts, faulty version of the curve is still the most common. Yann Giraud believes that this is due to the fact that it is most easily accessible to non-economists.[103]

The Laffer curve has been criticized for not considering that in most countries there is not a single tax rate, but progressive taxation. In addition, empirical research seems to indicate that the point of optimal tax returns is 71%, which is above the tax levels relevant in the political discussion in most countries. According to Roger Middleton, because of its academic shortcomings the Laffer curve remained a peculiarity of US politics, gaining traction with the tax cuts of the Reagan years. Even in the fiscally frugal Thatcher administration, economic advisers refrained from legitimizing their tax cuts with the "mistaken" Lafferism.[104] However, lately the Laffer curve has been used outside of the United States by market liberal think tanks, such as the Canadian Frazer Institute and the Swedish Timbro.[105] Sweden is one of the few countries—according to a study from 2005, the only country—where the Laffer curve suggests a lowering of tax rates could increase tax revenues.[106] However, also in countries with less extreme taxation, the fact that the Laffer curve promises more for everyone—taxpayers can keep more of their money while the government increases its revenues—have made policymakers and lobbyists fond of it.

Interestingly, Laffer has claimed that he did not invent the idea himself, but borrowed it from the medieval scholar Ibn Khaldun (1332–1406).[107] However, since the function concept was not yet developed in Ibn Khaldun's day, he did not have the diagrammatic tools to produce a curve similar to Laffer's.

[103] Giraud (2014).

[104] Middleton (2010, 412–418).

[105] Ferede (2019); Timbro (n.d.).

[106] Lundberg (2017); Heijman and van Ophem (2005). A tax cut on the highest incomes in 2020 seems to have resulted in increased tax revenues, according to Carlgren (2024).

[107] McCaffrey (2014).

Stock Charts: Icons of Capitalism

The stock chart is a common form of diagram in mature capitalist societies. The first stock charts appeared in the 1830s. However, the early charts usually depicted very long time series, and they only acquired their present role as tools for daily trading activities with the stock ticker, a machine for transferring stock market data electrically, which was invented by a telegraph operator in 1866. The stock ticker provided a continuous flow of information that could be represented in standardized charts.[108]

In a guide for stock traders from 1935, H.M. Gartley described the externalizing function of diagrams: "As in any other human endeavour, pictures more strikingly emphasize cold tabulations of facts or figures. Thus, charts relieve the memory of arduous work, and leave the mind free to contemplate the future."[109]

However, in contrast to other forms of statistical graphics, which have been the object of considerable academic research, the history of stock charts is largely unwritten. No overviews such as, for example, those Michael Friendly has written about the use of diagrams in public statistics and the sciences exist regarding the stock charts. Perhaps this is linked to the fact that stock charts have been practical tools for individual traders and investors, rather than enablers of scientific breakthroughs or advances in public policy. Trading based on analysis of price movements through charts, technical analysis, is also commonly considered less legitimate than long-term investing based on fundamental company value. In the words of a Japanese trading guide, "Speculation is the illegitimate child of capitalism."[110] Recently, prediction of price movements from charts has also become considered antiquated by "quants" who use computerized statistical calculations.[111] This is similar to how diagrammatic representations have periodically been seen as less "scientific" than symbolic mathematics by researchers and statisticians, for example, in the eighteenth and early twentieth centuries.

It is also considerably more difficult to do research about stock charts than about scientific diagrams. Publications that contain information about stock charts are seldom kept by research libraries, and are often rare and expensive on the secondhand market. Sometimes these publications

[108] Preda (2001); Lo and Hasanhodzic (2010, 72).
[109] Gartley (1935) 1981.
[110] Suminoe (1990, iii).
[111] Lo and Hasanhodzic (2010, viii).

contain some background information about the history of stock charts, but always brief and anecdotal, seldom based on reliable sources.

The above quoted work by Gartley from 1935 gives one of the clearest and best referenced overviews of the early history of stock charts. According to Gartley, line charts, that merely recorded closing prices, and bar charts, that displayed the whole range of trading, were both introduced by statistically inclined Wall Streeters in 1895 to 1900. However, they were preceded by figure charts, which were introduced around 1881. Figure charts illustrate price movements with lines, but only those above a certain threshold, such as one point. Thus, the X-axis is not time, but events: a new line is drawn whenever the price movement exceeds the threshold. As long as the price does not move enough, "time" seems to stand still on the chart.[112]

In the 1980s and 1990s, the Western world was introduced to Japanese candlestick charts (Fig. 6.13). An important person in this process was Steve Nison, who with the help of a translator read a vast amount of Japanese literature and in 1991 presented his findings in a book. According to Nison's first book, these charts were developed by the rice futures trader Honma Munehisha (1724–1803) from the town of Sakata. A candlestick chart displays the difference between the opening and closing price on a trading day through the length of a candle. If the price has been

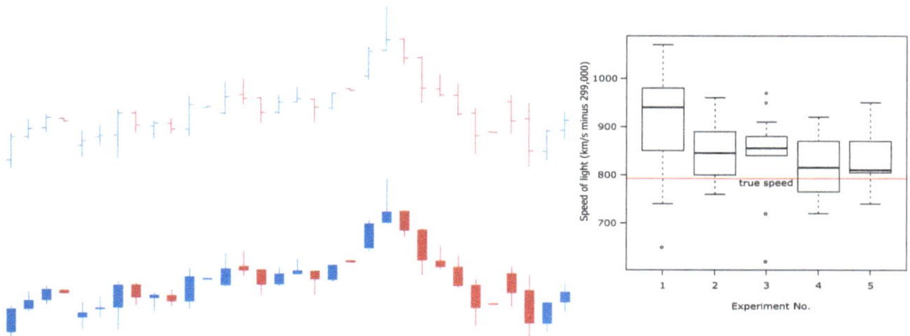

Fig. 6.13 A bar chart or OHCL chart (top left) and a candlestick chart (bottom left) that represent the same price data series. To the right a box plot. This particular box plot illustrates the result of different experiments for determining the speed of light. Drawn by author, box plot by Schutz

[112] Gartley (1935) 1981, 327.

rising, the candle is empty (or green), and the lower end represents the opening price and the upper the closing price. On a day with falling prices, the candle is filled (or red), and the upper limit represents the opening and the lower the closing price. In addition, the candlesticks are adorned with thin lines, in Japan called the hair and the tail, which represent the lowest and the highest prices reached during the day.[113] Visually, the candlestick diagrams with their "hair" are more or less identical to the box plots with "whiskers" introduced by John Tukey in 1970, and more broadly popularized by him in his 1977 *Exploratory Data Analysis*.[114] Their functions are also similar: to represent the range and distribution of data rather than mere averages or single points. However, in box plots the edges of the box indicate quartiles rather than openings and closes, which makes them useful for many different categories of data. The box plot had a prehistory before Tukey, as Mary Eleanor Spear in 1952 presented "range bars," which depicted quartiles in a similar fashion to Tukey's box plots. Range bars were also described in a US statistical journal back in 1948.[115]

Before candlestick diagrams spread to the West, the same data were represented in what is called a bar chart,[116] today often called an OHCL (Open-High-Close-Low) chart, where the length of a vertical line represents difference between highs and lows, and short horizontal lines left and right of the main line represent opening and closing prices. Although the information expressed by the two chart types are identical, the candlestick diagrams are better adapted for applying the Sakata rules, since the difference between opening and closing price and the pattern it forms from day to day is immediately visually available.

If Honma Munehisha had developed candlestick diagrams in the 1700s, he would have been far ahead of his time. In fact, it would totally undermine the thesis presented in this volume regarding the importance of analytical geometry and functions for the development of statistical graphics and time series charts. Functions were unknown in traditional Japanese geometry, and there was little Western influence on the country's mathematics in the eighteenth century.[117] Also in Europe, the earliest diagrams for data visualization had barely been drawn in Honma's day, and they

[113] Nison (1991).
[114] Tukey (1977); Wickham and Stryjewski (2011).
[115] Spear (1952, 164, 166); Haemer (1948).
[116] Not to be confused with the ordinary bar chart invented by William Playfair.
[117] D.E. Smith (1932, 538).

would not become common until the 1800s. However, there is reason to believe that the idea that Homma invented the candlestick diagrams is based on a misunderstanding. In a later book, Nison changed his narrative and suggested that the candlestick charts were probably introduced in the early part of the Meiji period (1868–1912), as the last step in a development beginning with line charts, via bar charts and anchor charts.[118] What Honma introduced was a set of rules for how to predict market movements based on patterns in the difference between open and closing prices, the so-called Sakata methods. There is no evidence suggesting that he would have drawn diagrams similar to modern candlestick charts in order to keep track of market history. None of the Sakata rules demand a longer historical record than a few days. It would not be inconceivable that the modern candlestick diagrams were developed by Meiji Japanese who combined influences from the contemporary Western Golden Age of statistical graphics with the domestic heritage of the Sakata rules.

However, if we look closer at editions of Honma Munehisha's works and comments on his methods from different time periods, we cannot find any Meiji period candlestick charts either. No books on Honma were published in the Edo period (1603–1867), but the diagrams in other publications from the period are similar to Buddhist *kewen* (Fig. 4.9), tree diagrams of lines connecting textual elements.[119] The first books on Honma's methods were published in the Meiji period. However, they contained similar tree diagrams, although Western influence was revealed by the trees being constructed of brackets, which were introduced in Japan during this period.[120]

A diagram reminiscent of a candlestick chart appeared in a 1932 book on the Tokyo stock exchange (Fig. 6.14). It depicts the yearly highs and lows of the Tokyo Stock Exchange since its inauguration in the eleventh year of Emperor Matsuhito's reign (1878).[121] The direction from left to right, the opposite of Japanese writing and, for example, Fig. 6.15, indicates that it is modeled after Western predecessors. The filled bars spanned the range of trading for the year, but unlike candlestick charts there were no hairs and tails and no difference in coloring indicating upward or downward price movement. The chart also reveals an extreme volatility

[118] Nison (1994, 14–18).
[119] See, for example, this work on tea ceremonies: Minami (1760).
[120] Honma (1898).
[121] Tōkyō kabushiki torihikisho (1932, 21).

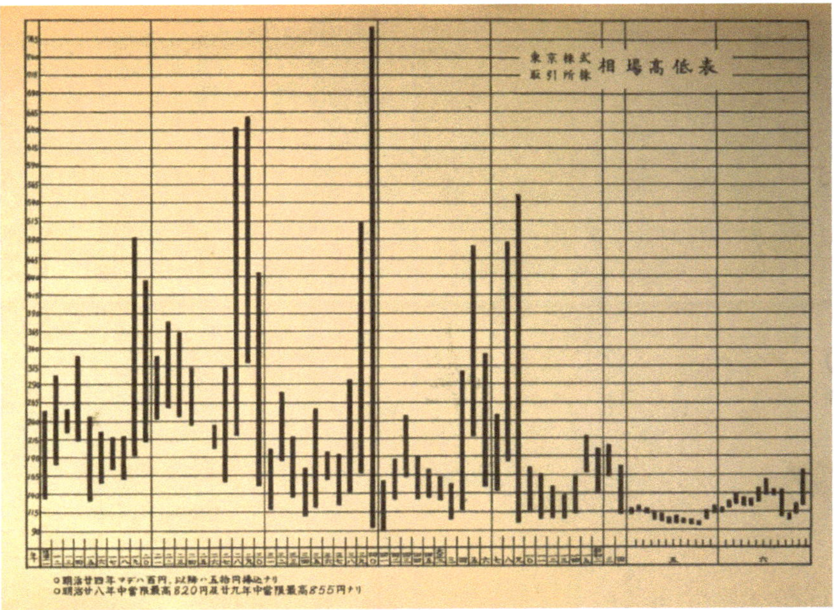

Fig. 6.14 Highs and lows of the Tokyo stock exchange from its inauguration until 1932. The data is presented yearly until the two last years, which are presented monthly. Tōkyō kabushiki torihikisho (1932, 21)

and absence of a long-term trend. In such conditions, it is not surprising that stock market actors turned their attention to short-term predictions based on market psychology, such as the Sakata rules, rather than to long-term investing.

A book published by the *Nihon Securities Journal* in 1969 seems to be the first that linked the Sakata rules to candlestick charts.[122] According to Saichirō Suminoe (Fig. 6.15), similar data was represented by anchor charts (a) in a book from the 1940s, and in between them yet another form (b) was used.[123] It seems likely that the fully developed candlestick charts (c) borrowed its form from range bars in US statistical publications.

[122] Nihon shōken shinbun-sha (1969).

[123] Arai (1950); Suminoe (1990, 88).

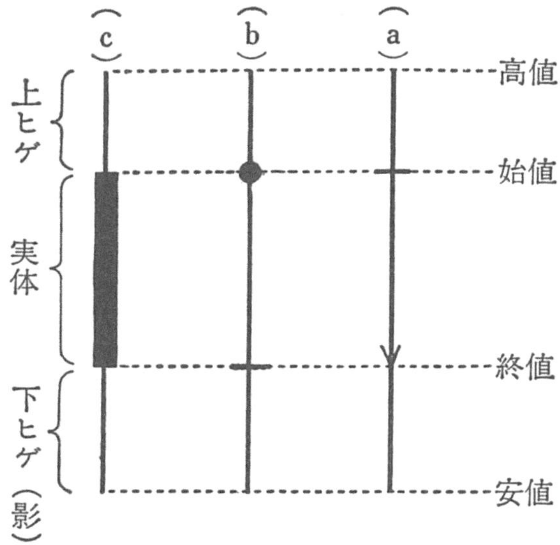

Fig. 6.15 To the left the candlestick chart presented in 1969 (c), to the right the anchor chart used in the 1940s (a). In the middle a variant used in the intermediary period (b). Suminoe (1990, 88)

That Japanese stock charts and other time series charts are influenced by Western diagramming is also reflected by the fact that time often flows from left to right, following the spatial agency bias of the Latin alphabet and not of Japanese writing.

The history and historiography of the candlestick charts illustrate several points in the general history of diagrams. First, it is one more example of how new diagrammatic forms have been introduced to illustrate older principles to new audiences. Second, it exemplifies transmission of diagrammatic forms forward and back between cultures in ways that are difficult to untangle. Finally, it also shows how successful diagrammatic forms can be endowed with "creation myths" that give them old and glorious histories than do not quite hold up under closer scrutiny.

The Sakata rules and other principles of technical analysis are based on the idea that there are recurring patterns in price changes, which are fundamentally based in the psychology of the actors on the market. Thus,

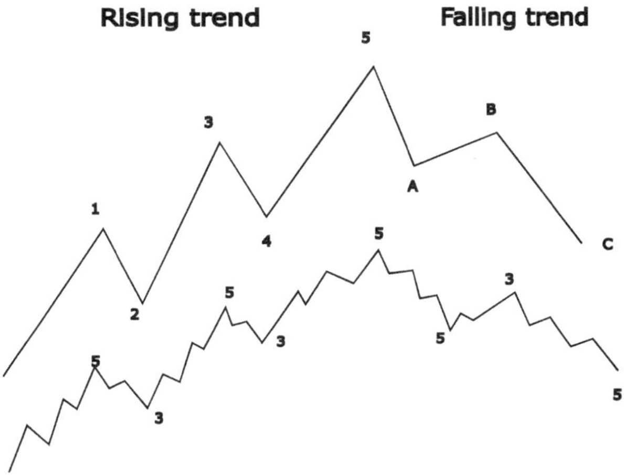

Fig. 6.16 According to Elliott wave theory, waves following trends can be divided into five smaller waves, while countertrends are divided into three smaller waves. This hierarchical categorization of thousands of small index movements conforms with the principles of folk taxonomies and respects the limits of human working memory. Drawn by author

diagrams such as candlestick charts are tools that enable traders to utilize their spatial cognition in order to analyze phenomena of mass psychology.

Another example of this is the Elliott wave theory. It was developed in the 1930s and 1940s by Ralph Nelson Elliott on the basis of the earlier Dow theory. The main principle of Elliott wave theory is that movements in stock indexes take place in five waves in one particular direction, which is then followed by a retraction of three waves. Each wave in the direction of the trend can be divided into five smaller waves, while each wave contrary to the trend consists of three smaller waves. These smaller waves might then be subdivided into even smaller waves, following the same principle, providing a hierarchy of waves from centuries down to seconds (Fig. 6.16).[124]

Elliott believed he had discovered something profound. In his later book *Nature's Law: The Secret of the Universe* from 1946, he proposed that

[124] Murphy (1999, 319–342).

the wave patterns he had identified were a fundamental law that could be used for predictions in all fields of nature and society.[125]

An alternative suggestion is that Elliott wave theory fills the same function in the analysis of time series charts as tree diagrams do in classification: it divides and subdivides chaotic and overwhelming amounts of information into a hierarchy of manageable chunks, not exceeding the capacity of working memory and thereby avoiding cognitive overload. Elliott wave theory and other methods of technical analysis do not immediately turn anyone into a successful trader, since they require a lot of experience. They are also open to interpretation, as two traders might reach different conclusions of how to interpret the candles or waves. For example, it is difficult to judge whether a recent retraction is just a shorter countertrend in a continuing trend or the beginning of a shift in the general trend. However, these methods are useful tools in the learning process of becoming a trader, since they remove the noise from the graphs. This helps traders focus on aspects of the diagrams that are simple and general enough to be grasped by the limited capacity of human cognition.

Diagrams of Social Hierarchies

In the Middle Ages, the *scala naturae,* Aristotle's hierarchical model of the natural world, was extended all the way up to God.[126] From the late Middle Ages, a similar diagram structure was used to divide human society into a vertical *Ständeordnung* of clergy, nobility, and commoners (Fig. 6.17).

The medieval and early modern diagrams of social stratification were qualitative analogies. However, with the emergence of data-driven science and statistics, social hierarchies also became quantified. In the nineteenth century, diagrams appeared that showed a stratification within the human race, based on normal distribution of hereditary traits. A pioneering work was Francis Galton's *Hereditary Genius* from 1869. Although Galton used a diagram of body length to illustrate the law of deviation, he presented his data on giftedness in tabular form. However, in 1895 the German anthropologist Otto Ammon produced a curve based on Galton's giftedness data (Fig. 6.18), and used it to argue against the social democratic

[125] Elliott (1994).

[126] Section "The Shape of Life" in Chap. 5.

Fig. 6.17 Tree-shaped *Ständeordnung* from the first half of the fifteenth century. The king in the center is Charles VII of France, who ruled 1422–1461. Unlike, for example, Fig. 2.2, this model places the Church—which commissioned the image—on top. From a copy of *L'arbre des batailles* by Honorat Bovet. Paris, Bibliothèque nationale de France, Bibliothèque de l'Arsenal, ms. 2695 folio 6v

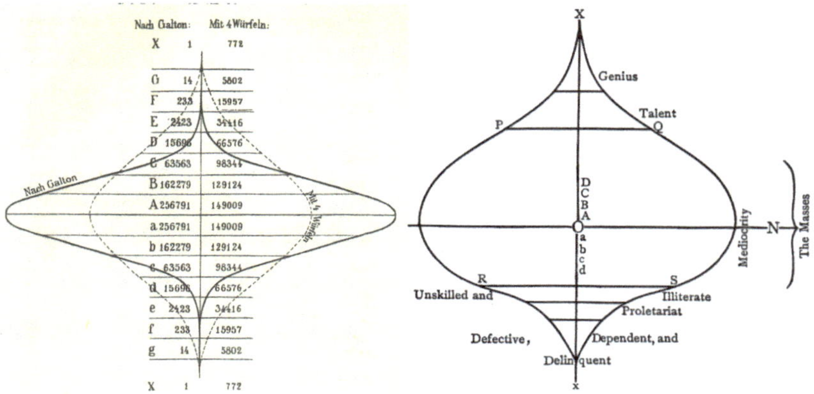

Fig. 6.18 Left: Galton's distribution of mental gifts according to Ammon (1895, 54). Right: This qualitative diagram of social value mimics the shape of normally distributed data. Sumner (1906, 40)

view of society.[127] Although most curves of normal distribution are horizontal, Ammon chose a double vertical curve overlaid on Galton's table.[128] Thereby, the diagram reflected the traditional vertical model of social stratification.

The US sociologist William Graham Sumner developed the diagram further (Fig. 6.18). He detached it from the original data to illustrate the general principle and linked it to social classes defined by their "social value." Although he admitted that social value was in part dependent on factors such as health and luck, he mainly considered it linked to mental power. Sumner recognized that social value was unmeasurable and could never be verified statistically, but he still assumed that it followed a curve of normal distribution, since that distribution "has been statistically verified so often, and for such a great variety of physical traits, that we may infer its truth for all traits of mind and character for which we have no units, and which we cannot therefore measure and statistically classify."[129]

The view that mental qualities were unmeasurable would soon be challenged. In the early twentieth century, intelligence tests developed by

[127] Galton (1869, 27–28); Ammon (1895, 83).

[128] Such double curves of distributions are today called violin plots. Hintze and Nelson (1998).

[129] Sumner (1906, 39–41).

Alfred Binet in France were put to use in the US schools by Henry Herbert Goddard. He argued that Binet's method for measuring intelligence was validated by the fact that the results followed a curve of normal distribution, which proved that the tests were "amazingly accurate."[130] Binet's and Goddard's tests were performed as interviews, but the deployment of IQ tests on a mass scale in the US Army during World War I spurred the invention of multiple choice questions.[131] Such questionnaires spread to other fields such as education, polling, and marketing, and they are probably responsible for more bar and pie charts than any other innovation in human history.

Sumner and Goddard's examples reveal that the bell curve of normal distribution was attributed with almost magical qualities, which made it capable both of confirming the accuracy of data and, where no data was available, of replacing it. The fascination has periodically recurred, as the book *The Bell Curve* by Richard J. Hernstein and Charles Murray from 1994 made a similar argument to that of Galton, Ammon, Sumner, and Goddard.[132] Biological interpretations of group differences in IQ tests have received heavy criticism for both political and statistical reasons. For example, it has been claimed that IQ tests, originally developed to identify individuals with low cognitive functions, are unsuitable for investigating high intelligence.[133] That the focus of intelligence testing after World War II shifted from the feeble minded to the gifted can be attributed to the damage that the atrocities before and during World War II had done to the eugenics movement.[134] However, the bell curve of intelligence was still commonly assumed to be a valid model for how hereditary intellectual traits were distributed across populations.

The stratification of society was thus perceived as a natural order based on scientific laws, which was spatialized with the help of the curve of normal distribution. Ammon warned that social democrats threatened this natural order and wanted to replace it with a new one based on their theories.[135] Indeed, socialists were crafting their own revolutionary diagrams of class stratification, with the intent of overthrowing it. In 1885, the Belgian Labor Party presented a poster with a "Pyramid a renverser," a pyramid to

[130] Zenderland (1998, 122).
[131] S. Murdoch (2007).
[132] Hernstein and Murray (1994).
[133] For the statistical arguments, see, for example, Taleb (2019).
[134] S. Murdoch (2007).
[135] Ammon (1895, 8).

be turned upside down. Nicolas Lokhoff from the union of Russian socialists produced a similar one the year after, and the most famous of them all, "Pyramid of the Capitalist System," was published in 1911 by the labor union Industrial Workers of the World (Fig. 2.2).[136] The iconographies of the three pyramids were adapted to the national settings, for example, with the Russian imperial eagle replaced by a capitalist bag of money in the US example. Notable is that while the early diagrams of normal distribution of talents as well as the Belgian social pyramid only referred to men, the Russian pyramid as well as that of the Industrial Workers of the World also included women and children. However, the basic diagrammatic shape of these representations of society was the same hierarchical stratification as in medieval models of society. As the diamond-shaped curves of normal distribution, the social pyramids divided the upper classes into several strata, although not based on hereditary genius, but on their capital of violence, superstitious reverence, and economic wealth. In the bell curve, the lower strata were also segmented, but in the social pyramids, everyone below the upper strata were united into one huge base of workers. These carried the whole pyramid and would, if they decided to straighten their backs, be able to overturn it.

Although the bell curve model of society is mostly associated with the political right and the social pyramid with the political left, these two diagrammatic representations of the social class structure are not mutually exclusive. Both Charles Darwin and Karl Marx criticized social Darwinism, but some socialists embraced the ideas of a biological foundation for classes. For example, Karl Pearson, Galton's disciple and successor both in the development of statistics and as head of the Eugenics Record Office, was a convinced socialist.[137] For him, a flatter class structure could only be achieved by eliminating the lowest classes through selective breeding.

The spatial metaphors used by eugenicists reveal that they believed that their science would allow humanity to move upward on the scale of nature, approaching regions that in the Middle Ages had been reserved for angels and the divine. The Russian biologist Nikolai Koltzov wrote in 1921 that eugenics aimed "to create . . . a higher type of human, the powerful king of nature and the creator of life." In 1923, Leon Trotsky claimed that through Bolshevism,

[136] Bendix (1978); Triantafillou (2019, 265–288).

[137] Colp (1974, 329–338); Krementsov (2018, 239–292); Pearl and MacKenzie (2018, 66–68).

> humankind, frozen Homo sapiens, will enter into radical reconstruction and will become—under its own fingers—an object of most complicated methods of artificial selection and psycho-physical training. . . Man will put forward a goal . . . to raise himself to a new level—to create a higher socio-biological type, an *[Ü]bermensch*, if you will.[138]

The diagrammatic model of a vertical social hierarchy was the same from Aristotle to Trotsky, and can even be found in later diagrams such as Fig. 6.7. However, this does not necessarily need to be the result of an uninterrupted historical legacy, as similar models, such as placing higher beings in the sky or depicting powerful individuals as higher than commoners, can be found also in other cultures. The universal tendency to place dominant actors and classes on top likely has its foundation in the primary metaphor "Control is Up" that is learned by all in early childhood through exposure to the force of gravity.[139]

However, some diagrams of social order produced in the 1920s (Fig. 6.19) put a twist on the hierarchical model of society by taking recourse to another primary metaphor, "Good is Bright/Bad is Dark."[140] In 1922, Herman Lundborg, a Swedish racial biologist, published a diagram of how the social structure changes in a degenerating population. Lundborg believed that the middle classes would diminish, and that the lower and upper strata of society would gradually be filled with degenerates, marked with black ink. The German Hans Günther, the favorite racial biologist of the Nazi Party, copied Lundborg's diagram in the 1924 edition of his *Rassenkunde*, but replaced it with a simplified model of his own design by 1928.[141] In contrast to Lundborg, who used four shades, Günther only used light and dark. In addition, Günther's diagram showed that degenerates could be found in all classes of society. In contrast to the early eugenics of, for example, Galton and Sumner, the fascist model of society was rotting from its core rather than from the bottom. This was useful for dictatorships such as Nazi Germany, since it could legitimize the persecution of anyone regardless of social status, not only the bottom strata of society. It also allows fascists from all social groups to consider themselves better than their adversaries, regardless of their social status. A peculiar difference between Lundborg's and Günther's models is that

[138] Quotes from Krementsov (2018, Chap. 5, para. 68).
[139] Lakoff and Johnson (1999, 53).
[140] Ortiz (2011).
[141] Lundborg (1922); H. Günther (1924, 390); (1928, 377).

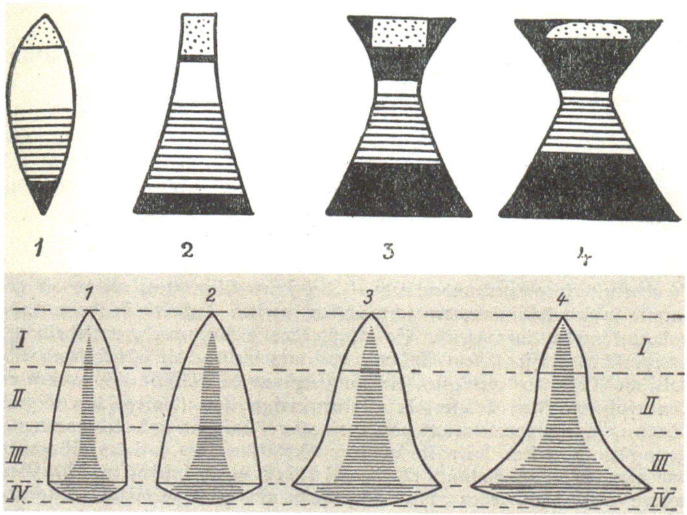

Fig. 6.19 Four stages of degeneration in industrial society according to Lundborg (top) and Günther (bottom). In Lundborg's diagram, white represents middle class including farmers, stripes represent workers, dots represent the culturally leading stratum, and black represents degenerates in the upper and lower classes. For Günther, I. represents the leading stratum, II. Middle Class and farmers, III. the workers, and IV. the lowest, unemployed stratum. The dark stripes represent the population with inferior genetic traits. Lundborg (1922, 21); Günther (1928, 377)

Lundborg seems to have been much more concerned with a swelling mass of degenerates in the upper social strata. It is possible that Günther's adaptation from 1928, which de-emphasized degeneration in the upper classes, was better suited to German politics in the late 1920s, when the Nazis began to seek alliances with traditional conservatives.

References

Ammon, Otto. 1895. *Die Gesellschaftsordnung und ihre natürlichen Grundlagen: Entwurf einer Sozial—Anthropologie zum Gebrauch für alle Gebildeten, die sich mit sozialen Fragen befassen* (The social order and its natural foundations: A sketch of a social anthropology for the use of all educated people who deal with social questions). Jena: Gustav Fischer.

Arai, Teishirō. 1950. *Sakata senjutsu shōkai.* (The Sakata method in detail). Tokyo: Kabushiki no kenkyū-sha.

Bellhouse, David R. 2023. *The flawed genius of William Playfair: The story of the Father of Statistical Graphics.* Toronto: University of Toronto Press.

Bender, John B., and Michael Marrinan. 2010. *The culture of diagram.* Stanford, CA: Stanford University Press.

Bendix, Reinhard. 1978. *Kings or people: Power and the mandate to rule.* Berkeley: University of California Press.

Beniger, James R., and Dorothy L. Robyn. 1978. Quantitative graphics in statistics: A brief history. *The American Statistician* 32 (1): 1–11.

Bent, Ian D., David W. Hughes, Robert C. Provine, Richard Rastall, Anne Kilmer, David Hiley, Janka Szendrei, Thomas B. Payne, Margaret Bent, and Geoffrey Chew. 2014. Notation. *Grove Music Online.* Oxford: Oxford University Press. https://www.oxfordmusiconline.com/grovemusic/display/10.1093/gmo/9781561592630.001.0001/omo-9781561592630-e-0000020114?rskey=PZblv0&result=3. Accessed 20 February 2023.

Bertin, Jacques. 1983. *Semiology of graphics: Diagrams, networks, maps.* Madison: University of Wisconsin Press.

Bigotti, Fabrizio, and David Taylor. 2017. The pulsilogium of Santorio: New light on technology and measurement in early modern medicine. *Societate şi politică* 11 (2): 53–113.

Blasius, Jörg, and Andreas Schmitz. 2015. History of correspondence analysis. In *International encyclopedia of the social and behavioral sciences*, ed. James D. Wright, 2nd ed., 6–11. Amsterdam: Elsevier.

Blaug, Mark, and Peter Lloyd. 2010. Introduction. In *Famous figures and diagrams in economics*, ed. Mark Blaug and Peter Lloyd, 1–23. Cheltenham: Edward Elgar.

Boberg, Johan. 2020. *Scientifically minded: Science, the subject and Kant's critical philosophy.* Uppsala: Uppsala University.

Börjesson, Mikael, Donald Broady, and Mikael Palme. 2020. 'In our school we have students of all sorts'. Mapping the space of elite education in a seemingly egalitarian system. In *Researching elites and power: Theory, methods, analyses*, ed. Francois Denord, Mikael Palme, and Bertrand Réau, 179–191. Cham: Springer Nature.

Bouras-Vallianatos, Petros. 2022. Diagrams in Greek medical manuscripts. In *The diagram paradigm: Cross cultural approaches*, ed. Jefferey F. Hamburger, David J. Roxburgh, and Linda Safran, 287–329. Washington, DC: Dumbarton Oaks.

Braun, Marta. 1992. *Picturing time: the work of Etienne-Jules Marey (1830–1904).* Chicago: University of Chicago Press.

Bullynck, Maarten. 2010. Johann Lambert's scientific tool kit: Exemplified by the measurement of humidity. *Science in Context* 23 (1): 65–89.

Burch, George E., and Nicholas P. DePasquale. 1990. *A history of electrocardiography*. San Francisco: Norman.

Campbell, Stephen R. 2004. On the concepts of quantity and quality in the history of Western thought. Online Submission. *The Annual Meeting of the American Educational Research Association*. https://eric.ed.gov/?id=ED495204

Carlgren, Fredrik. 2024. *Skattebaseffekter av värnskattens avskaffande*. Stockholm: Svenskt näringsliv.

Clagett, Marshall. 1959. *The science of mechanics in the Middle Ages*. Madison: University of Wisconsin Press.

Clagett, Marshall. 1968. *Nicolas Oresme and the geometry of qualities and motions*. Madison: University of Wisconsin Press.

Colp, Ralph Jr. 1974. The contacts between Karl Marx and Charles Darwin. *Journal of the History of Ideas* 35 (2): 329–338.

Costigan-Eaves, Patricia, and Michael Macdonald-Ross. 1990. William Playfair (1759–1823). *Statistical Science* 5 (3): 318–326.

Demos, Tim. 2013. Explanation of my ECG interpretation flow diagram. https://www.youtube.com/watch?v=Cy8RM4W8b_U. Accessed 3 January 2024.

Einthoven, Willem. 1896. *Onderzoekingen gedaan in het physiologisch laboratorium der universiteit te Leiden*. [Research conducted in the Physiological Laboratory of the University of Leiden] 2:2. Leiden: Van Doesburgh

Elliott, Ralph Nelson. 1994. Natures law: The secret of the universe. In *R.N. Elliott's masterworks: The definitive collection*, ed. Robert Rougelot Pretcher Jr., 213–308. Gainesville, Georgia: New Classics Library.

Euler, Leonhard. 1748. *Introductio in analysin infinitorum*. Lausanne: Marcum-Michaelem Bousquet & socios.

Euler, Leonhard. 2000. *Foundations of differential calculus*. Trans. J.D. Blanton. Cham: Springer.

Eves, Howard Whitley. 1990. *Foundations and fundamental concepts of mathematics*. 3rd ed. Boston: PWS-Kent.

Ferede, Ergete. 2019. *The revenue effects of tax rate increases on high-income earners*. Vancouver: Fraser Institute.

Floyer, John. 1707. *The physicians pulse watch*. London: Smith & Walford.

Franklin, John Curtis. 2002. Diatonic music in Greece: A reassessment of its antiquity. *Mnemosyne* 55 (6): 669–702.

Friendly, Michael. 2006. A brief history of data visualization. In *Handbook of computational statistics: Data visualization*, ed. Chun-houh Chen, Wolfgang Härdle, and Antony Unwin, vol. 3, 15–56. Heidelberg: Springer.

Friendly, Michael. 2008. The golden age of statistical graphics. *Statistical Science* 23 (4): 502–535.

Friendly, Michael, and Raymond Andrews. 2021. The radiant diagrams of Florence Nightingale. *Statistics and Operations Research Transactions* 45 (1): 1–18.

Friendly, Michael, and Howard Wainer. 2021. *A history of data visualization and graphic communication*. Cambridge, MA: Harvard University Press.
Funkhouser, H. Gray. 1937. Historical development of the graphical representation of statistical data. *Osiris* 3:269–404.
Galton, John. 1869. *Hereditary genius: An inquiry into its laws and consequences*. London: Macmillan.
Gartley, Harold M. (1935) 1981. *Profits in the stock market*. Washington, DC: Lambert-Gann.
Giraud, Yann. 2014. Legitimizing napkin drawing: The curious dispersion of Laffer curves, 1978 – 2008. In *Representation in Scientific Practice Revisited*, ed. Catelijne Coopmans, Janet Vertesi, Michael Lynch, and Stephen William Woolgar, 269–290. Cambridge, MA: MIT Press.
Günther, Siegmund. 1877. *Die Anfänge und Entwickelungsstadien des Coordinatenprincipes*. Nürnberg: Naturhistorischen Gesellschaft.
Günther, Hans. 1924. *Rassenkunde des deutschen Volkes*. 6, Umgearb. Aufl. München: Lehmann.
Günther, Hans. 1928. *Rassenkunde des deutschen Volkes*. 12., Umgearb. Aufl. München: Lehmann.
Haemer, Kenneth W. 1948. Range-bar charts. *The American Statistician* 2 (2): 23–23.
Haines, John. 2020. The visualization of music in the early Middle Ages: Three case studies. In *The visualization of knowledge in medieval and early Europe*, ed. Marcia Kupfer, Adam S. Cohen, and J. H. Chajes, 327–340. Turnhout: Brepolis.
Heijman, Wim J. M., and Johan A. C. van Ophem. 2005. Willingness to pay tax: The Laffer curve revisited for 12 OECD countries. *The Journal of Socio-Economics* 34 (5): 714–723.
Hernstein, Richard J., and Charles Murray. 1994. *The bell curve: Intelligence and class structure in American life*. New York: Free Press.
Hintze, Jerry L., and Ray D. Nelson. 1998. Violin Plots: A Box Plot-Density Trace Synergism. *The American Statistician* 52 (2): 181–184.
Hoff, Hebbel E., and L. A. Geddes. 1962. The beginnings of graphic recording. *Isis* 53 (3): 287–324.
Honma, Munehisha. 1898. *Sō Uchimoto Kansōkyūō Isho Honmasōkyū Cho*. (The Last Will and Testament of Munehisa Honma of Shonai). Sakata: Hayasaka Toyozō.
Humphrey, Thomas M. 1992. Marshallian cross diagrams and their uses before Alfred Marshall. *Federal Reserve Bank of Richmond Economic Review* 78 (2): 3–23.
Kant, Immanuel. (1768) 1992. Concerning the ultimate ground of the differentiation of directions in space. In *The Cambridge edition of the works of Immanuel Kant: Theoretical philosophy, 1755–1770*, ed. David Walford and Ralf Meerbote. Cambridge: Cambridge University Press.

Klein, Jacob. (1968) 1992. *Greek mathematical thought and the origin of algebra*. New York: Dover.

Krementsov, Nikolai. 2018. *With and without Galton: Vasilii Florinskii and the fate of eugenics in Russia*. Cambridge: Open Book Publishers.

Lakoff, George, and Mark Johnson. 1999. *Philosophy in the flesh: The embodied mind and its challenge to Western thought*. New York: Basic Books.

Lambert, Johann Heinrich. 1764. *Neues Organon oder Gedanken über die Erforschung und Bezeichnung des Wahren und dessen Unterscheidung vom Jrrthum und Schein*. Bd. 2. Leipzig: Wendler.

Lambert, Johann Heinrich. 1771. Essai d'Hygrométrie ou sur la mesure de l'humidité. In *Histoire de l'Académie Royale des Sciences et Belles-Lettres de Berlin. Année MDCCLXIX*, 68–127. Berlin: Libraires de la Cour & de l'Académie Royale.

Lo, Andre W., and Jasmina Hasanhodzic. 2010. *The evolution of technical analysis: Financial prediction from Babylonian tablets to Bloomberg terminals*. Hoboken, NJ: Bloomberg Press.

Lundberg, Jacob. 2017. *The Laffer curve for high incomes*. Uppsala: Uppsala University. Department of Economics Working Paper 2017:9. http://uu.diva-portal.org/smash/get/diva2:1148666/FULLTEXT01.pdf

Lundborg, Herman. 1922. *Degenerationsfaran och riktlinjer för dess förebyggande*. Stockholm: Norstedts.

Maclean, Ian. 2006. Diagrams in the defence of Galen: Medical uses of tables, squares, dichotomies, wheels and latitudes. In *Transmitting knowledge: Words, images, and instruments in early modern Europe*, ed. Sachiko Kusukawa and Ian Maclean. Oxford: Oxford University Press.

Magnello, M. Eileen. 2012. Victorian statistical graphics and the iconography of Florence Nightingale's polar area graph. *BSHM Bulletin: Journal of the British Society for the History of Mathematics* 27 (1): 13–37.

McCaffrey, Joe. 2014. Ibn Khaldun: The forgotten father of economics. *ISOR Journal of Business and Management* 16 (3): 1–3.

Middleton, Roger. 2010. The Laffer curve. In *Famous figures and diagrams in economics*, ed. Mark Blaug and Peter Lloyd, 412–418. Cheltenham: Edward Elgar.

Minami, Hideme. 1760. *Chaji-dan*. Kyoto: Nishimura Ichiroemon.

Muirhead, James Patrick. 1854. *The origin and progress of the mechanical inventions of James Watt: Illustrated by his correspondence with his friends and the specifications of his patents*. Vol. 1. London: Murray.

Murdoch, John E. 1984. *Album of science: Antiquity and the Middle Ages*. New York: Charles Scribbner's Sons.

Murdoch, Stephen. 2007. *IQ: A smart history of a failed idea*. London: Duckworth.

Murphy, John J. 1999. *Technical analysis of the financial markets: A comprehensive guide to trading methods and applications.* New York: New York Institute of Finance/Penguin.

Nightingale, Florence. 1859. *A contribution to the sanitary history of the British Army during the late war with Russia.* London: John W. Parker and Son.

Nihon shōken shinbun-sha [Nihon Securities Journal]. 1969. *'Sakata gohō' wa fūrinkazan.* [The "five Sakata rules" are fūrinkazan (wind, forest fire, mountain, a quote from Zun Tzu's art of war)]. Tokyo: Nihon shōken shinbun-sha.

Nison, Steve. 1991. *Japanese candlestick charting techniques: A contemporary guide to the ancient investment technique of the Far East.* New York: New York Institute of Finance.

Nison, Steve. 1994. *Beyond candlesticks: New Japanese charting techniques revealed.* New York: Wiley Finance.

Ortiz, Maria J. 2011. Primary metaphors and monomodal visual metaphors. *Journal of Pragmatics* 43 (6): 1568–1580.

Pearl, Judea, and Dana Mackenzie. 2018. *The book of why: The new science of cause and effect.* New York: Basic books.

Pearson, Karl. 1901. On lines and planes of closest fit to systems of points in space. *Philosophical Magazine and Journal of Science* 6 (2): 559–572.

Playfair, William. 1801. *The statistical breviary.* London: Bensley.

Plot, Robert. 1685. Observations of the wind, weather, and height of the mercury in the barometer, throughout the year 1684; taken in the musaeum ashmoleanum at Oxford, by Robert Plot, LLD. To which is prefixt a letter from him, to Dr. Martin Lister, F. off the R.S. concerning the use of this and such like historys of the weather. *Philosophical Transactions (1683–1775)* 15: 930–943.

Preda, Alex. 2001. The rise of the popular investor: Financial knowledge and investing in England and France, 1840–1880. *Sociological Quarterly* 42 (2): 205–232.

Pryer, Anthony. 2011. Notation. In *The Oxford companion to music.* Oxford: Oxford University Press., https://www.oxfordreference.com/display/10.1093/acref/9780199579037.001.0001/acref-9780199579037-e-4761?rskey=6GSQhH&result=5202. Accessed 8 August 2024.

Quetelet, Adolphe. 1846. *Lettres sur la théorie des probabilités, appliquée aux sciences morales et politiques.* Bruxelles: Hayez.

Rankin, Susan. 2022. *Sounding the word of God: Carolingian books for singers.* Notre Dame: University of Notre Dame Press.

Read, Sara. 2018. Sir John Floyer's pulse-watch. *Early modern medicine: A blog about bodies and medicine c. 1500–1780.* https://earlymodernmedicine.com/sir-john-floyers-pulse-watch/. Accessed 3 January 2024.

Risi, Vincenzo De. 2018. Analysis situs, the foundations of mathematics, and a geometry of space. In *The Oxford handbook of Leibniz*, ed. Maria Rosa Antognazza, 247–258. Oxford: Oxford University Press.

Sahoo, Santanu, Asit Subudhi, Manasa Dash, and Sukanta Sabut. 2020. Automatic classification of cardiac arrhythmias based on hybrid features and decision tree algorithm. *International Journal of Automation and Computing* 17 (4): 551–561.

Savage, Patrick E., Steven Brown, Emi Sakai, and Thomas E. Currie. 2015. Statistical universals reveal the structures and functions of human music. *Proceedings of the National Academy of Sciences* 112 (9): 8987–8992.

Schreiber, Peter. 2003. The beginnings of visualizing the invisible. *Prague Studies in the History of Science and Technology, new series* 7:157–168.

Scriba, Christoph J., Peter Schreiber, and Jana Schreiber. 2015. *5000 years of geometry: Mathematics in history and culture*. 15th ed. Basel: Springer.

Smith, D. Eugene. 1932. *History of Mathematics.* Volume I. Boston: Ginn & Company.

Spear, Mary Eleanor. 1952. *Charting statistics.* New York: McGraw-Hill.

Spear, Mary Eleanor. 1969. *Practical charting techniques.* New York: McGraw-Hill.

Spelke, Elizabeth S. 2011. Natural number and natural geometry. In *Space, time and number in the brain: Searching for the foundations of mathematical though*, ed. Stanislas Dehaene and Elizabeth Brannon, 287–317. London: Academic Press.

Spence, Ian. 2005. No humble pie: The origins and usage of a statistical chart. *Journal of Educational and Behavioral Statistics* 30 (4): 353–368.

Spence, Ian, and Howard Wainer. 2017. William Playfair and the invention of statistical graphs. In *Information design: Research and practice*, ed. Alison Black, Paul Luna, Ole Lund, and Sue Walker, 43–60. London: Routledge.

Stefanou, Maria Ioanna. 2020. The footprints of neuroscience in Alexandria during the 3rd-century BC: Herophilus and Erasistratus. *Journal of Medical Biography* 28 (4): 186–194.

Suminoe, Saichirō. 1990. *Nihon Keisenshi: Nihon Tekunikaru Anarisuto Kyōkai* (History of technical analysis in Japan), 10th print. Tōkyō: Nihon Keizai Shinbunsha.

Sumner, William Graham. 1906. *Folkways: A study of the sociological importance of usages, manners, customs, mores, and morals.* Boston: Ginn.

Taleb, Nassim N. 2019. Fooled by correlation: Common misinterpretations in social 'science.' Academia Online. https://www.academia.edu/39797871/Fooled_by_Correlation_Common_Misinterpretations_in_Social_Science. Accessed 3 January 2024.

Tang, Anson C. Yan. 2012. Review of traditional Chinese medicine pulse diagnosis quantification: Complementary therapies for the contemporary healthcare. In *Complementary therapies for the contemporary healthcare*, ed. Marcelo Saadand Roberta De Medeiros, 61–80. Rijeka: InTech.

Taruskin, Richard. 2005. *The Oxford history of Western music, vol. 1: The earliest notations to the sixteenth century.* Oxford: Oxford University Press.

Theocharis, Reghinos D. 1983. *Early developments in mathematical economics.* 2nd ed. London: Macmillan.

Tilling, Laura. 1975. Early experimental graphs. *The British Journal for the History of Science* 8 (3): 193–213.

Timbro. n.d. Lafferkurvan. *Timbro.se.* https://timbro.se/skattekoll/lafferkurvan/. Accessed 15 February 2022.

Tōkyō kabushiki torihikisho. 1932. *Tōkyō kabushiki torihikisho* (Tokyo stock exchange). Tōkyō: Tōkyō kabushiki torihikisho chōsa-ka.

Triantafillou, Eric. 2019. To make what is vertical horizontal: Picturing social domination. *Journal of Visual Culture* 18 (3): 265–288.

Truesdell, Clifford. (1972) 2007. Leonhard Euler, supreme geometer. In *The genius of Euler: Reflections on his life and work,* ed. William Dunham, 13–41. Washington, DC: Mathematical Association of America.

Tufte, Edward R. 2001. *The visual display of quantitative information.* 2nd ed. Cheshire: Graphics Press.

Tufte, Edward R. 2006. *Beautiful evidence.* Cheshire: Graphics Press.

Tukey, John W. 1977. *Exploratory data analysis.* Reading, MA: Addison-Wesley.

Verboon, Annemieke R. 2010. *Lines of thought: Diagrammatic representation and the scientific texts of the Arts Faculty, 1200–1500.* Dissertation. Lieden: Leiden University Scholarly Publications.

Wallace, Anthony F. C. 1961. On being just complicated enough. *Proceedings of the National Academy of Science* 47 (4): 458–464.

Wickham, Hadley, and Lisa Stryjewski. 2011. 40 years of boxplots. Hadley Wickham's academic portfolio. https://vita.had.co.nz/papers/boxplots.pdf. Accessed 3 January 2024.

Youschkevitch, Adolph P. 1976. The concept of function up to the middle of the 19th century. *Archive for history of exact sciences* 16 (1): 37–85.

Zenderland, Leila. 1998. *Measuring minds: Henry Herbert Goddard and the origins of American intelligence testing.* Cambridge: Cambridge University Press.

CHAPTER 7

Diagramming Flows and Causality

Not all modern diagrams are preoccupied with representing quantitative data. Some are still used as spatial representations of abstract theories, connections, and categorizations, as was the case before the quantitative revolution. In their modern forms they have sometimes incorporated quantitative elements. In these new diagrams, the connecting elements not only link items in static trees of categorization, but dynamically illustrate change, the flow of time, and causality.

As human worldviews have changed, so have our spatial representations of the world. However, this is not a one-way process, since the diagrams we use to understand the world also have repercussions for how we perceive it. For example, Section "Timelines: Spatial Representations of Time" discusses how the invention of the timeline as a diagram of history inspired the idea of traveling backward in time, something which seems not to have been contemplated before the second half of the 1800s. Section "Flow Charts: From Liquids to Paper and Back Again" focuses on flow charts and their historical trajectory of increased abstraction: from maps of physical flow of liquid substances via more metaphoric flow in industrial production to maps of abstract logical processes in computer programs. Section "Causal Path Diagrams and Fuzzy Cognitive Maps" describes how diagrammatic thinking can be combined with symbolic mathematics in order to investigate causality, one of the hardest problems

J. Holmén, *A History of Diagrams*, Palgrave Studies in the History of Science and Technology,
https://doi.org/10.1007/978-3-031-99691-7_7

in science. It also discusses how philosophically incompatible theories can share striking structural similarities when presented in diagram form.

Timelines: Spatial Representations of Time

The history of the timeline and other diagrammatic representations of time have been described by Rosenberg and Grafton in their *Cartographies of Time*. They show that the timeline in its present form is not older than the mid-eighteenth century, but that it was preceded by a rich variety of other cartographies of time, such as different types of maps.[1] Although space is a metaphor for time in most languages, the exact formulation of these metaphors varies.[2] Therefore, the historical diversity in visuospatial representations of time should not be surprising. However, it can be argued that the victory of the timeline among the many competing spatial models of time might be due to its simple and straightforward foundation in primary metaphors and innate associations. However, which spatial analogies we use in our thought process have consequences for how we think. For example, the adoption of the timeline as a spatial analogy for time seems to have enabled the development of the notion of time travel.

Travel forward in time, for example, by protagonists falling asleep for centuries, has an old history in mythology and folklore.[3] Every day, we all take part in forward time travel at a leisurely pace, 60 minutes per hour. In contrast, traveling backward in time was unheard of until the late nineteenth century. One exception that confirms this rule is Samuel Madden's *Memoirs of the Twentieth Century* from 1733, which is sometimes described as an early example of travel back in time. However, in that story only some sources from the future traveled back to the writer's age, which enabled him to write the history of the future. The documents are simply handed to Madden by his *genius*, and it is not described how they ended up in his time.[4] It was a literary convention of the day to pretend that fictional stories were based on actual events and sources, and the documents from the future filled that purpose. For all practical purposes, Madden made an intellectual journey to the future rather than back in time.

[1] Rosenberg and Grafton (2010).
[2] See Chap. 2, Section "The Neurology of Spatial and Abstract Thinking".
[3] Jones and Ormond (2015, 10–11).
[4] Madden (1733).

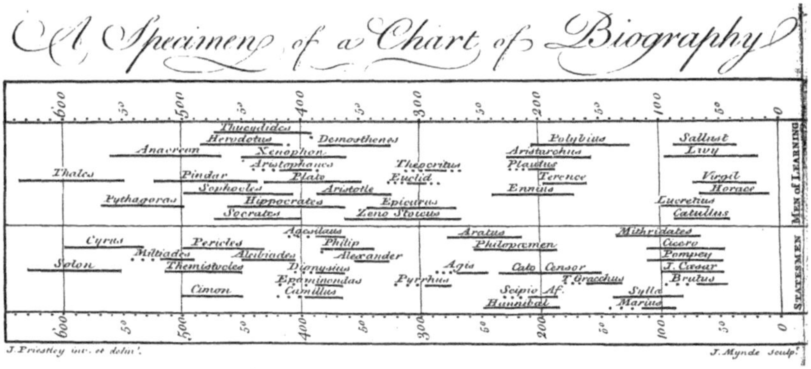

Fig. 7.1 A chart from Priestley's *Description of a Chart of Biography*, 1764, where he explained the principles behind the poster-sized chart of biography that he published the following year. Omissions such as Augustus and errors such as the short lifespan of Livy were corrected in later editions. (Priestley 1764)

Madden's story is the only known description of backward time travel written before the timeline had become the standard model for understanding time. Although the timeline's breakthrough was preceded by a gradual development from the layout of early medieval chronicles, it is generally considered that it was invented by Joseph Priestley (1733–1804) in the 1760s. He used horizontal lines to delimit the lives of historical persons, spanning from their birth to the left to their death to the right, using dots where dates were uncertain (Fig. 7.1).[5]

Other attempts at visualizing chronology were made around the same time, and some of them seem to have had the potential to inspire notions of time travel. In 1753, Jacques Barbeu-Dubourg published a 54-meter-long scroll with the chronology of world history. It was viewed with the help of a "chronographic machine" that allowed you to crank the scrolls, and thereby move forward and back in history. If this cranking had become a common pastime, it is possible that it would have inspired notions of machines that allowed movement forward and back in time. However, Barbeu-Dubourg's invention did not become a commercial success, and few came in contact with it.[6] The moving diagram in the chronographic

[5] Priestley (1764); Rosenberg and Grafton (2010, 19).
[6] Rosenberg and Grafton (2010, 112).

machine had a good foundation in the primary metaphor "Time is Motion,"[7] but the timeline, based on the primary metaphor "Linear Scales are Paths" and the innate association between line length and number, was much simpler to produce and distribute and could be included on ordinary book pages.

Therefore, the timelines of Joseph Priestley became widely used, and were described by contemporaries as an essential part of a gentleman's library. Before Priestly, several attempts had been made to use cartographic methods to illustrate chronology. However, with Priestley it became accepted that the line rather than the conventional map was the best analogy for time.[8]

Priestley carefully explained his lines as analogies for time, as this was new to his readers. The influence of his graphs is evident from the fact that after him, according to Rosenberg and Grafton, "most readers simply assumed the analogy between historical time and measured graphic space." They attribute the success of Priestley's timeline to it being useful, intuitive, and resonant with the Enlightenment's linear view of history.[9] The introduction of the timeline illustrates that although a new diagram form is intuitive, in the sense that it resonates with cognitive principles, a learning process is still needed. However, in such cases the learning process is much faster than regarding, for example, Raymond Llull's challenging diagrams.[10]

In the eighteenth century, advances in chronometry made a global synchronized time possible. In the course of the nineteenth century, factory schedules, railroad timetables, and long-distance telegraph communication made Americans and Europeans internalize the idea of a uniform fabric of time.[11] Jones and Ormond have suggested that fictional stories about time travel appeared in the late nineteenth century, when technology, travel, and shifting ideas about the nature of work encouraged the conception of time as a segmented, measured phenomenon.[12] It could be argued that diagrams in the form of timelines were one of the factors that contributed to the idea of traveling backward in time. The spatial analogy made it easier to fathom the idea of moving in time, just as one might in

[7] Lakoff and Johnson (1999, 52).
[8] Rosenberg and Grafton (2010, 123, 126).
[9] Rosenberg and Grafton (2010, 126, 140).
[10] For Llull, see Chap. 4, Section "Llullism: Incomprehensible Originality".
[11] Rosenberg and Grafton (2010).
[12] Jones and Ormond (2015, 13).

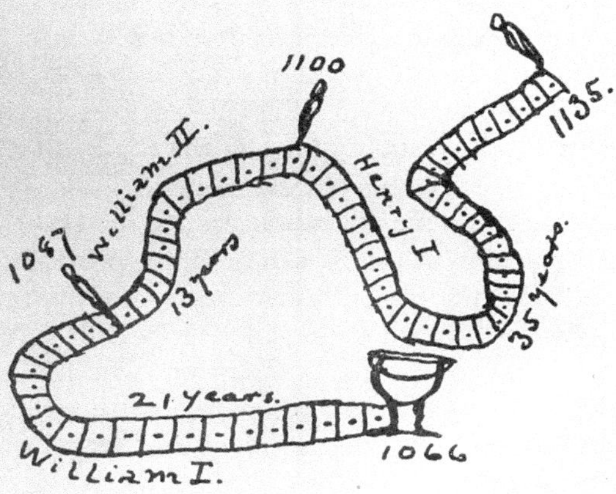

FIG. 2

Fig. 7.2 Twain's roadway was quite straight, but he depicted it as a serpent to fit the paper. Each foot represented one year of history. (Twain 1914)

space. This idea is confirmed by the important role diagrams and spatial analogies have in the early history of backward time travel in fiction.

One of the first novels featuring backward time travel was Mark Twain's *A Connecticut Yankee in King Arthur's Court* from 1889.[13] In 1883, Twain made a game for his children on the road to his summer house. Similar to how knowledge is connected to space along the songlines of oral cultures, the reigns of English monarchs and other historical events were marked with stakes (Fig. 7.2). No less than 817 years of history were represented by as many feet of roadway, all of which could be seen from their porch. Walking up and down the road, Twain's children learned important historical dates, which their governess without success had tried to teach them all summer. As Twain often forgot books and other articles lying around outdoors, he found the division of the roadway into historical periods useful, since he could ask his children to fetch his belongings from the right reign.[14] In 1884, Twain filed for a patent for a history board game inspired by the outdoor activities from the year before.[15]

[13] Twain (1889).
[14] Twain (1914).
[15] Rosenberg and Grafton (2010, 198); Rosenberg (2013).

It is likely that Twain's practice in 1883 of sending his children back to the reigns of medieval kings to fetch forgotten books primed him for sending a Yankee to the court of King Arthur a few years later. In fact, although the book was not published until 1889, it was conceived in December 1884,[16] the year after the history game on the road and the same year as the patent application for the history board game.

Several other time-traveling stories appeared in the 1880s. The first work of fiction using a machine to go back in time was probably *The Clock that Went Backward*, a short story published anonymously in the New York newspaper *The Sun* by Edward Page Mitchell in 1881.[17] When the clock went backward, so did time: "The hands were whirling around the dial from right to left with inconceivable rapidity. In this whirl we ourselves seemed to be borne along. Eternities seemed to contract into minutes while lifetimes were thrown off at every tick."[18] That a clock going backward is described as moving from right to left is not self-evident: Between 9 and 3, the hands of a clock do indeed move from left to right, but from 3 to 9 the opposite is the case. It is likely that the identification of time moving backward with moving from right to left in space stems from the analogy of the timeline, where time usually flows from left to right, as Latin script.

Another early work of fiction featuring backward time travel is the Spanish novel *El Anacronópete* from 1888 by Enrique Gaspar. It contained the first description ever of a purpose-built time machine.[19] The idea of a time machine was also used by the most renown early author of time travel stories, Herbert George Wells. In his works, the space–time analogy is made obviously clear. In Wells's short story *The Chronic Argonauts* from 1888, the inventor of the time machine, Dr. Nebogipfel, has a "diagram-covered bench" in his workshop. He also explains his machine to Reverend Cook by making an analogy between extension in time and extension in space:

> "When we take up this new light of a fourth dimension and reexamine our physical science in its illumination," continued Nebogipfel, after a pause, "we find ourselves no longer limited by hopeless restriction to a certain beat of time—to our own generation. Locomotion along lines of duration—chronic navigation comes within the range, first, of geometrical theory, and

[16] Baetzhold (2018, 445).
[17] Mitchel (1881).
[18] Mitchel (1881, 5).
[19] Ayala (1998).

> then of practical mechanics. There was a time when men could only move horizontally and in their appointed country … Then man burst his bidimensional limits, and invaded the third dimension, soaring with Montgolfier into the clouds, and sinking with a diving bell into the purple treasure-caves of the waters. And now another step, and the hidden past and unknown future are before us. We stand upon a mountain summit with the plains of the ages spread below."[20]

Although there were some forerunners in the 1700s,[21] the idea that time is a fourth dimension was presented in *Nature* in 1885 by someone using the initial S., which has been speculated to be an acquaintance of Wells. Paul Nahin claims that Wells's novel *The Time Machine* from 1895 made the idea of time as a fourth dimension enter the popular mind.[22] However, we can see that the concept was familiar to Wells already in 1888. In fact, in the older short story, the spatial analogy and diagrammatic thinking underlying time travel was even more obvious. When the story's time-traveling genius, Nebogipfel, struggled to be believed, he reached for the ultimate tool of explanation:

> "Come and look at these diagrams," he said, and then with elaborate simplicity he commenced to explain again to the clergyman the new quadridimensional geometry. Insensibly Cook's aversion passed away, and seeming impossibility grew possible, now that such tangible things as diagrams and models could be brought forward in evidence.[23]

In addition to the timeline, other diagrammatic models of time have been suggested in science fiction. One argument against the idea of time travel is that if a time machine had ever been invented, history would be full of tourists. In part to counter this argument, several science fiction writers have portrayed time as a spiral, where hops in time can only be performed between nearby coils. Our time might be in a distant region in space–time, which would explain why we have not been reached by travelers from other times.[24]

After Priestley's introduction of timelines in 1765, the exposure to spatial representations of time gradually increased. In the 1800s, the growth

[20] Wells (1888).
[21] For example, Immanuel Kant. Boberg (2020, 144).
[22] Nahin (2001, 141).
[23] Wells (1888).
[24] Nahin (2001).

of statistical graphics accelerated this process, since the horizontal axis of time series diagrams is essentially a timeline. The second half of the nineteenth century, when backward time travel was introduced as a literary genre, has been described as the golden age of statistical graphics.[25] Just as the diagrammatic practices of scholastic division and subdivisions might have created a habitus out of which gothic architecture could form,[26] tinkering with timelines in the nineteenth century might have made movement back and forth in time a mental habit that allowed the birth of time travel science fiction.

However, in order to explain the concentrated outburst of such stories that took place within a relatively narrow period in the 1880s, we need a supporting explanation that can be pinpointed more exactly in time. One such factor might have been the introduction of unified time zones, which was driven by the need to synchronize railway timetables, and culminated in the International Meridian Conference in 1884.[27] This development heightened the public interest in problems related to time, and could have triggered authors to realize the potential of the space–time analogy that had accumulated for a century under the influence of timelines.

That the idea of travel backward in time was inspired by analogical thinking based on movement backward on a timeline complies with the invariance hypothesis, the idea that metaphorical mappings preserve the cognitive topology and inferential structure of the source domain.[28] If we can move forward and back in space, and we use space as a metaphor for time, the preserved referential structure induces us to believe that travel back in time is possible. However, this analogy seems to lead to a false inference. In contrast to what is the case with the spatial dimensions, experience suggests that time has a direction.

Flow Charts: From Liquids to Paper and Back Again

The earliest predecessors of flow charts are illustrations of flowing water. Many early civilizations developed around rivers. Channels were constructed in order to provide irrigation, transportation, and protection from floods, and these channel systems were commonly depicted in

[25] Chap. 6, Section "Statistical Graphics".
[26] See Chap. 4, Section "The Scholastic Diagram Habitus".
[27] Schivelbusch (1978); Galison (2003, 98–155).
[28] See Lakoff (1990).

diagrammatic form. Maps of irrigation systems on Mesopotamian clay tablets were highly stylized; the canals were outlined with simple lines reminiscent of a modern flow chart. This simplicity was partly imposed by the medium, the clay tablet, and the reasons for drawing these maps might have been entirely different than for modern flow charts. For example, some irrigation maps were used to calculate the labor needs for constructing the canals.[29] Thus, on such maps the actual flow of water was secondary. On later drawings and printed maps, the depictions of rivers and channels were often more figurative and naturalistic, although, for example, Leonardo da Vinci's sketches of flowing water also had a stylized diagrammatic element.[30]

Flow charts have their foundations as spatial analogies in the primary metaphors "Organization is Physical Structure" and "Linear Scales are Paths." However, this does not mean that all aspects of them are intuitively understandable to everyone, since they also incorporate a large number of learned conventions and symbols. For example, it was not until the seventeenth century that arrows were introduced to indicate the direction of waterflow in diagrams and maps.[31] Before that, pointing fingers often filled the same function as arrows do today, such as indicating directions or pointing out details on flow charts as well as other types of diagrams.[32]

During the golden age of information graphics in the second half of the nineteenth century, maps were developed that indicated, for example, the flow of trade between different countries. Some of the most famous of these flow maps were constructed by Charles-Joseph Minard, such as his map from 1869 depicting how Napoleon's *Grand Armée* gradually withered away as it flowed to Moscow and back again in the winter of 1812.[33] Minard's career encompassed work with both actual and abstract flows, since he had a background as an engineer involved in the inspection and construction of waterways.[34]

In the late nineteenth century, when the second Industrial Revolution was transforming manufacturing, flow charts were used to illustrate

[29] Wheat (2013).
[30] See, for example, Gombrich (1976, 41–48).
[31] Gombrich (1990, 28).
[32] See, for example, the fifteenth century *Seligenstädter Lateinpädagogik*.
[33] Friendly and Wainer (2021, 167–170, 235–239).
[34] Rendgen (2018, 11).

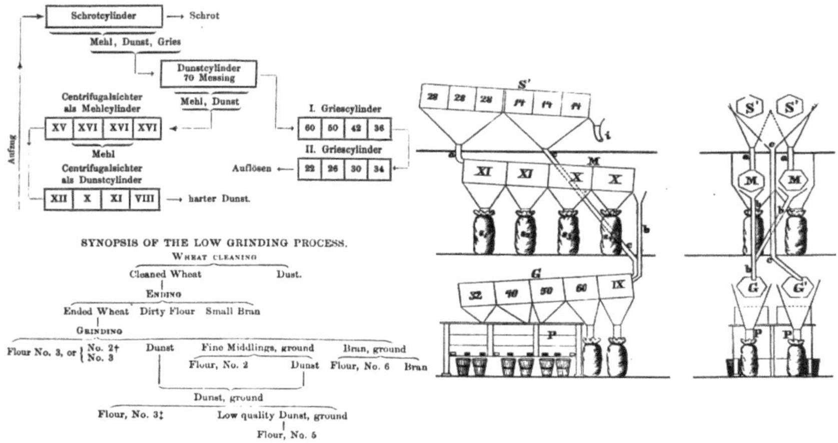

Fig. 7.3 Illustrations from Kick's book on flour manufacture combines several older diagrammatic traditions that anticipated the flow chart. Top left: Kick 1878, 329; Right: Kick 1878, 324, Bottom left: Kick and Powles 1888, 207 (English translation)

industrial processes involving flowing liquids or granular substances.[35] An early publication on the subject, by Friedrich Kick from 1878, used a wide range of traditional forms of illustration, such as printed bracketed diagrams and figurative illustrations of pipes connecting sacks of grain (Fig. 7.3). These diagrams were called *schema*, in the English translation "scheme" or "synopsis."[36] Thus, both Kick's diagrams and the concepts he used to describe them were rooted in an older and general tradition of diagramming.

However, in a book on the subject from 1912, Percy Amos's *Process of Flour Manufacture*, the flow chart had evolved into a distinct type of diagram described by a designated concept, "flow sheets." Amos claimed that "correctly worked out, a flow sheet is a scientific scheme, based upon well-tried formulae and capable of close reasoning." Amos displayed several different types of charts, some describing details of the flow between

[35] Morris and Gotel (2006).
[36] Kick (1878); Kick and Powles (1888).

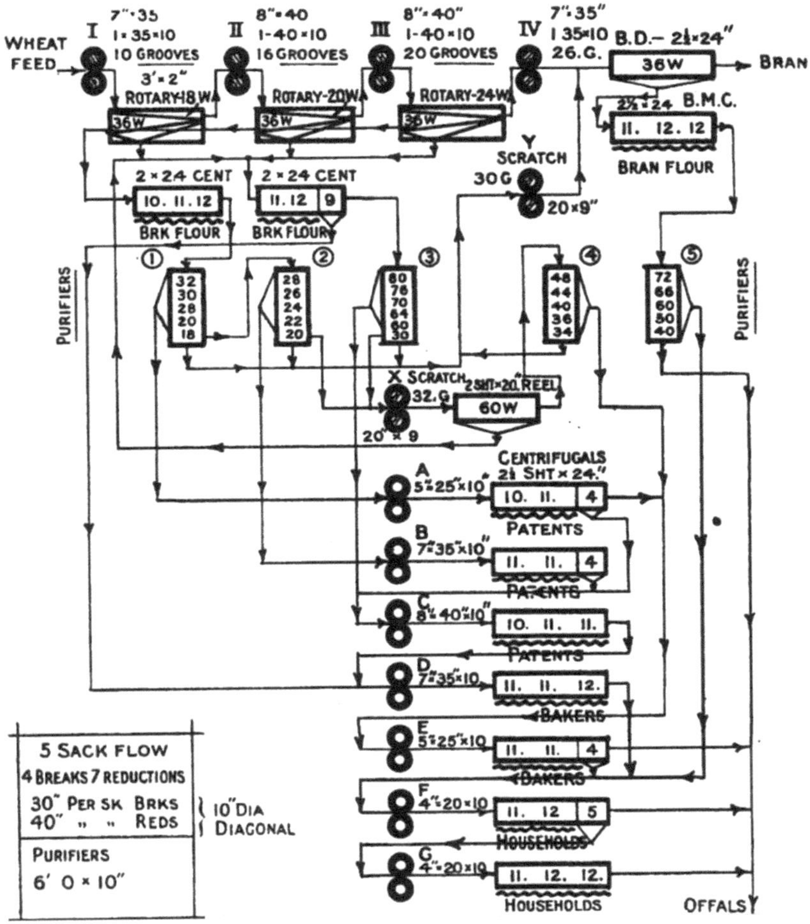

Fig. 7.4 The complete flow sheet of a five-sack plant, an overview of the flour manufacturing process. (Amos 1912, 174)

particular pieces of machinery while others were higher-level overviews of the entire production process (Fig. 7.4).[37]

[37] Amos (1912, 171–183, quote on 175).

In addition to flow processes in production, the second industrial revolution was also characterized by increased electrification. Diagrams that described in detail the wiring of electric appliances, such as those for radio transmission, were in the early 1900s called "circuit diagrams." By the mid-1920s, higher-level "block schematics" or "schematic diagrams" appeared. They depicted the general function of electrical systems without detailed circuit notation.[38]

The realization that flow sheets were useful in streamlining production soon spread from the manufacture of literally flowing products to other parts of industry. In 1921, the married couple Frank Bunker Gilbreth and Lillian Moller Gilbreth, pioneers in industrial engineering management, introduced what they called process charts to the American Society of Mechanical Engineering. These charts were intended for streamlining industrial processes, for example, by eliminating unnecessary steps in the production (Fig. 7.5).[39]

From the early 1930s, industrial engineer Allan H. Mogensen spread the use of the Gilbreths' tools at his Work Simplification Conferences in Lake Placid, New York. Two Mogensen graduates from 1944, Art Spinanger and Ben S. Graham, applied and developed the methods further. Spinanger introduced the Deliberate Methods Change Program at Procter and Gamble, while Graham, a director at Standard Register Industrial, adapted flow process charts to information processing. In 1947, the American Society of Mechanical Engineering adopted a standardized symbol system for flow process charts, based on the Gilbreths' work. The system involved symbols for Operations (doing work), Transportation, Inspection, and Storage/Delay. Graham added two symbols adapted to information processing, representing the creation and addition/altering of information.[40]

One of the newest branches of engineering at the time was the emerging computer industry. In 1946, Herman Goldstine and John Van Neumann suggested that "flow diagrams" with defined symbolic notation should be used for what is today known as software engineering. Diagramming became increasingly popular in the development of computer programs and algorithms. In the ISO standard of 1985, the number of symbols used in the flow charts had increased from the original 5

[38] See, for example, Pratt (1915); Morris and Gotel (2006, 134–135).
[39] Gilbreth and Gilbreth (1921, 38–41).
[40] Graham (2004, 2–4).

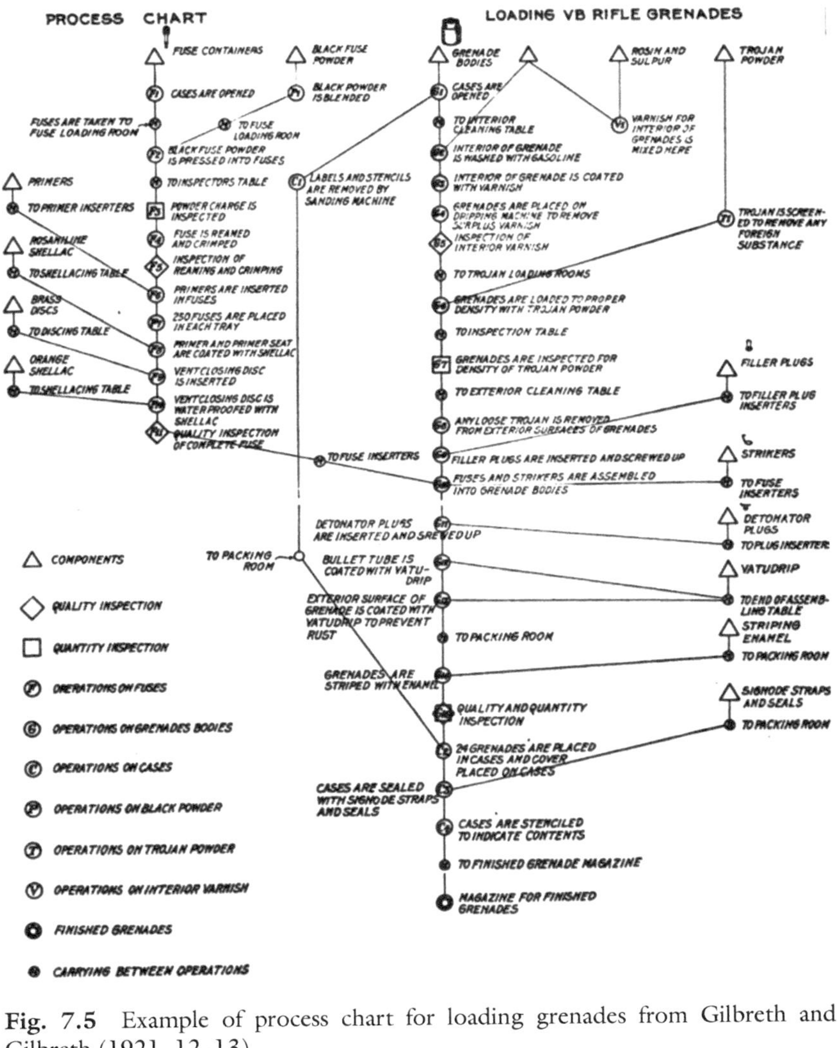

Fig. 7.5 Example of process chart for loading grenades from Gilbreth and Gilbreth (1921, 12–13)

(operation, test, flow arrow, stop, start) to 40. As the tools for programming evolved and higher-level programming languages were developed, the use of flow charts moved from detailed constructions of algorithms to broader concerns of software engineering.[41]

In Japan, Kaoru Ishikawa (1915–1989), a central figure behind quality initiatives in manufacturing, considered flow charts one of the key tools of quality control. However, he also invented the cause-and-effect diagram, now often called the Ishikawa Diagram (Fig. 7.6). Ishikawa first developed these diagrams in the summer of 1943 while explaining to engineers at the Kawasaki Steel works how to sort and relate various factors in the production process. The diagrams were further popularized through the quality control circles for Japanese factory workers that Ishikawa initiated. In the 1960s, his textbooks for these circles began to be translated into English.[42]

The Ishikawa diagrams are also called fishbone diagrams, since they sometimes might have that shape. However, the only diagrams that look like fishbones in Ishikawa's books are the cautionary examples he presents under the heading "Poor Cause and Effect Diagrams":

> A cause generally consists of many complex elements. Therefore, cause-and-effect diagrams usually turn out to be rather complicated, like the one in figure [7.6, bottom]. If it turns out looking like the one in figure [7.6, top], it means that either your knowledge of the manufacturing process is too shallow or the diagram is too generalized.[43]

However, in the more complicated diagram, according to Ishikawa, none of the arrows is directly affected by more than five other arrows, as the theories on magical numbers and cognitive load would predict. In contrast, in the too generalized diagram the outcome is affected by 15 direct possible causes. This means that in the first step in the process of finding a cause, the seemingly simpler diagram is more difficult to use. Thus, the main difference between the diagrams is not their level of complexity, but that the good one makes better use of hierarchical categorization.

Simple circular flow diagrams were used in economic literature from the late nineteenth century to illustrate the circulation of money. It has

[41] Morris and Gotel (2006); Zobel (2004, 117).
[42] Ishikawa (1998, i, 29).
[43] Ishikawa (1998, 28–29).

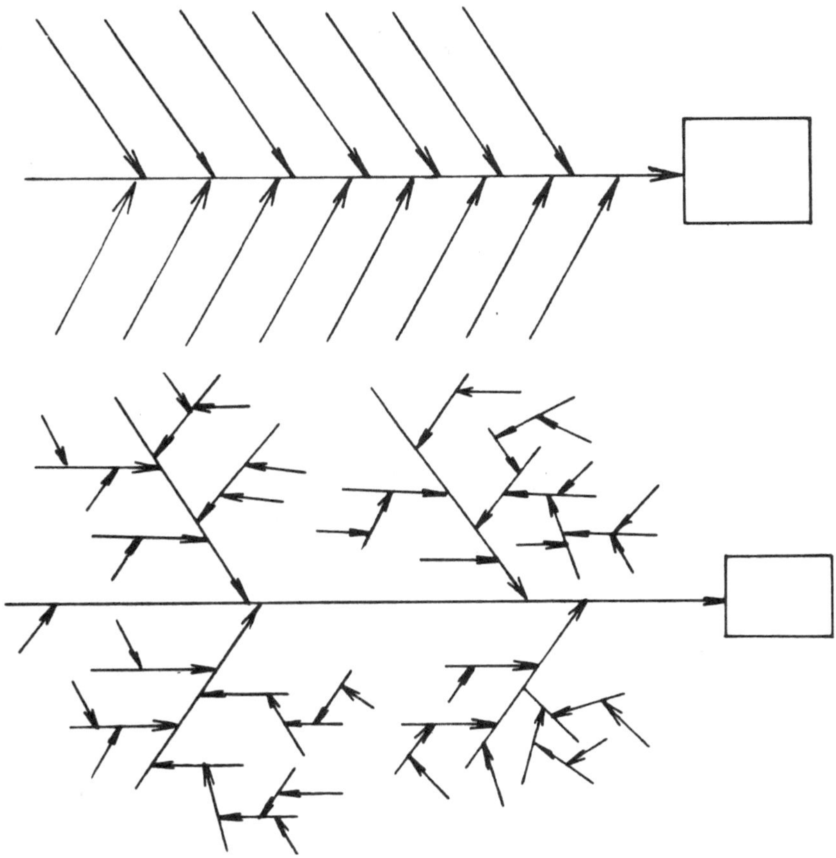

Fig. 7.6 A cause-and-effect diagram that is too simplified (top) and a good one that is relatively complicated (bottom) but makes better use of hierarchical categorization. (Ishikawa 1998, 28–29)

been argued that Quesnay's *Tableau Economique* from 1759, which illustrated money flows between landowners, farmers, and artisans, functioned as an early flow chart. However, Quesnay's intention was that the table should be used as a calculating device.[44]

[44] Backhouse and Giraud (2010).

An example of the opposite, a teaching aid that surprisingly could be used as a calculating device, was the Phillips machine or MONIAC, invented in 1949. The machine simulated a nation's economy with the help of tanks connected through pipes and valves. Colored water flowed between the tanks in order to illustrate processes such as taxation and investments. Two machines could be connected in order to simulate the interacting economies of two nations. Although constructed for pedagogical purposes, the mechanism proved accurate enough to function as an analog computer and was capable of simulating national economies within an error margin of 4%. The visualization capabilities of MONIAC were far ahead of early computers.[45]

Other computing devices have been either mechanical or electronic. The MONIAC is an anomaly, and it has been debated why Phillips chose this mechanism despite its lower accuracy. In all likelihood, Phillips was not familiar with electronic computers when he began his construction of the machine. However, Doron Swade believes that, despite better accuracy, early electronic computers with output of punched holes would not have fulfilled Phillips's pedagogical aims.[46] If we understand the MONIAC not as a computer but as a moving flow chart, Phillips choice of a fluid mechanism becomes intuitive.

The flow chart became fully developed relatively late. In part, this is because it builds on several other diagram types, such as the timeline, and general diagrammatic conventions, such as arrows, which did not appear until the 1700s. Another reason is that the complex flows in industries, electrical circuits, or computers that it is used to depict did not begin to appear before the second Industrial Revolution in the late nineteenth century. However, although many types of flow charts are associated with specialized and formalized conventions, their fundamental structure is still intuitively understandable due to their resonance with primary metaphors such as "Linear Scales are Paths."

Causal Path Diagrams and Fuzzy Cognitive Maps

Today, representations of different types of networks are common forms of diagrams. These figures are in mathematics called graphs. Unlike trees, graphs do not represent hierarchies. This means that they can contain

[45] Bissell (2007).
[46] Swade (2000).

loops and that two nodes can be connected by several lines. The theory behind graphs was not fully developed until the 1900s. However, many diagrams that fulfill the criteria for graphs were produced earlier. For example, several diagrams labeled "trees" in the Middle Ages were graphs, as some branches reconnected to each other. One example is the Kabbalistic trees found in Jewish mysticism. They are believed to have developed under the influence of not very treelike "trees" of Porphyry similar to Fig. 4.5.[47] The first person to study such forms mathematically was Leonhard Euler in his 1736 paper on the seven bridges in Königsberg.[48]

One type of graph, called causal graphs, causal path diagrams, or causal Bayesian networks,[49] can be used to investigate causality, something which has been very difficult to achieve with traditional statistical methods. As we saw in a previous chapter, Francis Galton set out on his investigation of human heredity in search of the causes of genius.[50] However, what he found was more complex, such as correlation, variance, and regression to the mean. As a consequence, he and his followers dove into mathematical statistics, developing tools for investigating correlation rather than causation. Particularly for Galton's collaborator and successor Karl Pearson, the search for correlation became synonymous with scientific practice, and questions of causation were deemed unscientific.[51]

In 1920, Sewall Wright published the first causal path diagrams, which attempted to determine the influence of hereditary, developmental, and environmental factors on the coat colors of guinea pigs. He faced fierce attacks from statisticians, who with references to Galton and Pearson claimed that his method was philosophically flawed. However, Wright had found a genuine method for how to weigh the influence of different causes in cases where the causes are known and a data set is available. According to the computer scientist and philosopher Judea Pearl, Wright's unique contribution was that he combined qualitative information in arrow diagrams with quantitative data.[52]

Midcentury, the causal path methods were adopted by sociologists and economists. However, in both cases their potential was eventually

[47] See, for example, Chajes (2020).

[48] Biggs et al. (1986).

[49] Another related term is directed acyclic graphs (DAGs), a graphical notation often used to construct causal path diagrams. Digitale et al. (2022).

[50] Chap. 6. Section "Diagrams of Social Hierarchies".

[51] Pearl and Mackenzie (2018, 66–68).

[52] Pearl and Mackenzie (2018, 72–80).

curtailed. Sociologists called the diagrams "structural equation modeling" (SEM). When they in the 1970s began to use a software package that automated the calculations, some practitioners of SEM became passive software users and let the analysis of the diagrams fade into the background. According to Pearl, they applied SEM as a rote statistical method without looking under the hood to actively construct path diagrams of causation. Economists never really used the path diagrams, but appropriated the algebraic part under the name "simultaneous equation models." However, in contrast to Wright, economists did not construct diagrams that enabled them to analyze causality.[53]

What made the causal path diagrams susceptible to criticism was their subjectivity. They were not a statistical model that could be applied to any data set, but individual models built on researchers' expertise in their field. According to Pearl, we must have a mental model of the world in order to speak of causality.[54] He believes that in the future, causal models have an even more important role to play in Big Data. Large-scale data mining has until now been crunching databases for correlation, but diagrammatic thinking about causality is needed:

> In some fields our knowledge may be in such an embryonic state that we have no clue how to begin drawing a model of the world. But Big Data will not solve this problem. The most important part of the answer must come from such a model, whether sketched by us or hypothesized and fine-tuned by machines.[55]

According to Pearl, causal modeling might also be crucial in the field of artificial intelligence. Recent development in AI has leaned heavily on deep learning by training neural networks on Big Data sets. This method has not been transparent, since it is impossible to tell why it works or not, and it does not incorporate any explicit representation of the environment in which the AI operates. However, in order to create AI with agency, it will be necessary for these machines to create causal models of their experiences.[56]

[53] Pearl and Mackenzie (2018, 86).
[54] Pearl and Mackenzie (2018, 88).
[55] Pearl and Mackenzie (2018, 349–358, quote on 352).
[56] Pearl and Mackenzie (2018, 358–370).

While the causal path diagrams have evolved from statistics, the functionally similar fuzzy cognitive maps have roots in political science and logic. In the 1970s, the political scientist Robert Axelrod transformed policy texts into cognitive maps reminiscent of flow charts, where arrows marked with plus or minus indicated whether a circumstance increased or decreased the likelihood of a certain decision. In 1986, Bart Kosko combined Axelrod's cognitive maps with fuzzy logic, which the mathematician Lotfi Zadeh had developed starting in the 1960s. Unlike in binary logic, fuzzy statements do not need to be either true or false but can take values in between. Kosko claims that the mind imposes a mental viewing screen on our perception of the world which consists of time and space, but also of causality, continuity, connectedness, and probability. He is especially critical of probability, which, unlike fuzziness, he does not believe exists outside our minds.[57] In contrast, causal path diagrams are based on statistical methods with their roots in probability theory. This philosophical difference might help explain why causal path diagrams and fuzzy cognitive maps are viewed as disconnected fields, although the underlying diagrams are similar (Fig. 7.7).

The main structural difference between fuzzy cognitive maps and causal path diagrams is that the latter are unidirectional, since they are used to investigate cause and effect. However, a development of them, the causal loop diagrams, also allow for feedback loops.[58] Thus, there is even less of a difference between the causal loop diagrams and fuzzy cognitive maps. In fact, sometimes structural equation modeling is even combined with fuzzy cognitive maps.[59]

The case with the fuzzy cognitive maps and the causal path diagrams illustrates how similar diagrammatic models and forms of diagrammatic thinking can be arrived at from entirely contradictory theoretical positions. The reason for this is in all likelihood that the constructors of the diagrams, despite their philosophical differences, not only share the same primary metaphors and innate understanding of space, but also the same cultural conventions related to diagramming. The basic diagrammatic structures of their thoughts are the same, although they are verbalized differently in different academic discourses.

[57] Holmén (2020).
[58] Maruyama (1963); Richardson (1999, 200–204).
[59] For example, by Singh et al. (2016).

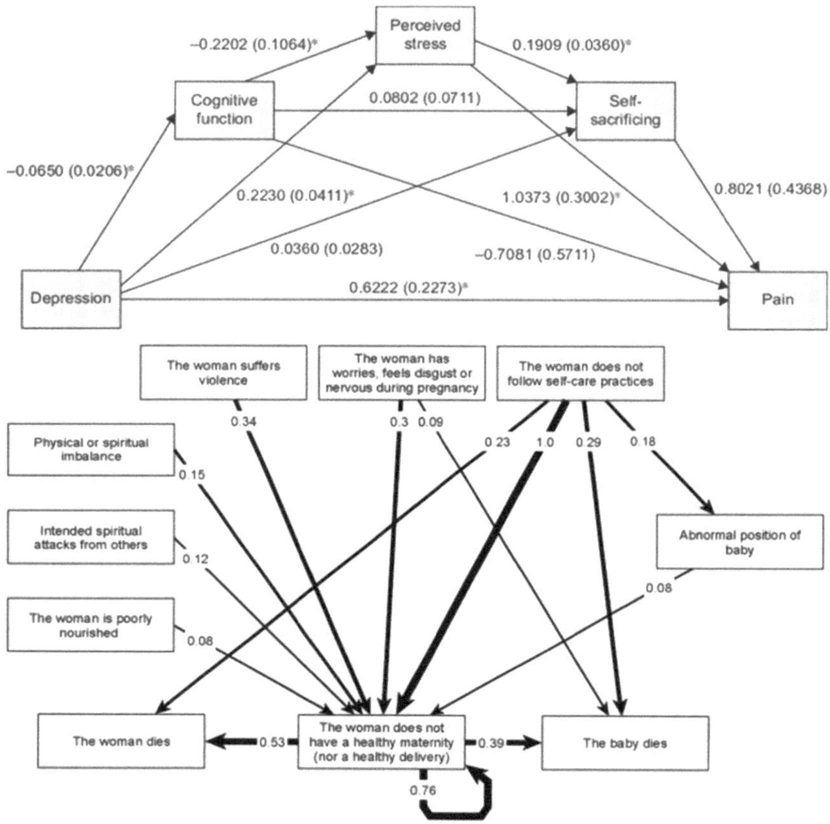

Fig. 7.7 Top: Causal path diagram. The weights before the parenthesis are path coefficients that express possible causal linkages. Wongpakaran et al. 2016, 1856. Bottom. Fuzzy cognitive map. The weights refer to fuzziness. (Sarmiento et al. 2020, 10)

References

Amos, Percy A. 1912. *Process of flour manufacture*. London: Longmans.

Backhouse, Roger E., and Yann Giraud. 2010. Circular flow diagrams. In *Famous figures and diagrams in economics*, ed. Mark Blaug and Peter Lloyd, 221–229. Cheltenham: Edward Elgar.

Baetzhold, Howard G. 2018. 'Well, My Book Is Written—Let It Go…': The making of a Connecticut Yankee in King Arthur's court. In *A Connecticut Yankee in King Arthur's court: A Norton critical edition. An Authoritative Text. Contexts. Criticism*, ed. Henry B. Wonham, 445–477. New York: Norton.

Biggs, Norman E., Keith Lloyd, and Robin J. Wilson. 1986. *Graph theory 1736–1936*. Oxford: Oxford University Press.

Bissell, Chris C. 2007. The Moniac: A hydromechanical analog computer of the 1950s. *IEEE: Control Systems Magazine* 27 (1): 69–74.

Boberg, Johan. 2020. *Scientifically minded: Science, the subject and Kant's critical philosophy*. Uppsala: Uppsala University.

Chajes, J. H. 2020. The kabbalistic tree. In *The visualization of knowledge in medieval and early Europe*, ed. Marcia Kupfer, Adam S. Cohen, and J. H. Chajes, 449–473. Turnhout: Brepolis.

Digitale, Jean C., Jeffrey N. Martin, and Medellena Maria Glymour. 2022. Tutorial on directed acyclic graphs. *Journal of Clinical Epidemiology* 142:264–267.

Friendly, Michael, and Howard Wainer. 2021. *A history of data visualization and graphic communication*. Cambridge, MA: Harvard University Press.

Galison, Peter. 2003. *Einstein's clocks, Poincaré's maps*. London: Sceptre.

Gilbreth, Frank Bunker, and Lillian Moller Gilbreth. 1921. *Process charts*. New York: American Society of Mechanical Engineers.

Gombrich, Ernst. 1976. *The heritage of Apelles*. Oxford: Phaidon.

Gombrich, Ernst. 1990. Pictorial instructions. In *Images and understanding: Thoughts about images. Ideas about understanding*, ed. Horace Barlow, Colin Blakemore, and Miranda Weston-Smith, 26–45. Cambridge: Cambridge University Press.

Graham, Ben B. 2004. *Detail process charting: Speaking the language of process*. Hoboken, NJ: Wiley.

Holmén, Janne. 2020. Time and space in time and space: Mapping the conceptual history of mental maps and historical consciousness. *Contributions to the History of Concepts* 15 (2): 105–129.

Ishikawa, Kaoru. 1998. *Guide to quality control*. 2nd rev. ed., 14th printing. Tokyo: Asian Productivity Organization.

Jones, Matthew, and John Ormond. 2015. Introduction. In *Time travel in popular media: Essays on film, television, literature and video games*, ed. Matthew Jones and John Ormond, 5–18. Jefferson, NC: McFarland.

Kick, Friedrich. 1878. *Die Mehlfabrikation: Ein Lehrbuch des Mühlenbetriebes*. 2. ergänzte und revidirte Aufl. Leipzig: Felix.

Kick, Freidrich, and Henry Handley Pridham Powles. 1888. *Flour manufacture: With supplement, recent progress in flour manufacture*. London: Lockwood.

Lakoff, George. 1990. The invariance hypothesis: Is abstract reason based on image schemas? *Cognitive Linguistics* 1 (1): 39–74.

Lakoff, George, and Mark Johnson. 1999. *Philosophy in the flesh: The embodied mind and its challenge to Western thought*. New York: Basic Books.
Madden, Samuel. 1733. *Memoirs of the twentieth century*. London: Osborn.
Maruyama, Magoroh. 1963. The second cybernetics: Deviation-amplifying mutual causal processes. *American scientist* 51 (2): 164–179.
Mitchel, Edward Page. 1881. The clock that went backward. *The [New York] Sun*, 18 September 1881.
Morris, Stephen J., and O. C. Z. Gotel. 2006. Flow diagrams: Rise and fall of the first software engineering notation. In *Diagrams 2006: Diagrammatic representation and inference*, ed. Dave Barker-Plummer, Richard Cox, and Nik Swoboda, 130–144. Berlin: Springer.
Nahin, Paul J. 2001. *Time machines: Time travel in physics, metaphysics, and science fiction*, 2nd ed., second corrected printing. New York: Springer.
Pearl, Judea, and Dana Mackenzie. 2018. *The book of why: The new science of cause and effect*. New York: Basic books.
Pratt, H. 1915. Long range reception with combined crystal detector and audion amplifier. *Proceedings of the Institute of Radio Engineers* 3 (2): 173–178.
Priestley, Joseph. 1764. *A description of a chart of biography*. Warrington.
Rendgen, Sandra. 2018. *The Minard system: The complete statistical graphics of Charles-Joseph Minard*. New York: Princeton Architectural Press.
Richardson, George P. 1999. *Feedback thought in social science and systems theory*. Waltham, MA: Pegasus communications.
Rosenberg, Daniel. 2013. The Pleasures of Memory. *Time OnLine*. https://time-online.uoregon.edu/twain/pleasures.php#page-1. Accessed 3 January 2024.
Rosenberg, Daniel, and Anthony Grafton. 2010. *Cartographies of time: A history of the timeline*. New York: Princeton Architectural Press.
Sarmiento, Ivan, Sergio Paredes-Solís, David Loutfi, Anna Dion, Anne Cockcroft, and Neil Andersson. 2020. Fuzzy cognitive mapping and soft models of indigenous knowledge on maternal health in Guerrero, Mexico. *BMC medical research methodology* 20(1): 1–16.
Schivelbusch, Wolfgang. 1978. Railroad space and railroad time. *New German Critique* 14 (14): 31–40.
Singh, Manpreet, Levi Monteiro Martins, Patrick Joanis, and Vijay K. Mago. 2016. Building a cardiovascular disease predictive model using structural equation model and fuzzy cognitive map. In *2016 IEEE International Conference on Fuzzy Systems (FUZZ-IEEE)*, 1377–1382. IEEE.
Swade, Doron. 2000. The Phillips Machine and the history of computing. In *A.W.H. Phillips: Collected works in contemporary perspective*, ed. R. Leeson, 120–126. Cambridge: Cambridge University Press.
Twain, Mark. 1889. *A Connecticut Yankee in King Arthur's court*. New York: Harper & Brothers.

Twain, Mark. 1914. How to make history dates stick. *Harper's Monthly Magazine,* December.

Wells, Herbert George. 1888. *The chronic argonauts.* Telelib.com https://www.telelib.com/authors/W/WellsHerbertGeorge/prose/chronicargonauts/index.html. Accessed 30 July 2024. Originally published in *The Science Schools Journal,* April, May, and June 1888.

Wheat, Elizabeth Ruth Josie. 2013. *Terrestrial cartography in ancient Mesopotamia.* Birmingham, AL: University of Birmingham.

Wongpakaran, Tinakon, Nahathai Wongpakaran, Sitthinant Tanchakvaranont, Putipong Bookkamana, Manee Pinyopornpanish, Kamonporn Wannarit, Sirina Satthapisit, Daochompu Nakawiro, Thanita Hiranyatheb, and Kulvadee Thongpibul. 2016. Depression and pain: Testing of serial multiple mediators. *Neuropsychiatric disease and treatment* 12:1849–1860.

Zobel, Justin. 2004. *Writing for computer science.* 2nd ed. London: Springer.

CHAPTER 8

Diagramming the Political Space

Even the highly abstract dimensions of politics can be made understandable—and be manipulated—through the construction of spatial models and diagrams. An early example was the analogy between politics and cosmological systems. Since the French Revolution, the left–right scale has gradually become the dominant spatial representation of the main axis in politics (Chap. 8, Section "The Rise of the Left–Right Political Scale"). However, the history of left–right scale in twentieth-century Sweden reveals how two contradictory spatial models can coexist simultaneously, until circumstances finally force one model to dominate over the other (Chap. 8, Section "The Inversion of the Swedish Political Scale"). Revolutionary ideologies, such as communism, feminism, and LGBTQ, have, despite their radical political outlook, made use of traditional diagrammatic forms and principles in order to convey their messages and views of the political situation (Chap. 8, Section "Revolutionary Diagrams: Breaking the Barrier"). Often, they have done so by constructing diagrams that visualizes the political barriers that they are attempting to break through, which encourages views of the political field as a dichotomy rather than as a continuum.

J. Holmén, *A History of Diagrams*, Palgrave Studies in the History of Science and Technology,
https://doi.org/10.1007/978-3-031-99691-7_8

The Rise of the Left–Right Political Scale

During the Renaissance, the heavenly spheres were used as analogies not only for musical harmonies[1] but also for a harmonious social and political order. For example, in his 1588 *Sphaera civitatis*, John Case constructed a diagram of Queen Elisabeth's political and moral virtues based on the Ptolemaic cosmological system.[2] From the late seventeenth century, the natural balance codified in the Newtonian laws became an appealing analogy for the political system. Political Newtonianism provided a language in which arguments as well as counterarguments could be framed, as both action and inaction could be justified by the need to restore or maintain "balance" in the political system.[3] For example, in the British colonies in North America, the Newtonian system became a source of analogies to the consequences of disturbing the old political system by trying to break away from the English monarchy.[4] Later, federalism was likened to the heliocentric systems. For example, on early US coins, the states were depicted as circles orbiting congress, which was portrayed as a sun.[5]

When an official flag for the United States was accepted, the union was depicted as a constellation rather than as a solar system. A constellation with several stars of equal magnitude was considered a more republican representation, as a dominating sun was associated with monarchy.[6]

Still, the age of parliamentarian democracy needed a spatial model that did not only convey harmony, but could also visualize the ideological differences between competing political actors. The left–right spectrum is commonly said to have its roots in the parliament of the French Revolution, where radicals were placed to the left and conservatives to the right. However, the gradual incorporation of this dichotomy in the political vocabulary spanned over centuries. In England, references to the left and right sides of the parliament were made at least as early as 1672, then referring to government and opposition. It is unclear whether the English example inspired the use of the term in revolutionary France, but it is known that some parliamentary practices were copied from England.[7]

[1] See Chap. 5, Section "The Musical Prelude to Diagrams of Scientific Systems".
[2] Shalev (2018, 194–195).
[3] Shalev (2018); Striner (1995).
[4] Striner (1995, 599–600).
[5] Shalev (2018, 190–192).
[6] Shalev (2018, 201–202).
[7] Gauchet (1996, 241–242).

Left and right were first used in the Third Estate of the French Estates General on 8 May 1789. Delegates Count Mirabeau and Baron Malouet had made rival proposals on the combination of the orders, and a vote was conducted by asking the supporters of Malouet's proposals to gather to the right, and those of Mirabeau's more radical proposal to the left. It is possible that this contributed to the formation of the left as the home for radical reformists. Another explanation is that the priests continued to sit to the right in the National Assembly, which first convened July 9, 1789, as they had done in the old estates, thereby providing the right side with a conservative core. From late August, observers noted that there was a spatial division in the National Assembly. According to several sources, it was the radicals who first seated themselves to the left, thereby pushing the others to the right.[8]

In the following years, the terms "left," "right," and "center" were used referring to radicals, conservatives, and moderates, reflecting their seating in the National Assembly. However, their usage was limited to reports from proceedings in the parliament, and there is no indication that people identified as being right or left because of their political views. The spatial division of the parliament was considered problematic and a cause of conflict, and therefore it was decided in 1795 that seating should be rearranged by lots every month.[9]

Left and right did not become a permanent political division until the French Restoration. When monarchy was restored after the Napoleonic Wars, the ultra-royalists came to dominate the chamber and decided to take a seat to its right. According to Marcel Gauchet, the years 1815 to 1820 became crucial in establishing the ultimate sense of the political terms left and right.[10] Still, the colors white and red were the most commonly used symbols for the main political factions in France throughout the nineteenth century. However, left and right gradually gained strength from the 1870s and came to dominate from the early 1900s. The process culminated at the election of 1906, where left and right were mobilized as signs of identity. Gauchet points out that if there had only been two major parties in France, there would not have been any lexical need for left and right, but the terms caught on since both left and right were divided. Through their contrast, red and white were perfect symbols for opposed

[8] Gauchet (1996, 243–244).
[9] Gauchet (1996, 245–247).
[10] Gauchet (1996, 247–248).

camps, but the pair left and right "was an amazing marriage of radical simplification with the possibility of broadening nuance" that " allows one to conceive simultaneously of both radical opposition and a continuous, infinitely subdividable spectrum."[11] Left and right were better suited to describe the fractured political landscape that came to dominate most countries with the advent of democracy.

In the National Assembly of revolutionary France, the division of the parliament had been qualified by notions such as "the left extremity of the left part." During the Restoration, the positioning in the chamber became an even more elaborate spectrum, where delegates seated themselves with great attention to nuances in their positions vis-à-vis the government and the political extremes.[12] It is the spatial metaphor that makes the left–right axis such a versatile tool for describing political nuances. Distance in political opinion could be described as distance in space by journalists present in the chamber. Readers could, using their spatial cognition, intuitively infer what it meant that delegate x had a political view slightly to the left of delegate y, but to the right of delegate z, without being closely familiar with the political questions.

That the left–right pair can be understood as both radical opposition and a spectrum is related to its roots in two different primary metaphors. Either left and right can, through the primary metaphor "Categories are Containers," be understood as two separate categories, or they can through "Linear Scales are Paths" be understood as two directions on a scale. Thus, left and right can function as verbal cues to two different spatial models of the political system, which we can switch between. As will be discussed in Chap. 8, "Section Revolutionary Diagrams: Breaking the Barrier", we gravitate toward the dichotomous interpretation in times of increased stress.

According to Laponce, the left–right political concept did not spread outside of France until the twentieth century. Its use was pioneered by the left. In the 1920s, the terms were used by socialists and communists in the Soviet Union, Britain, France, and the United States. Still today, left-wing supporters are more likely to describe themselves as left than supporters of the right are to identify as right.[13] However, there are exceptions, such as the Nordic countries, where "right" has even been included in the names

[11] Gauchet (1996, 253–257, 261, quotes on 264 and 249).
[12] Gauchet (1996, 245, 249).
[13] Laponce (1981, 55).

of the parties. The oldest of them, the first Danish *Højre* (right) party, was founded already in 1848.[14]

Diagrammatic representations of the left–right spectrum have evolved from the physical seating of representatives in parliaments via maps and sketches of this seating to more abstract diagrams. During the French Revolution, a theater was remodeled into a permanent seat for the National Assembly. This semicircular architecture came to dominate parliamentary buildings except for in the British Commonwealth, where government and opposition are facing each other.[15] In France, it became common to represent the political positions of the delegates in plans outlaying their habitual seating in the chamber (Fig. 8.1).[16] Schematic drawings of this semicircular architecture have been the most common spatial representations of multiparty politics.

Generally, spatial representations of the political field are confined to the left–right dimension, although it is widely accepted that this does not reflect the multidimensionality of political differences. If two or three dimensions were used, it would be possible to graphically represent more aspects of party politics. However, of the three available dimensions, left–right is the most neutral. If political parties were represented in a system that distinguished them on the over–under and in front–behind dimensions, chances are that it would not be favored by supporters of the under/behind parties.[17] The relative neutrality of the left–right spectrum makes it difficult to replace it with other spatial models, regardless of their descriptive potential. However, as the following chapter illustrates, not even the horizontal dimension of politics is completely value-neutral.

The Inversion of the Swedish Political Scale

Jean A. Laponce has pointed out that the left–right scale is charged with affective values. In all larger cultural areas, with the possible exception of China, right has been associated with the divine, the dominating, and privileged, and left with challenges to that order. According to Laponce, the modern political left–right scale has developed from this older idea. It

[14] Holmén (2024).
[15] Goodsell (1988).
[16] Gauchet (1996, 245).
[17] This was also the case with the GAL-TAN scale which became popular in the lead-up to the 2016 Swedish parliamentary election, but was criticized by the parties placed below. Holmén (2024).

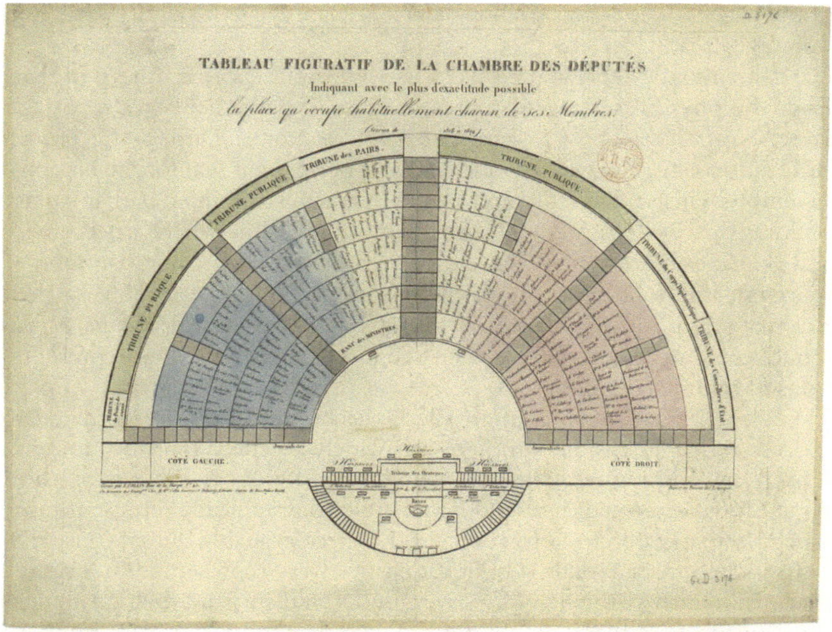

Fig. 8.1 Plan of the seating in the French Chamber of Deputies from 1818, engraved by the cartographer E. Collin. This is the first image of this type that I have been able to find. Bibliothèque nationale de France, Tableau figuratif de la Chambre des députés. (Session de 1818–1819)

can be described as a collapsed societal pyramid, which at the advent of democracy toppled over to the right, so that parties that represent the old upper classes are now associated with the right. In the words of Laponce, "[t]he horizontal left/right classification, born from a revolution that sought to deny the very notion of verticality, became a mirror where up and down were reflected." In Laponce's opinion, it was no coincidence that the pyramid fell to the right, but the positively charged poles on the two axes—up and right—constitute a semantic cluster, a group of words that are intimately associated with each other.[18]

[18] Laponce (1981, 8). This chapter is based on Holmén (2024).

This connection between up and right is, however, not self-evident under all circumstances. According to the principle of spatial agency bias, a literate culture that uses the Latin alphabet, where the direction of reading is from left to right, is likely to follow that directionality in its spatial analogies. As a consequence, powerful and agentic actors are more often placed to the left in diagrams, as action takes place from left to right.[19]

The fact that vertical series are usually directed from top to bottom and horizontal from left to right suggests that left and up can become interchangeable. Based on this, it could be assumed that the social pyramid would fall to the left, which would make the first element in the vertical order the first in the horizontal. That would contradict Laponce's hypothesis that up and right form a strong cluster. An illustrative example of how these two opposing principles have interacted with each other can be found in Sweden. There, parliamentarians are not seated according to political party but home region. This means that spatial representations of the party system cannot be constructed based on the actual seating of parliamentarians, as is the case in many other countries. This has left the floor open for the two contradictory cognitive principles, the "semantic cluster" and the "spatial agency bias," to perform an intricate dance with some surprising turns. This is evident from how the spatial representation of the political spectrum in Swedish school textbooks has developed in the last century.

In the early 1900s, the Swedish political parties were only presented in a textual narrative. The texts began by describing the political right, the conservatives, and then continued with the liberals followed by the socialists. Beginning in the interwar period, election results were presented in tables, and soon also in data visualizations. The spatial order from the early texts was maintained in the tables: the political right was spatially placed first, that is on top or to the left in tables. When the first diagrams of the parties' political ideologies appeared in textbooks in the 1970s (Fig. 8.2, top), they inherited the spatial order of this inverted right–left scale from the texts and tables, placing the political right to the spatial left.[20]

Text, tables, and data visualizations primarily strived to present the parties' strength in voter support and number of parliamentarians. The parties' spatial order was of less importance and might have been entirely random. However, some kind of order always finds its way into cultural

[19] Tversky et al. (1991); Maass and Schubert (2011, 4); Chatterjee (2011).

[20] See Holmén (2024).

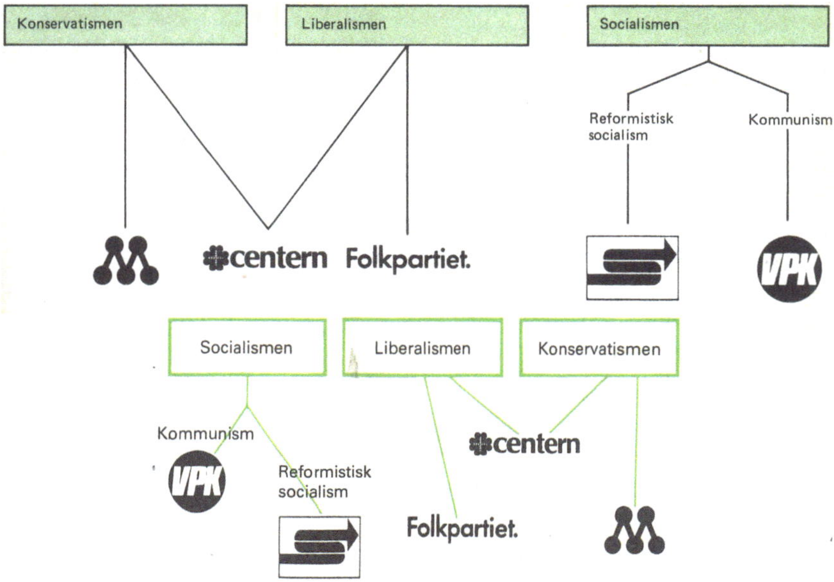

Fig. 8.2 In Sweden, the introduction of diagrams of the political system flipped the political scale, from having placed the radical "left" to the right, to placing them to the left. This is illustrated by two editions from the same textbook. (Liljequist et al. 1978, 37 (top) and 1985, 212 (bottom))

constructions, also when it is not consciously strived for.[21] This made it possible for several spatial analogies—the timeline, the social pyramid, and political affinity—to combine more or less subconsciously in a way that gave birth to the inverted right–left scale.

Probably, the conservatives were mentioned first since they were seen as the oldest political power, defenders of an established order challenged first by liberals and then by socialists. This interpretation is supported by the fact that the agrarian party, which in Sweden was late to organize politically, in the 1920s and 1930s was placed last in text and tables, although they were ideologically close to the conservatives that were placed first. New parties founded in the second half of the twentieth

[21] Several studies indicate that humans are unable to randomize. Wagenaar (1972); Figurska et al. (2008).

century were also initially placed last, until they found their ideological place on the left–right scale.[22]

To place the political right (the upper classes) first (on top or to the left) resonated with views from class society and from the even older feudal order of estates. Sweden also experienced a gradual expansion of suffrage from the mid-nineteenth century through the removal of income barriers, which meant that conservative voters were granted suffrage first and socialist voters last. Conservatives from the upper classes also dominated the first chamber of the Swedish parliament. In all likelihood, several of these logics interacted in establishing a political scale with conservatives first, that is to the left.

Laponce's claim that "up" and "right" form a semantic cluster has to be nuanced in light of the Swedish example. The parties placed on top have always also been placed to the left, both in the older inverted right–left scale and later in the regular left–right scale. The habit formed by reading and writing the Latin alphabet seem to have created a strong connection between left and first. Since first means up in horizontal tables, there is also a strong indirect connection between left and up. Thus, spatially the parties that represented the old upper classes were placed to the left, although semantically they were called right. The long period during which the right-wing parties were placed to the left also seems to have left traces in Swedish semantics. A Google search reveals that in English "left–right political spectrum," with the word left placed to the left, is about nine times more common than "right–left political spectrum." However, in Swedish, "höger-vänsterskala," with the word *höger* (right) placed to the left, is around three times more common than the reverse.[23]

The inverted right–left axis was not turned around until the mid-1980s. This followed the appearance of diagrams of the ideological roots of the political parties in the 1970s. Until then, the parties had been presented in texts or tables of election results, with the political right placed to the spatial left. The first diagrams followed the same spatial model, but they were flipped over in new editions (Fig. 8.2). It is possible that students who came in contact with the political field for the first time through the diagrams, and were not habituated with the inverted right–left scale,

[22] Holmén (2024).

[23] "Left-right political spectrum": 35,700, "right-left political spectrum": 3,830, "höger-vänsterskala": 13,400, "vänster-högerskala": 4,610. Searches performed on 11 November 2022.

perceived it as strange that the right should be placed to the left, and that might have created pressure for change. According to Julio Santiago et al., two contradictory spatial models, such as the spatial agency bias which placed the political representative of the upper classes to the left and the semantic cluster which associated them with right, can coexist in long-term memory. However, when they are simultaneously recalled to short-term memory, for example, through a diagram of political ideologies where parties described as "right" politically are placed to the left spatially, some form of congruence is forced upon the model.[24] In this case, congruence was achieved by flipping the diagrams over.

In addition to the left–right axis, a new dimension was added to the political diagrams in the 1980s: the industrial–agrarian or economic–ecological. It was based on research in political science from the late 1970s, and came in vogue because of the 1980 referendum on nuclear energy, which split the political field across party lines. The construction of such diagrams involved a greater element of conscious spatial thinking. This probably contributed to a questioning of habitual spatial conventions in favor of reflection on what was the most logically and didactically appealing form of representation. For a while, the inverted right–left scale continued to be used in text and tables after it had been replaced by the regular left–right scale in diagrams. However, from the mid-1990s, the political left has in civics textbooks consequently been placed first or to the left in texts and tables as well as in diagrams. After a period when the narrative structure of the texts had conditioned the spatial structure of tables and diagrams, there was a transitional period from the mid-1980s to the mid-1990s during which the new spatial order of the regular left–right scale spread from diagrams to other forms of representation. By the early 2000s, the regular left–right scale had also spread to other parts of the public sphere, such as the election-night watch parties on national television.[25]

That diagrams of the parties' ideological positions began to appear in the 1970s and 1980s might be related to the fact that it was during this period that the voters began to choose. Still in the 1960s, voting was to a large degree class based, in Sweden as well as in other Western societies, and the share of convinced party supporters was high. This changed in the following decades. The change was swifter in Sweden than in many other

[24] Santiago et al. (2011, 81).
[25] Holmén (2024).

countries because of rapid social change and an advanced individualization.[26] Voters began to find their way to the ballot boxes in a new way, from having been led there by a guide—the party they were more or less born into—to navigate there with their own map. The diagrams of the political field that began to appear in civics textbooks at this time functioned as navigational tools for wayfinding in the new political environment. Earlier, the Swedish party system consisted of a small number of old parties with well-known profiles. That new parties appeared, and that all parties began to shift positions in order to compete for the favors of the more mobile voters, also contributed to the demand for a diagrammatic overview of the political field.[27]

There has been a debate among linguists and neurologists regarding whether the spatial layout of diagrams are mere visualizations of verbal spatial metaphors.[28] The history of the Swedish inverted political scale suggests that this is not the case. Since the standard diagrammatic spatial representation of the political system for a long period directly contradicted the dominant verbal spatial metaphor, the former can hardly be considered a visualization of the latter.

Revolutionary Diagrams: Breaking the Barrier

Revolutionary movements are not content to act inside the framework of the existing social and political order. They want to break through the barriers of the present society in order to reach a better reality located on the other side. Consequently, diagrams constructed by revolutionaries often contain representations of borderlines: between revolutionaries and reactionaries, between the bad past and the promised future, or between the ideals of the oppressors and those of the oppressed.

For example, an early Soviet propaganda poster (Fig. 8.3) illustrates how revolutionary workers break through the defense lines of the reactionaries in the revolutions of 1905 and 1917 on their way to world revolution.[29]

The poster is figurative and filled with metaphors and symbols, but it also contains a strong diagrammatic element of barrier lines being broken.

[26] Rose and McAllister (1986); Ekengren Oscarsson (2018).
[27] Holmén (2024).
[28] Ortiz (2011).
[29] Zitser (2018, 254).

Fig. 8.3 "The Way to the Russian Revolution and the World Revolution." A figurative propaganda poster with strong diagrammatic elements where several spatial analogies are combined: spatial agency bias from left to right, the primary metaphor "Control is Up," a timeline from left to right, and, primarily, a distinct frontline that has to be crossed in order to achieve the political goal. (Zitser 2018, figure 23)

In the same collection of 30 early Soviet propaganda posters there is another with a similar diagrammatic structure, where a Soviet tank from the upper left overruns a barrier in order to crush a capitalist.[30] The direction from the upper left in the images means that spatial agency bias from left to right is combined with the primary metaphor "Control is Up" in order to portray the Soviet troops as dominant and agentic. This is a contrast to propaganda posters depicting pre-revolutionary society, such as social pyramids with workers at their base (Fig. 2.2).

The diagrammatic aspect of a dividing border is more clearly discernible in non-figurative diagrams with a similar basic diagrammatic structure. Feminist theory experienced a period of rapid development in the late 1960s and early 1970s. Some feminists, such as Shulamith Firestone and Ti-Grace Atkinson, believed that they needed to complement their lectures and texts with diagrams in order to explain their novel ideas. In the words of Atkinson, "[e]ach of the 35 charts saved me pages of explanation. … Relationships can be taken at a glance that simply could not be explained adequately in any other way."[31] Both Firestone's and Atkinson's diagrams ventured into the field of art. Firestone, herself an art major, constructed a diagram for her 1970 book *The Dialectic of Sex*, which was sold as a poster in connection to the book's republication in 2015. Atkinson worked in collaboration with an artist, Barbara Nassim, in constructing the diagrams.[32] Atkinson's and Firestone's diagrams are purely abstract without any figurative elements and combine text with horizontal and vertical lines as well as diagram elements such as arrows, brackets (in the case of Firestone), and texture patterns (in the case of Atkinson).

Ti-Grace Atkinson's diagrams were originally produced for a lecture in 1971, but were also published in book form in 1974 together with explanatory text. In a series of strategy charts (Fig. 8.4), clearly inspired by military maps, she illustrated how the oppressed feminists could stage a successful rebellion against the oppressive patriarchy. She also tried to figure out the role of lesbians in the revolution. The diagrams were drawn in a period when Atkinson went from viewing lesbians as a neutral buffer between the oppressors and the oppressed to considering them an important asset for feminism.[33]

[30] Zitser (2018, figure 11).
[31] Atkinson (1974, ccxxxvii).
[32] McBean (2021).
[33] Atkinson (1974).

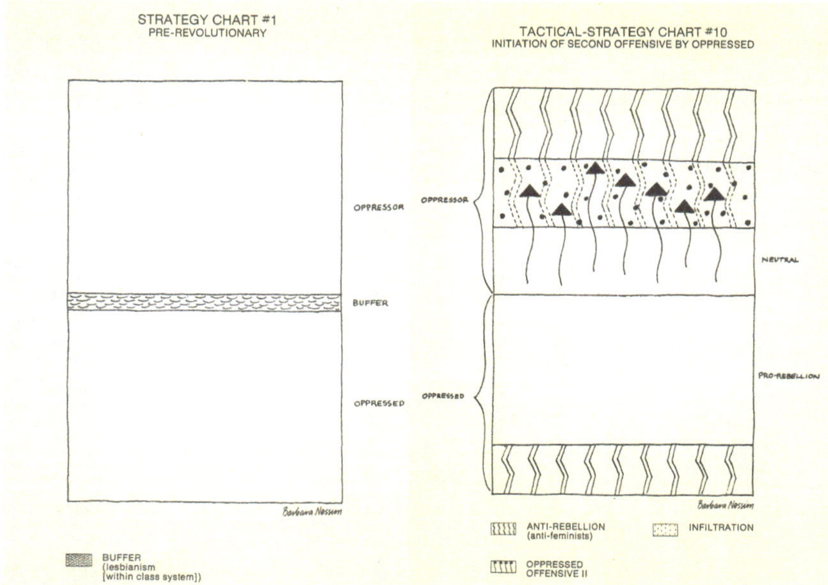

Fig. 8.4 Atkinson's strategy charts for feminist rebellion were inspired by military charts and their frontlines. (Atkinson 1974, 141, 173)

Atkinson's first strategy chart is the simplest diagram imaginable, in its dichotomous division between good and evil comparable to the Pythagorean and Christian ypsilon (Y).[34] The oppressors are here placed on top with the oppressed below, separated by a thin buffer. Such a simple diagram would probably not have been developed on its own, but its function was to describe the starting point for a series of diagrams where the following ones became more complicated. Atkinson's diagrams are essentially one dimensional, since the horizontal dimension, which often represents time in diagrams, is not carrying any information. Time is in her case instead represented by the succession of diagrams.

Firestone's diagram, on the other hand, is more complex and two-dimensional, with the horizontal axis as a timeline (Fig. 8.5). Her book discussed the dialectics of sex, class, and culture, which was supposed to lead to sexual, economic, and cultural revolutions. The accompanying

[34] Chap. 4, Section "Tree Diagrams".

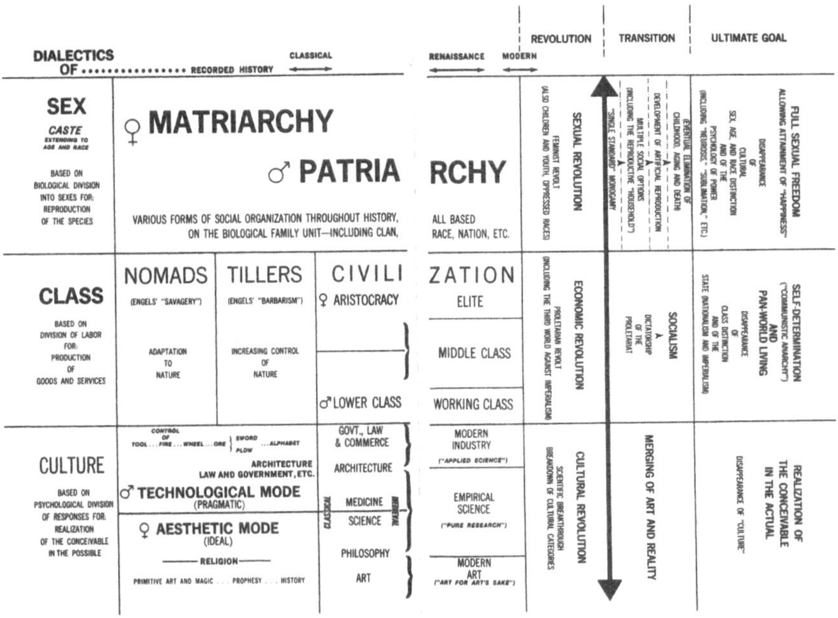

Fig. 8.5 "3D revolution. Dialectic of Sex." The three parallel timelines of sex, class, and culture merge when the borderline representing the revolution is broken. (Firestone [(1970) 2015, 243–244])

diagram was structured as three parallel timelines charting the developments in these fields from before the beginning of recorded history until the ultimate goal. The top left corner, containing the sexual and economic fields until the revolution, was based on Friedrich Engels's *Origin of the Family, Private Property and the State*. Firestone's largest personal contribution was to map the cultural development. The cultural revolution would be characterized by a breakdown of the modern cultural categories such as applied science, pure science, and art for art's sake in what Firestone described as a merger of art and reality. From that perspective, her diagram can be seen as a bridge between art and science. According to Firestone, science, philosophy, and art had been united in the classical and medieval periods.[35]

[35] Firestone [(1970) 2015, 6, 160–161, 172–173].

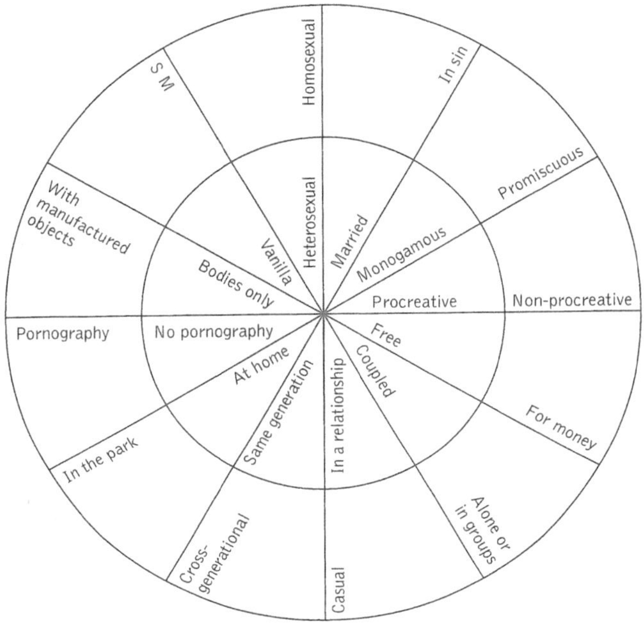

Fig. 8.6 "The charmed circle." Good practices inhabit the space inside the borderline of the inner circle while bad, abnormal practices dwell in the outer limits. (Rubin [(1984) 2008, 293])

In 1984, Gayle Rubin published the article "Thinking Sex: Notes towards a Radical Theory of the Politics of Sexuality." There, she used a diagram of two concentric circles to explain the dichotomy between accepted and forbidden sexual practices (Fig. 8.6). In this diagram, the inner "charmed circle" is constituted by "Good, Normal, Natural, Blessed Sexuality," while "the outer limits" are populated by "Bad, Abnormal, Unnatural, Damned Sexuality." In each sector of the circle, blessed and damned practices are in conflict with each other, such as heterosexual–homosexual and same generation–cross generational.[36] This dichotomous concentric diagram, where the sharp borderline is circular, was relatively static and left little room for revolution.

[36] G. Rubin [(1984) 2008, 293].

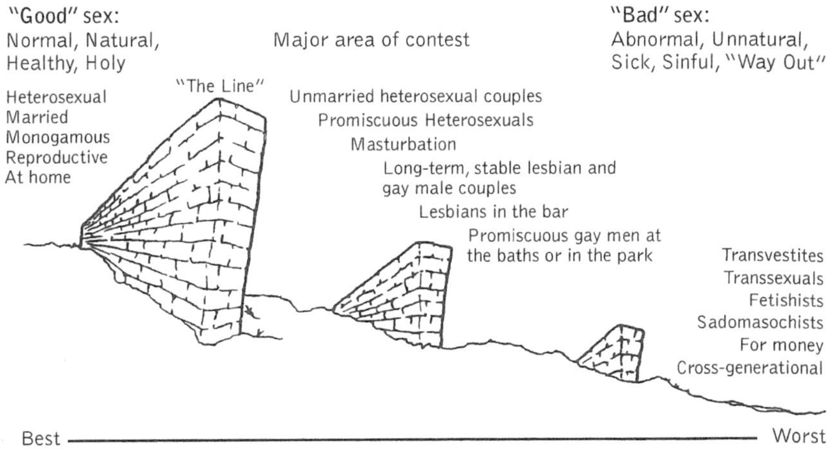

Fig. 8.7 A series of walls illustrate how the frontier of the acceptable is moving. (Rubin [(1984) 2008, 294])

However, Rubin complemented it with a second linear spectrum, which illustrated that bad sexual practices can be bad to varying degrees, with some eventually crossing the border to the legitimate (Fig. 8.7). Rubin raised protective walls at different positions along the spectrum, in order to illustrate the "domino theory of sexual peril: the fear that if anything is permitted to cross this erotic DMZ, the barrier against scary sex will crumble and something unspeakable will skitter across." According to Rubin, most systems of judgment—religious, psychological, feminist, or socialist—attempt to determine on which side of the barrier particular sexual acts should be placed. In the last decades, some behavior close to the border had inched across it, such as unmarried couples, monogamous homosexuality, and masturbation.[37] Thus, the diagram is a scale rather than a dichotomy, and represents more of a reformist than a revolutionary view of development.

The roots of the human propensity for dichotomous thinking and the creation of dualities have been explained in many ways, for example, as a reflection of the alterations between night and day, of the difference

[37] G. Rubin (2008, 294).

between self and others, or of the two sexes.[38] It can, however, be considered a consequence of our ability to detect borderlines that, as other parts of our spatial cognition, has been transformed to abstract realms. A distinct mental borderline establishes a dichotomy by dividing the world into two categories.[39] The analogical transfer from the spatial to the abstract domain is made obvious in some of the figures (Figs. 8.3 and 8.7), which allows terrain and ideas to merge on an axis from left to right, which in some instances also constitutes a timeline (Figs. 8.3, 8.5, and 8.7). While Atkinson's diagrams are clearly influenced by military maps, Rubin's charmed circle seems to have been modeled on medieval maps where dragons and cyclops dwelled in the fringes. It is founded on a spatial metaphor that equates center with good and periphery with evil.[40] Although a table with two columns or opposing terms on each side of a straight line could have performed the same simple classificatory function, the circular shape added an emotional dimension.

Our propensity to think in binary opposites is influenced by external factors. According to Laponce, individuals or cultures under stress are more likely to use dichotomies than continuums. For example, left and right in politics can become perceived as polarized opposites, rather than as two directions on a continuous spectrum.[41] Laponce's observation might be related to a general tendency to switch between systems for spatial orientation depending on stress levels, which was observed by Edward Tolman in 1947. He found that rats under stress or under strong motivational forces were more likely to develop narrow strip maps of the maze in which they were tested than rats allowed to wander freely. Tolman believed these findings could be generalized to humans and explain the racial and Cold War tensions of his time.[42] Later psychological research also indicates that under stress the brain uses a faster but less exact system for spatial orientation, and that this is related to a general tendency to shift between fast intuitive and slow thoughtful systems of cognition depending on situation.[43]

As our mental apparatus for spatial cognition has been appropriated for abstract tasks, the tendency for crude dichotomization during stress has

[38] Laponce (1981, 23).
[39] The line as an archetype for political thinking is discussed in Laponce (1975).
[40] For the metaphor center-periphery, see Deane (1995).
[41] Laponce (1981, 16).
[42] Tolman (1947; Chap. 5).
[43] Starcke and Brand (2012); Epstein et al. (1996, 390); Kahneman (2003, 697).

entered the social and political realms. A 1976 study of revolutionary leaders revealed that those with the most polarized rhetoric were most successful in carrying through a revolution, although more complex messages were needed after revolution was accomplished.[44] In order to induce social and political change, it might therefore be beneficial to combine polarizing rhetoric, that incites motivation for dichotomous thinking, with diagrams that visualizes the polarizing barriers that the revolution intends to break. Studies indicate that barriers in the environment fragmentize the signals from the grid cells that construct our perception of a continuous space.[45] When the same system for spatial cognition is used as an analogy for abstract domains, dichotomous political diagrams with barriers encourage interpretations of political polarization rather than of a continuous political spectrum. The human propensity for cognitive polarization and simplification in stressful situations might contribute to driving social change by encouraging revolutionary action. Thereby, it is an example of how cognitive principles, although themselves constant and unchanging in a historical timeframe, might help explain historical change and development.

References

Atkinson, Ti-Grace. 1974. *Amazon Odyssey: The first collection of writings by the political pioneer of the women's movement.* New York: Links Books.

Chatterjee, Anjan. 2011. Directional asymmetries in cognition: What is left to write about? In *Spatial Dimensions of Social Thought*, ed. Anne Maas and Thomas W. Schubert, 189–210. Berlin: De Gruyter Mouton.

Deane, Paul D. 1995. Metaphors of center and periphery in Yeats's *The Second Coming*. *Journal of Pragmatics* 24 (6): 627–642.

Derdikman, Dori, Jonathan R. Whitlock, Albert Tsao, Marianne Fyhn, Torkel Hafting, May-Britt Moser, and Edvard I. Moser. 2009. Fragmentation of grid cell maps in a multicompartment environment. *Nature Neuroscience* 12 (10): 1325–1332.

Ekengren Oscarsson, Henrik. 2018. *Flytande väljare 2018.* Gothenburg: University of Gothenburg.

Epstein, Seymour, Rosemary Pacini, Veronika Denes-Raj, and Harriet Heier. 1996. Individual differences in intuitive–experiential and analytical–rational thinking styles. *Journal of Personality and Social Psychology* 71 (2): 390–405.

[44] Suedfeld and Rank (1976).

[45] Derdikman et al. (2009); He and Brown (2019).

Figurska, Małgorzata, Maciej Stańczyk, and Kamil Kulesza. 2008. Humans cannot consciously generate random numbers sequences: Polemic study. *Medical hypotheses* 70 (1): 182–185.

Firestone, Shulamit. (1970) 2015. *The dialectic of sex: The case for a feminist revolution*. London: Verso.

Gauchet, Marcel. 1996. Right and left. In *Realms of memory: The construction of the French past. I. Conflicts and divisions*, ed. Pierre Nora, 241–298. New York: Columbia University Press.

Goodsell, Charles T. 1988. The architecture of parliaments: Legislative houses and political culture. *British Journal of Political Science* 18 (3): 287–302.

He, Qiliang, and Thackery I. Brown. 2019. Environmental barriers disrupt grid-like representations in humans during navigation. *Current Biology* 29 (6): 2718–2722.

Holmén, Janne, Ann-Sofie Jägerskog, Daniel Schumann, and Malin Tväråna. 2024. Connecting the nation: Diagrams of unity and diversity in civics textbooks. In *Diagrams in civic education: Visuospatial models of society in textbooks and teaching*, ed. Janne Holmén, Ann-Sofie Jägerskog, Daniel Schumann, and Malin Tväråna, 11–59. Palgrave Macmillan.

Kahneman, Daniel. 2003. A perspective on judgment and choice: Mapping bounded rationality. *American Psychologist* 58 (9): 697–720.

Laponce, Jean A. 1975. Spatial archetypes and political perceptions. *The American Political Science Review* 69 (1): 11–20.

Laponce, Jean A. 1981. *Left and right: The topography of political perceptions*. Toronto: University of Toronto Press.

Liljequist, Kurt, Svenåke Björkryd, Sture Andersson, and Lennart Walles. 1978. *Världen—vårt samhälle: Samhällskunskap för gymnasieskolans 3- och 4-åriga linjer*. Uppsala: Almqvist and Wiksell.

Liljequist, Kurt, Anders Altersten, Svenåke Björkryd, Hans Olof Hanselid, and Reidar Larsson. 1985. *Världen vårt samhälle: Samhällskunskap för gymnasieskolan*. Uppsala: Almqvist and Wiksell.

McBean, Sam. 2021. Feminist diagrams. *Feminist Theory* 22 (2): 206–225.

Ortiz, Maria J. 2011. Primary metaphors and monomodal visual metaphors. *Journal of Pragmatics* 43 (6): 1568–1580.

Rose, Richard, and Ian McAllister. 1986. *Voters begin to choose: From closed-class to open elections in Britain*. London: Sage.

Rubin, Gayle. 2008 (1984). Thinking sex: Notes towards a radical theory of the politics of sexuality. In *The Routledge critical and cultural theory reader*, eds. Neil Badmington and Julia Thomas, 281–323. London: Routledge.

Santiago, Julio, Antonio Román, and Marc Ouellet. 2011. Flexible foundations of abstract thought. In *Spatial dimensions of social thought*, ed. Anne Maas and Thomas W. Schubert, 39–108. Berlin: De Gruyter Mouton.

Schubert, Thomas W., Sven Waldzus, and Beate Seibt. 2011. More than a metaphor: How the understanding of power is grounded in experience. In *Spatial dimensions of social thought*, ed. Anne Maas and Thomas W. Schubert, 153–185. Berlin: De Gruyter Mouton.

Shalev, Eran. 2018. Our federal sun: Planetary politics before the civil war. *American Political Thought* 7 (2): 189–215.

Starcke, Katrin, and Matthias Brand. 2012. Decision-making under stress: A selective review. *Neuroscience & Biobehavioral Reviews* 36 (4): 1228–1248.

Striner, Richard. 1995. Political Newtonianism: The cosmic model of politics in Europe and America. *The William and Mary Quarterly* 52 (4): 583–608.

Suedfeld, Peter, and A. Dennis Rank. 1976. Revolutionary leaders: Long-term success as a function of changes in conceptual complexity. *Journal of Personality and Social Psychology* 34 (2): 169–178.

Tolman, Edward C. 1947. Cognitive maps in rats and men. *Psychological Review* 55 (4): 189–208.

Tversky, Barbara, Sol Kugelmass, and Atalia Winter. 1991. Cross-cultural and developmental trends in graphic productions. *Cognitive Psychology* 23 (4): 515–557.

Wagenaar, Willem A. 1972. Generation of random sequences by human subjects: A critical survey of literature. *Psychological Bulletin* 77 (1): 65–72.

Zitser, Ernest A. 2018. Raphael Lemkin and the Soviet propaganda poster collection at Duke university library. *Slavic and East European Information Resources* 19 (3–4): 242–262.

CHAPTER 9

Balancing Complexity and Simplification

Diagrams generally strive for a simplification of data and theories by presenting information in a format compatible with the limitations of human cognition. However, diagrams can for different reasons sometimes become overly complex, while in other cases the quest for simplicity might lead too far. In the age of information technology, it has become easier to generate extremely complex diagrams with the help of specialized data visualization software, to the detriment of the viewers. In some instances, overcomplexity is instead used as a conscious communication strategy. This is illustrated in Chap. 9, Section "Labyrinthic Chartjunk: Implication through Cognitive Overload", which discusses examples of "Labyrinthic chartjunk" such as political art, East German attempts to discredit the West German political system, and charts used by the QAnon movement to weave together conspiracy theories. Chapter 9, Section "Diagrammatic Reasoning in Quantum Theory", on diagrams in quantum physics, comments the discussion of whether Feynman diagrams should be interpreted as images or symbolic calculation devices. It also analyzes a new diagrammatic system that employs spatial and embodied cognition to simplify learning of quantum processes. The final chapter of the book before the conclusion, Chap. 9, Section "String Theory: Post-Modern Platonism", is dedicated to string theory, a proposed theory of everything where small strings are supposed to be the foundation of all other phenomena in the universe. Since string theory has evolved into a large research field despite

J. Holmén, *A History of Diagrams*, Palgrave Studies in the History of Science and Technology,
https://doi.org/10.1007/978-3-031-99691-7_9

little connection to empirical evidence, it illustrates important questions regarding the limitations of diagrammatic reasoning. The chapter includes a discussion of the pros and cons of our tendency to favor models of the world that appeal to the limitations and preferences of our cognition.

Labyrinthic Chartjunk: Implication Through Cognitive Overload

The digital revolution has enabled the construction of computer-generated network diagrams. These can often compress enormous amounts of data into a single diagram, such as diagrams of the structure of the internet.[1] However, often these diagrams are so overloaded with information that it is impossible to make any inferences from them—if there are any important patterns, they are hidden in the general chaos where everything seems to be connected to everything.

Such graphs might be considered the result of toying around with technology without a clear purpose of what information or interpretations the diagrams are supposed to transmit. Complex machine-made graphs displaying all available information might also be perceived as more objective and scientific than human-made simplified graphs, since no subjective middle step of selection is involved. They might appear to be what Peirce called indexes, signs directly linked to datasets, rather than human interpretations. Yet, since our mind cannot directly handle the complexity of such diagrams, the task of simplification and interpretation is not eliminated, but moved from the constructor to the observer of the diagram.

However, sometimes diagrams drawn by humans can reach a level of complexity that, although simpler than its computer-generated peers, still overloads the capacity of our working memory. One example is the huge and complex network diagrams that have been constructed as art. As the architecture of the Middle Ages reflected the diagrammatic structure of scholastic logic, art in the information age has sometimes adopted the form of data visualization. Some artists have created entirely abstract network diagrams for their aesthetic qualities, without basing them on any data.[2] Other artists visualize actual information. The most famous representative of that genre is probably Mark Lombardi, who in the second half of the 1990s penciled intricate diagrams of political and financial

[1] See, for example, Lima (2011).

[2] See, for example, Lima (2011, 223–244).

connections between different actors. Works such as *George W. Bush, Harken Energy, and Jackson Stephens, ca 1979–90* displayed crimes and conspiracies by analyzing the networks connecting politicians, criminals, and businessmen. Although these diagrams were artwork, they mapped real connections and could be used for analysis of the networks in question. It has been claimed that, following the 9/11 attacks, the FBI consulted Lombardi's drawing *BCCI-ICIC & FAB 1972–91* (4th version) for clues to the network of Osama bin Laden. Lombardi was influenced by the minimalist ideal of graphic design expressed by Edward R. Tufte. Another inspiration for Lombardi was Hans Haacke, a German artist active in the United States, who in 1971 portrayed shady business in the New York real estate sector.[3]

Later examples are found in the QAnon movement, which has since its emergence in the 2010s spread complicated conspiracy theories with the help of information visualizations.[4] A series of particularly elaborated diagrams have been produced by Dylan Louis Monroe, a former fashion designer and design school graduate who is running the website DeepStateMappingProject.com (Fig. 9.1).[5]

Although a firm believer in the QAnon cause, Monroe has admitted that the aesthetics sometimes takes priority over the information: "'When there's a gap somewhere I might not have anything to put there, but I'll add something in just to fill the gap to almost create a pattern.'" The artist Maura Brewer contrasts the art of Lombardi and Monroe to Tufte's ideal that every choice in the visualization should be made in the service of information, and that clarity should be achieved by eliminating "chartjunk." Brewer claims that Lombardi's and Monroe's graphs fail to provide coherent explanations, since they display the "undisciplined ambiguity" that Tufte despises. However, she sees the intent of these works as something entirely different than explanation: to express the anxiety of their makers. According to Brewer, these diagrams are "maps of the individual consumed by the system, swallowed up, lost in a sea of information—diagrams that are as ambiguous and undisciplined as reality itself."[6] In a similar psychological explanation, Matthew Hannah claims that the diagrams of the QAnon movement contribute to the members' apophenia, a

[3] Hobbs (2004).
[4] Hannah (2021).
[5] Monroe (n.d.).
[6] Brewer (2023).

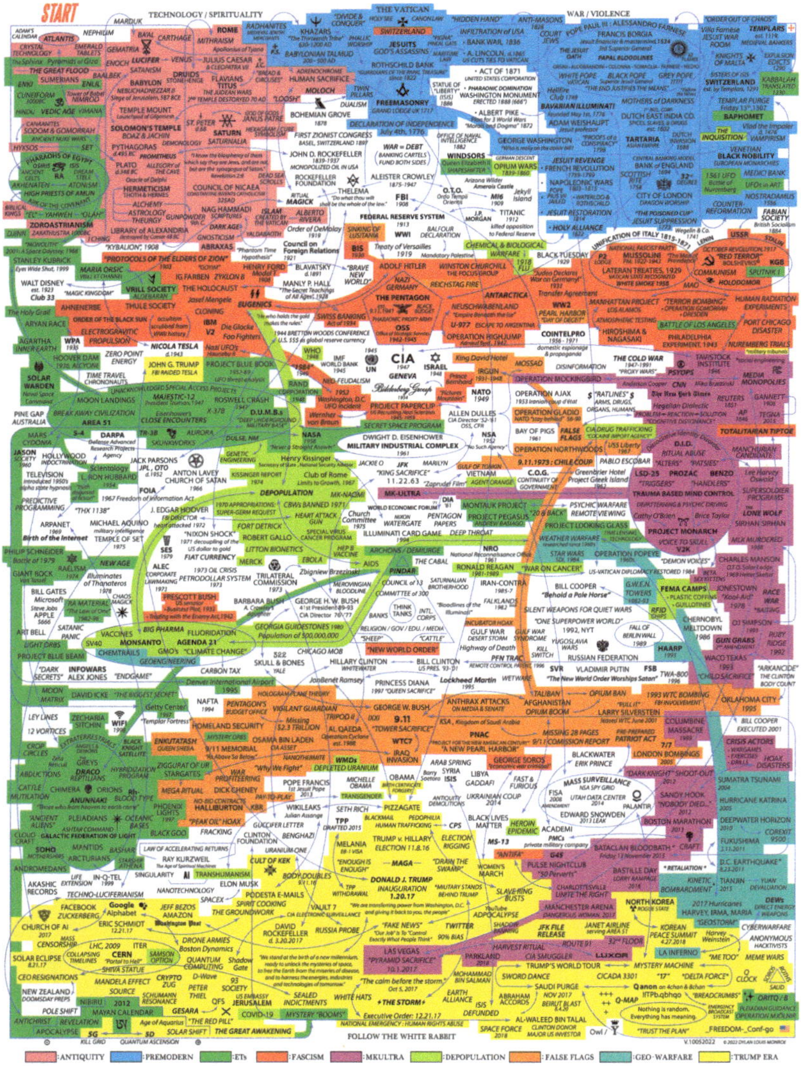

Fig. 9.1 This diagram of interwoven conspiracy theories within the QAnon movement shows an overwhelming number of connections. Dylan Louis Monroe. *Q-KEY*. 2018. Digital. 8.5 × 11 in. Deep State Mapping Project

psychological condition described by the schizophrenia expert Claus Konrad as an unmotivated seeing of connections accompanied by a feeling of abnormal meaningfulness.[7]

The ambiguity created by network diagrams with an overload of connections does indeed create a feeling that something is wrong in the system. However, that might be a conscious propaganda strategy or artistic choice rather than a reflection of the artist's mental state. For example, in civics textbooks from the German Democratic Republic during the Cold War, diagrams were drawn to illustrate, and discredit, the political system in West Germany. In Fig. 9.2, the West German parliament is a small box placed in the bottom right corner of the diagram. If we combine the primary metaphor "Control is Up" with the spatial agency bias of placing more powerful actors to the left in cultures writing from left to right, we can see that bottom right is the least powerful position in the whole diagram. The box on top of the diagram, representing the committee for German industry, is not centered but placed to the left, underscoring its power through spatial agency bias. In contrast, diagrams of the GDR's political system generally placed the parliament in the top left position of power.[8] Thus, the diagram constructor seems to have used space consciously for presenting their political message.

Complexity was used in the same calculated way. The weakness of the West German parliament vis-à-vis business interests is conveyed by a superfluously high number of links to the black box labeled "Forms of Influence." The web of connections alludes to a complicated plot through nefarious connections, in a strategy similar to that of Lombardi, Haacke, and Monroe. It is not impossible that the GDR textbook was inspired by the system-critical art of Haacke, since his work received some attention in the very years when the textbook was written. They are also visually similar, with crossing straight lines. Later versions of the same textbook diagram returned to a more simple, traditional textbook layout without unnecessary complexity.[9] The diagram's constructors seem to have carefully used principles of spatial cognition in order to achieve an optimal propaganda effect. The overcomplexity is applied only in part of a diagram that is otherwise clearly and logically structured. These methods are not

[7] Hannah (2021); Conrad (1958).

[8] For example, Redlow et al. (1971, 418). See Holmén et al. (2024) for more analysis of textbooks diagrams from the GDR and nine other countries.

[9] See, for example, Heuer (1979, 68); Paff (1983, 60).

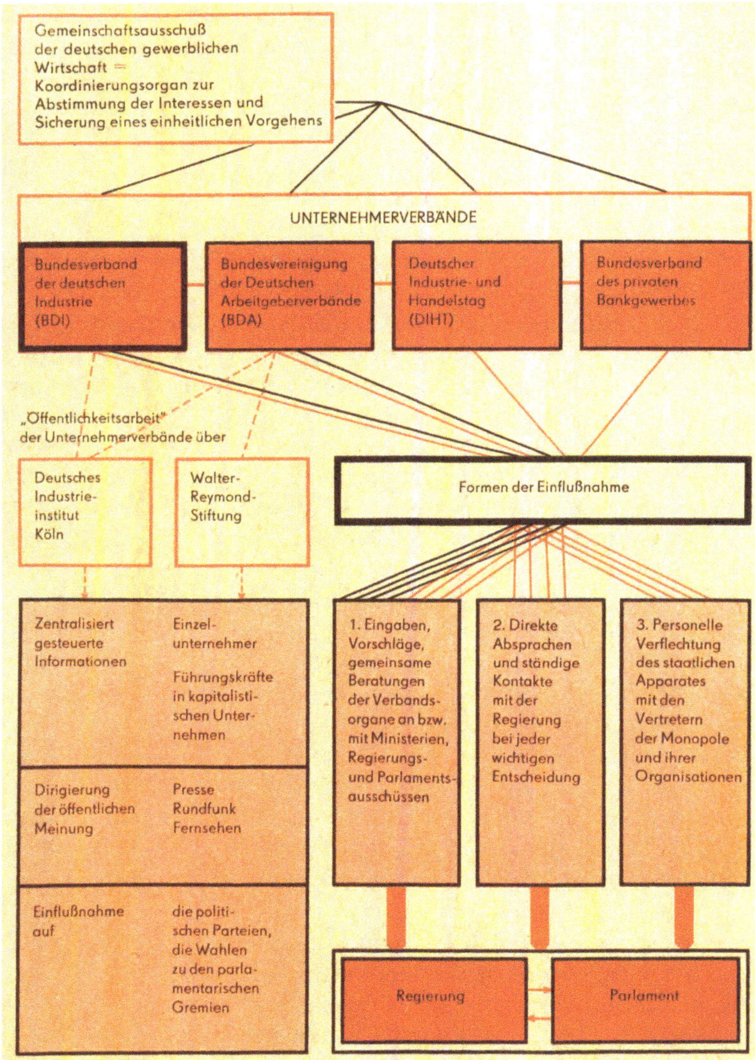

Fig. 9.2 In this East German diagram, the influence of business over politics in West Germany is emphasized by a superfluous amount of connections. (Heuer et al. 1973, 68)

unique to one-party states with large propaganda apparatuses. Jessica Hullman and Nick Diakopoulos have described how similar strategies of "visual noise" are used in present US politics by actors seeking to discredit the policies of their opponents.[10]

We can get a better understanding of this complex, overloaded, and conspiratorial genre of diagrams by comparing them to labyrinths rather than to ordinary diagrams. Mathematically, both labyrinths and network diagrams belong to the class of graphs.[11] However, they are constructed for entirely different purposes. The first labyrinthic shapes are found in southern Europe dating from around 4000 years ago, although they have older Neolithic precursors in the form of spirals and meanders. As many of the early spirals and labyrinths are simple rock carvings, it is difficult to determine their symbolism. However, some of them are found on the entrances to graves. That others have been "converted" to Christianity by later graffiti in the form of crosses suggests that they have been associated with religious meaning. These early European labyrinths have a distinct shape called the classical labyrinth.[12]

Since then, labyrinths, including classical ones, have appeared on all continents. Whether they have spread from a common Mediterranean origin or been invented independently is debated. The classical labyrinths found in Native American art are probably influenced by Spanish missionaries, and, in some cases, by unscrupulous antiques dealers who have "improved" Olmec pottery. However, other labyrinthine forms are found in pre-Columbian America, for example, among the Nasca lines in Peru.[13] Thus, it seems like labyrinths have been invented independently at least twice. The repeated occurrence of the labyrinth and the ease with which its classical form has spread around the world indicate that labyrinthic shapes are particularly appealing to aspects of the human mind. While grid patterns and hierarchical tree structures with a limited number of branches have a universal function as diagrams that order knowledge, confusing labyrinthic patterns function as metaphors for the unknown, complex, and mystic aspects of existence.

Of the Peircean triad diagram, metaphor, and image, it is the metaphorical aspect that dominates in labyrinths. In her study of ancient and

[10] Hullman and Diakopoulos (2011).
[11] Fraenkel (1970).
[12] Saward (2003, 36–39).
[13] Saward (2003, 68–79).

medieval labyrinths, Penelope Reed Doob identified three basic metaphorical functions of labyrinths: as signs of complex artistry, as signs of impenetrability or inexplicability, or as signs of a difficult process.[14] These functions of ancient labyrinths can also be applied to modern labyrinthic chartjunk found in artwork, propaganda, or conspiracy visualizations that are too complex to function as ordinary diagrams. They are not attempting to meaningfully clarify the structure of information as a traditional diagram. Instead, by causing a cognitive overload, they function as a labyrinthic metaphor for the complexity of the universe.

Diagrammatic Reasoning in Quantum Theory

Quantum physics is notoriously difficult and counterintuitive. Some diagrammatic representations were developed early on, such as the Feynman diagrams introduced in the late 1940s (Fig. 9.3), which illustrate interaction of subatomic particles. Similar diagrams were drawn earlier by Gregor Wentzel in 1938 and Ernst Stückelberg in 1941. However, Feynman's diagrams were superior in the sense that they were not mere illustrations of an idea but could also be used as aids for calculations, which has made them widespread tools in physics.[15]

There has been a philosophy of science debate regarding whether the Feynman diagrams should be considered a form of mathematical symbolic notation or pictorial representations of particle trajectories—or both.[16] However, this debate has overlooked the semiotic definition of diagrams as a category of signs in their own right, distinct from both images and symbols. In addition, the discussion on why the Feynman diagram has been so successful has touched upon explanations such as its similarity to real particle trajectories and its power of visualization.[17] Also this debate would be enlightened by the semiotic perspective of diagrams as icons that are similar to their objects not by exactly depicting them but by highlighting some of their most important traits.

With the exception of Feynman diagrams, research within quantum physics has predominantly relied on symbolic notation. However, in the

[14] Doob (2019, 64–91).

[15] Feynman (1949); Mehra (1994, 213, 272, 289–294, 577); Stückelberg (1941); Wentzel (1938).

[16] See, for example, Meynell (2008).

[17] Wüthrich (2010, 7–10).

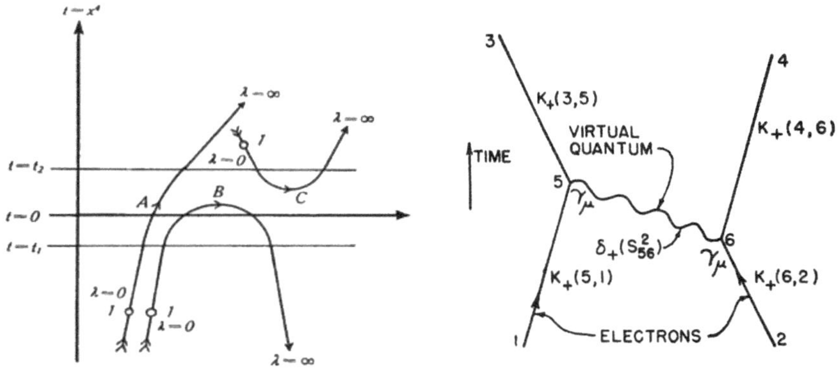

Fig. 9.3 While Stückelberg's diagram (left) was an illustration of his view on particle interactions, Feynman's diagram (right) can also be used for practical calculations. That particles seem to travel backward in time in some of these diagrams has contributed to the reluctance for considering them depictions of actual particles. In Feynman diagrams, straight lines are electrons (or positrons if they travel back in time) and wavy lines are photons (quanta) that the electrons exchange with each other. (Stückelberg 1941, 390; Feynman 1949, 772)

first decade of the twenty-first century a new school of researchers have emerged who argue that quantum theory is best presented and understood in diagrammatic form. This development has been summarized in a textbook by Bob Coecke and Aleks Kissinger.[18] Their diagrams are an extension of the notation for abstract tensor systems introduced by Roger Penrose in 1971.[19] However, they also recognize influence from other disciplines where flow charts have been used. The style of the new quantum physics diagrams developed in the first decade of the twenty-first century reveals that much of the early work was done at computer science departments.[20]

The textbook on quantum diagrams begins with a general description of diagrammatic reasoning using charts where "wires" represent flow of data or "stuff" between "boxes" that represent processes, a good definition of flow charts.[21] Coecke and Kissinger are so convinced by the

[18] Coecke and Kissinger (2017, 6).
[19] Penrose (1971).
[20] Coecke and Kissinger (2017, xiv, 16–17).
[21] Coecke and Kissinger (2017, 9–10, 29–30).

advantages of diagrammatic reasoning that they claim their 800-page book on quantum theory is accessible to readers without physics or computer science background, as long as they know some basic high school mathematics.[22]

The diagrams of Coecke and Kissinger not only utilize visuospatial layouts to explain quantum physics, but also appeal to sensorimotor manipulations (Fig. 9.4). Quantum theory has proven difficult to grasp because of its defiance of our accustomed principles for how objects should behave in space. However, the new diagrammatic representations can be described as an attempt to make quantum physics understandable by simplifying it into a familiar shape, which makes it possible to literally get a grip of the problems at hand through embodied cognition.

Coecke and Kissinger compare their diagrammatic system to a high-level language for computer programming, while they describe traditional symbolic notation for quantum theory as "strings of complex numbers," comparable to the strings of binary numbers in low-level machine code. They claim that the complexity of the older symbolic notation was the reason it took so long to discover basic features of quantum theory, such as 60 years for quantum teleportation. In diagrammatic form, by contrast, quantum teleportation is "staring you in the face." They also note that their diagrammatic language for quantum theory has found applications for modeling meaning in natural language, for doing proofs in formal logic, in control theory, and in modeling electric circuits, and predict that diagrammatic reasoning will become increasingly important in the sciences in general.[23]

While Coecke and Kissinger have undoubtedly developed a system that efficiently utilizes spatial cognition in the teaching of complex quantum processes, they might be overstating the fundamental novelty of their approach. The strings of complex numbers they mention probably refer to the matrices that are often used in quantum physics. However, these are used to represent vectors, which can, whether or not diagrams are used, undoubtedly be thought of in spatial terms.[24] And as this volume illus-

[22] Coecke and Kissinger (2017, 19).

[23] Coecke and Kissinger (2017, 9–10, 29–30).

[24] For the use of matrices to represent vectors, see, for example, this relatively old quantum theory textbook: Bates (1961).

Thanks to the clever shorthand for the transpose, it now looks as if we can slide boxes across the ∪-shaped wires and ∩-shaped wires:

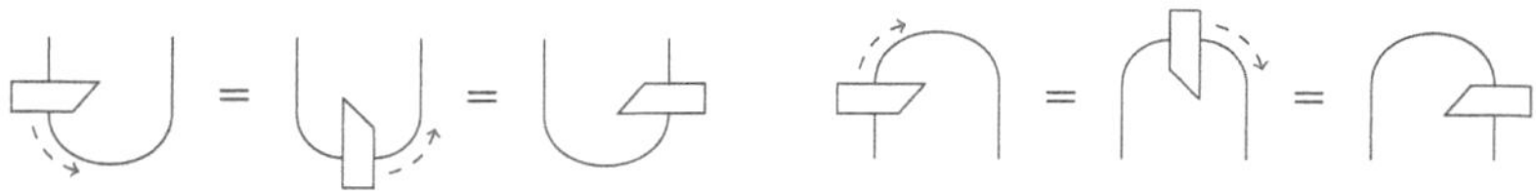

Thus we can slide boxes around on wires like beads on a necklace. For example, this is a valid equation:

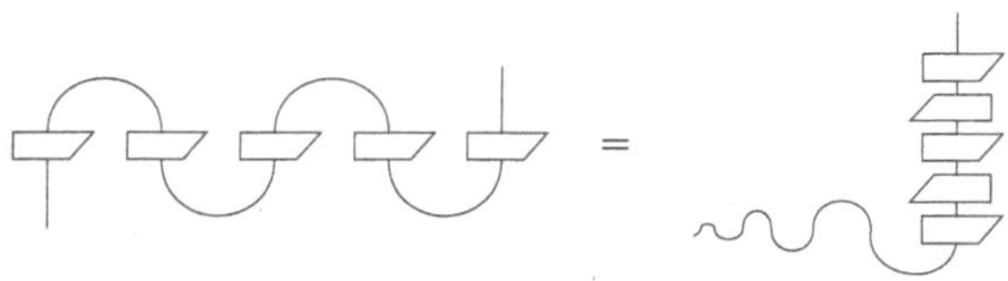

This is also very easy to remember. Just think about what would happen if we took the definition of the transpose (4.15) and yanked on the wires:

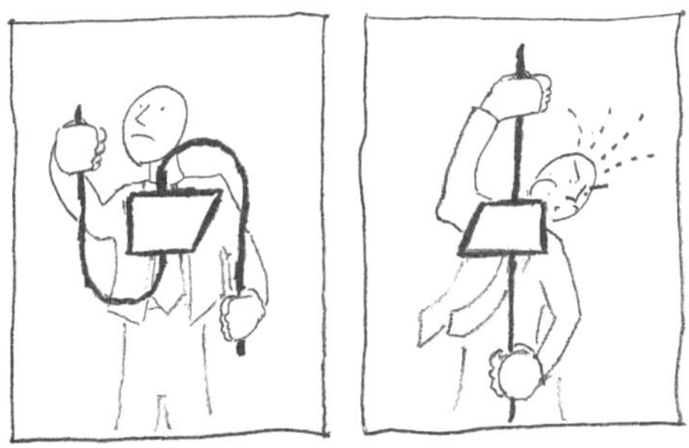

Fig. 9.4 Embodied cognition applied to quantum physics equations. (Coecke and Kissinger 2017, 98)

trates, diagrammatic reasoning has throughout the ages had a central place in the sciences. However, an increased awareness of the power of spatial cognition, combined with conscious efforts to optimize it, such as those of Coecke and Kissinger, might bring additional advantages.

String Theory: Post-Modern Platonism

String theory has been an established branch of theoretical physics for half a century. By 2016, Joseph Conlon estimated that about 20,000 research papers had been published in the field.[25] This means that it is extremely difficult to get an overview of all variants of string theory. Unlike, for example, the quantum theory discussed in the previous chapter, string theory is completely unverified by experimental observations, and neither has it produced any predictions that are likely to be possible to verify within the near future. In a way, string theory could be described as advanced mathematics of small strings that are attached to branes. It is unquestionable that string theory has made real contributions to mathematics. If they were merely viewed as geometrical diagrams, these strings and branes would have been left out of this book, as other examples of pure mathematical geometry. However, string theory has the ambition to represent the fundamental elements that constitute our universe—and possibly other universes as well. Thereby, it becomes an interesting example of the possibilities, limitations, and risks associated with diagrammatic reasoning.

String theory originates in the early 1970s, when several physicists independently of each other suggested that elementary particles should be understood as strings. These strings stretched when they gained energy and contracted when it was lost. Their various states of vibration corresponded to the multitude of different particles that had been discovered in the preceding decades (Fig. 9.5).[26]

This strange "zoo of particles" was unsatisfactory to quite a few physicists. Compared to many other diagrams, it is still quite simple, and forms a cognitively palatable hierarchy of dichotomies and tripartite divisions. However, the recent history of science had led scientists to believe that for each level of nature they revealed, they would encounter less and less complexity. The complexity of chemical compounds had been explained by the elements on the periodic table, and their building blocks turned out to be even simpler: protons, neutrons, and electrons. Against that background, it was a disappointment that the building blocks of neutrons and protons were so complicated. Thus, they welcomed a theory that could simplify

[25] Conlon (2016, xi).
[26] Smolin (2006, 103).

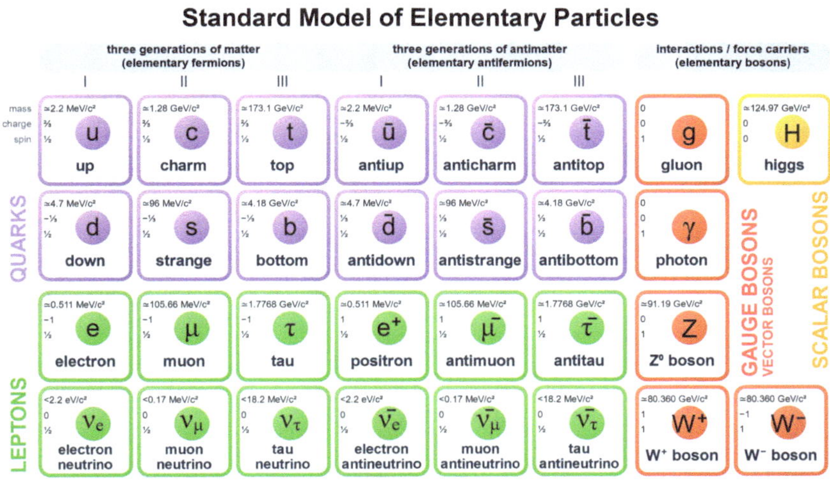

Fig. 9.5 The standard model's complexity has made many physicists search for a simpler and more "beautiful" alternative. (Drawn by Cush, based on 2018 data from the Particle Data Group)

the complex diagram in Fig. 9.5 by replacing it with a single string-shaped particle.

In 2010, the mathematician Shing-Tung Yau, who has made important contributions to string theory, likened the methods of string theorists to Plato's attempt to deductively conclude that the universe consisted of four elements, each constituted by a perfect geometrical body:

> Plato correctly surmised that symmetry ought to lie at the heart of any credible description of nature. For if we are ever to produce a real theory of everything—in which all the forces are unified and all the constituents obey a handful (or two) of rules we'll need to uncover the underlying symmetry, the simplifying principles from which everything else springs.[27]

To Yau, Plato's idea that the world we see is a mere reflection of an underlying geometry that we might not see related closely to his own work with the Calabi-Yau manifolds (Fig. 9.6). According to developments of string theory from the mid-1980s, in every point in our

[27] Yau and Nadis (2010, xix).

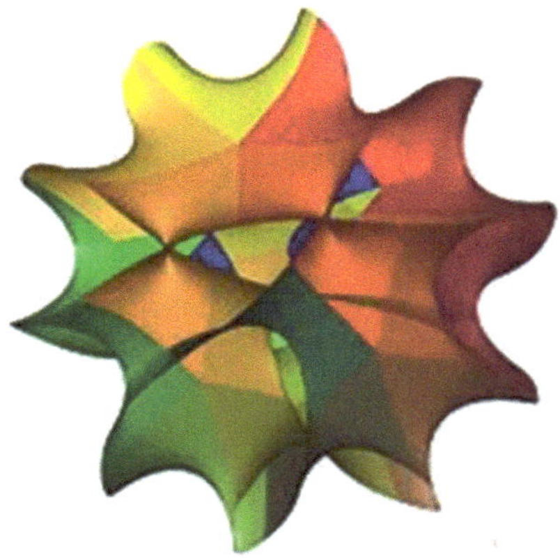

Fig. 9.6 Two-dimensional cross section of a six-dimensional Calabi-Yau manifold. Yau likens it to the solids that were the cosmos' fundamental building blocks in Plato's philosophy (Fig. 4.1). Generated by Andrew J. Hanson, Indiana University

four-dimensional space–time, there is a wrapped up six-dimensional Calabi-Yau manifold. Because of the wrapping and small scale, these dimensions are not visible to us.[28]

Yau ended his 2010 book by stating that "Plato used the solids named after him . . . in much the same way that string theory relies on Calabi-Yau manifolds, though we hope the results will be better this time."[29] In 2010, there was still hope that the Large Hadron Collider (LHC) would find evidence of super-symmetry partners, which are predicted by some versions of string theory. Yau even hoped that higher dimensions could manifest as particles in the collider.[30] However, despite being operational since 2008, the LHC has revealed nothing of the sort. Through the discovery of the Higgs boson, the LHC has confirmed the ugly, messy, standard model that string theory was supposed to replace, but besides that, not much has been found.[31]

In addition, string theory has not been able to fulfill the hopes of simplifying physics. In order to comply with observations, string theory has been modified to such an extent that it contains more constants than the theories it is replacing. Although it got rid of the zoo of particles, it created zoos of dimensions (usually 10) and of universes, of which ours is

[28] Yau and Nadis (2010, xix, 138).
[29] Yau and Nadis (2010, 326).
[30] Yau and Nadis (2010, 135, 231–235).
[31] Hossenfelder (2018, 5, 13).

only one of many variants. Theoretical physicists have also developed a zoo of different string theories, and so far, there is no way of finding out which of them that would be correct. Since string theory has not been verified by any experiments, its appeal rests entirely on its mathematical properties. For 20 years, there has been a debate in theoretical physics and philosophy of science regarding whether the methods of string theory can be considered science at all. Critics such as Håkon Enger claim that string theory does not correspond to any ideals of what constitutes accepted scientific practices.[32]

However, string theory's defenders, such as Richard Dawid, claim that string theory is an example of a new paradigm of theory assessment. According to this paradigm, theories can be validated by an absence of alternatives, by using mathematics that have worked before and by discovering unexpected connections.[33] Similarly, Joseph Conlon defends string theory by claiming that although it does not fulfill the scientific gold standard of experimental verifiability, it achieves the silver standard of making predictions for theoretical problems whose answers are not known in advance. According to him, the fact that string theory has proven to be of value in mathematics and in fields of physics that it was not intended for, such as quantum gravity, proves that it is not an "idle argument."[34] This line of thought rests on the tacit assumption that there is a common underlying mathematical truth behind different aspects of reality.

Since experimental evidence is lacking, the debate for and against string theory has been connected to the concept of beauty.[35] Some critics, such as Peter Woit, argue that the supposedly beautiful string theory is not very beautiful at all, in the sense of mathematical simplicity found in the theories of, for example, Newton, Einstein, or Dirac.[36] The beauty of string theory seems to lie in the simple diagrammatic image of a vibrating string underlying all phenomena in the universe; when one attempt to express this idea in mathematical symbolism, ugly complications arise.

Sabine Hossenfelder has challenged the concept of beauty on a more fundamental level. In her book *Lost in Math*, with the original title *Das hässliche Universum* (The ugly universe), she claimed that the search for

[32] Enger (2003).
[33] Dawid (2015).
[34] Conlon (2016, 234).
[35] Conlon (2016, 221).
[36] Woit (2006, 198–207).

beauty in theoretical physics in the second half of the twentieth century has been misguided. She believes that one of the root causes of the problem is that heroes such as Dirac and Einstein toward the end of their careers began to advocate beauty in theory building, in reaction to the supposed ugliness of developments in quantum physics. However, Hossenfelder claims that this was not how Dirac and Einstein actually conducted science in their earlier more productive years. On the contrary, they achieved their greatest accomplishments while working in intimate interaction with experimental data.[37]

It can be argued that a good theory, which is perceived as beautiful, has to be easy to accommodate by the human cognitive apparatus, with its limited capabilities. Candidates for good theories should therefore be simple and preferably easy to visualize, which makes it possible for us to grasp them with our spatial cognition. The case of string theory illustrates that if only the underlying diagrammatic visualization of the theory is simple and elegant enough, the theory can gain tremendous traction, despite convoluted mathematics and a lack of experimental evidence. Hossenfelder gives a similar example from the nineteenth century, when "vortex theory" was supposed to be a beautiful underlying explanation to the recently discovered but ugly laws of electromagnetism.[38]

One of the criteria for a beautiful theory in physics is that it only contains numbers that are close to one, such as 2, 3, ½, or 1/3. Very small or very large numbers are called "fine-tuned" and are supposed to be a sign that the theories have been tampered with to forcedly fit the data. Another criterion is symmetry, which means that a theory contains patterns or repetitions, and that some factors remain constant when others change. This allows it to be summarized in an elegantly economic way.[39] One example is the speed of light, which is constant in all systems.

The symmetry concept had its major breakthrough in physics with the mathematician Hermann Weyl, who in 1928 used group theory to analyze phenomena in quantum physics.[40] Group theory is still an active research field within physics, where symmetry has a well-defined meaning derived from the geometrical origin of the methods used. However, in 1952, Weyl published a popularized book on symmetry in nature, which contributed

[37] Hossenfelder (2018, 21–30).
[38] Hossenfelder (2018, 30–31).
[39] Hossenfelder (2018, 25, 37).
[40] Weyl (1928, 220–220, 236–243).

to a wide proliferation of the concept. Within a year of its publication, Weyl's book had inspired the double helix interpretation of DNA.[41] However, when the symmetry concept spread from a narrower mathematical environment, its meaning also became more generalized. In his 1952 book, Weyl equated it with beauty and harmony.[42] He referred back to Kepler's attempt to fit the orbits of the planets into Platonic solids (Fig. 4.1), and claimed that

> [w]e still share his belief in a mathematical harmony of the universe. It has withstood the test of ever widening experience. But we no longer seek this harmony in static forms like the regular solids, but in dynamic laws.[43]

It is perhaps telling that Yau and Weil, who have drawn parallels between modern physics and Platonic solids, are both primarily mathematicians rather than physicists.

It has been argued that the human sense of beauty is linked to the underlying structure of laws of nature, since the sense of beauty has provided an evolutionary benefit in finding true explanations. Hossenfelder's counterargument to this is that humans have not been exposed to, for example, elementary particle physics during our evolutionary history, and therefore it is no guarantee that the beauty criteria we have developed in the everyday world will be effective in other fields.[44]

The reasons symmetry is perceived as beautiful have been discussed within cognitive archaeology. Explaining why stone knappers 500,000 years ago chose to make their hand axes symmetrical, Wynn and Coolidge argue that

> [s]ymmetry is an example of a "good Gestalt"—a regular form that by itself elicits pleasure. The vertebrate visual cortex has cell groups that are attuned to symmetrical patterns in the visual field, especially vertical or horizontal bilateral symmetries. When these cell groups activate, they send neural impulses forward in the brain to cells in the pleasure/reward system. There are good evolutionary reasons for this. Food (e.g., fruit) and danger (e.g., lions) have symmetrical features, and rapid detection had a selective payoff.

[41] Weyl (1952); Darvas (2007, vii).
[42] Weyl (1952, 3).
[43] Weyl (1952, 77).
[44] Hossenfelder (2018, 151).

> Similarly, a symmetrical object, including a tool, is more pleasing to the eye than an amorphous lump.[45]

That the smallest building blocks of the universe would follow the same basic spatial layout as food and predators is possible, but not self-evident. The case of string theory highlights the limitations associated with the simplification provided by diagrammatic thinking. First, it is far from certain that the fundamental laws of the universe will satisfy the constraints that the limitations of our cognitive capabilities put on what is to be considered a good, beautiful theory. Nothing prevents reality from being more complex or uglier. Second, even if it would be the case that true theories are also simple and beautiful, we could still not from that deduce that any simple and beautiful theory is also true. And, there is no guarantee that we can identify a true theory among a multitude of simple and beautiful theories by evaluating which of them is the most simple and beautiful. Third, we can only reach certainty regarding whether the criterion of beauty and simplicity was a useful one after the fact, when we have found and verified an underlying "theory of everything." However, given the human propensity for simple, beautiful theories, there is a high likelihood that a complicated theory of everything would not be recognized as such, but that the hunt for a better, underlying theory would continue until a form was found simple enough to be grasped by human consciousness with its limited working memory and its propensity for spatial cognition.

A theory is not a scale model of the world, but an interface between observational data and the human consciousness, where important aspects of the world are outlined, diagrammed, in a sufficiently simple way to be made comprehensible to us. Understood as such an interface, the theory must fulfill constraints provided both by the world in the form of correspondence with observations and by the human consciousness in the form of diagrammatic simplicity. Since all theories are affected by both these constraints, similarities that we find among older theories, such as beauty and symmetry, might stem from any of them or from both. I would argue that the history of diagrams, when viewed from the perspective of its cognitive foundations, illustrates that the beauty criterion does stem from human consciousness and its dependence on spatial analogies and simplification. Of course, this does not exclude the possibility that beauty could

[45] Wynn and Coolidge (2022, 54).

also reflect a fundamental symmetry in nature. However, the suspicion that beauty has leaked into earlier theories from human diagrammatic thinking undermines the reasoning that since earlier theories have been beautiful, new ones will necessarily meet the same criteria, and that we could therefore find the fundamental theory of nature by searching for beauty.

In fact, the opposite argument could also be made: since we have a cognitive apparatus adapted to rapidly detect symmetrical shapes and consider them beautiful, we have relatively early in our history identified the theories and laws of nature that adhere to this aesthetic ideal. The same applies to our preference for simplicity and small numbers. Chances are therefore that, when the beautifully symmetric and simple theories have all been discovered, the remaining undiscovered ones might be ugly, messy, and complicated.

References

Bates, D. R. 1961. *Quantum theory, vol. 1: Elements.* New York: Academic Press.

Brewer, Maura. 2023. Chartjunk. *X-traonline.* 13 March. https://www.x-traonline.org/online/chartjunk. Accessed 25 May 2023.

Coecke, Bob, and Aleks Kissinger. 2017. *Picturing quantum processes: A first course in quantum theory and diagrammatic reasoning.* Cambridge: Cambridge University Press.

Conlon, Joseph. 2016. *Why string theory?* Boca Raton, FL: CRC Press.

Conrad, Klaus. 1958. *Die beginnende Schizophrenie: Versuch einer Gestaltanalyse des Wahns.* Stuttgart: Georg Thieme Verlag.

Darvas, György. 2007. *Symmetry: Cultural-historical and ontological aspects of science-arts relations: The natural and man-made world in an interdisciplinary approach.* Basel: Birkhäuser.

Dawid, Richard. 2015. *String theory and the scientific method.* Cambridge: Cambridge University Press.

Doob, Penelope Reed. 2019. *Idea of the labyrinth from classical antiquity through the Middle Ages.* Ithaca, NY: Cornell University Press.

Enger, Håkon. 2003. String theory and the scientific method. Håkon Enger homepage. https://henger.org/~henger/publ/mnvit-essay.pdf. Accessed 29 July 2024.

Feynman, Richard P. 1949. Space-time approach to quantum electrodynamics. *Physical Review* 76 (6): 769–789.

Fraenkel, Aviezri S. 1970. Economic traversal of labyrinths. *Mathematics Magazine* 43 (3): 125–130.

Hannah, Matthew N. 2021. A conspiracy of data: Qanon, social media, and information visualization. *Social Media + Society* 7(3). https://doi.org/10.1177/20563051211036064.

Heuer, Kurt, Lothar Schimmelpfennig, and Erwin Swillus. 1973. *Staatsbürgerkunde: Volk und Wissen. Klasse 9.* Berlin: Volkseigener Verlag.

Heuer, Kurt, Ruth Swillus-Baumann, Lothar Diersch, Christina Muller, and Erwin Swillus. 1979. *Staatsbürgerkunde. Volk und Wissen. Lehrbuch für Klasse 9.* Berlin: Volkseigener Verlag.

Hobbs, Robert. 2004. *Mark Lombardi: Global networks.* In *Mark Lombardi: Global networks*, ed. Robert Carleton Hobbs, 11–49. New York: Independent Curators International.

Holmén, Janne, Ann-Sofie Jägerskog, Daniel Schumann, and Malin Tväråna. 2024. Connecting the nation: Diagrams of unity and diversity in civics textbooks. In *Diagrams in civic education: Visuospatial models of society in textbooks and teaching*, ed. Janne Holmén, Ann-Sofie Jägerskog, Daniel Schumann, and Malin Tväråna, 11–59. Palgrave Macmillan.

Hossenfelder, Sabine. 2018. *Lost in math: How beauty leads physics astray.* New York: Basic Books.

Hullman, Jessica, and Nick Diakopoulos. 2011. Visualization rhetoric: Framing effects in narrative visualization. *IEEE Transactions on Visualization and Computer Graphics* 17 (12): 2231–2240.

Lima, Manuel. 2011. *Visual complexity: Mapping patterns of information.* New York: Princeton Architectural Press.

Mehra, Jagdish. 1994. *The beat of a different drum: The life and science of Richard Philips Feynman.* Oxford: Clarendon Press.

Meynell, Letitia. 2008. Why Feynman diagrams represent. *International Studies in the Philosophy of Science* 22 (1): 39–59.

Monroe, Dylan Louis. n.d. Deepstatemappingproject.com https://deepstatemappingproject.com/. Accessed 4 January 2024.

Paff, Werner, Martin Döbler, Christina Müller, Horst Riechert, Günter Schippel, and Georg Warkotsch 1983. *Staatsbürgerkunde: Volk und Wissen. Klasse 9.* [Civics: People and Knowledge. Grade 9.] Berlin: Volkseigener Verlag.

Penrose, Roger. 1971. Applications of negative dimensional tensors. In *Combinatorial mathematics and its applications: Proceedings of a conference held at the mathematical institute, Oxford, from 7–10 July, 1969*, ed. Dominic J. A. Welsh, 221–244. London: Academic Press.

Redlow, Götz, Helmut Frommknecht, Matthäus Klein, Eberhard Fromm, Herbert Hörz, Gottfried Stiehler, and Harald Wessel. 1971. *Einführung in den dialektischen und historischen Materialismus. Lehrbuch für Staatsbürgerkunde Klassen 11 und 12* [Introduction to dialectical and historical materialism: Textbook for civics classes 11 and 12]. Berlin: Dietz Verlag.

Saward, Jeff. 2003. *Labyrinths and mazes: The definitive guide to ancient and modern traditions*. London: Gaia Books.
Smolin, Lee. 2006. *The trouble with physics: The rise of string theory, the fall of a science, and what comes next*. Boston: Houghton Mifflin.
Stückelberg, Ernst C. G. 1941. Remarque a propos de la création de paires de particules en théorie de relativité. *Helvetica Physica Acta* 14 (372): 588–594.
Wentzel, Gregor. 1938. Schwere Elektronen und Theorien der Kernvorgänge [Heavy electrons and theories of nuclear processes]. *Die Naturwissenschaften* 26 (18): 273–279.
Weyl, Hermann. 1928. *Gruppentheorie und Quantenmechanik*. Leipzig: Hirzel.
Weyl, Hermann. 1952. *Symmetry*. Princeton, NJ: Princeton University Press.
Woit, Peter. 2006. *Not even wrong: The failure of string theory and the continuing challenge to unify the laws of physics*. London: Jonathan Cape.
Wüthrich, Adrian. 2010. *The genesis of Feynman diagrams*. Dordrecht: Springer.
Wynn, Thomas Grant, and Frederick L. Coolidge. 2022. *An introduction to evolutionary cognitive archaeology*. New York: Routledge.
Yau, Shing-Tung, and Steven J. Nadis. 2010. *The shape of inner space: String theory and the geometry of the universe's hidden dimensions*. New York: Basic Books.

CHAPTER 10

Conclusions

In this concluding part of the book, Chap. 10, Section "The History of Diagrams and the Waves of Knowledge" discusses the relationship between diagrams and the great waves of knowledge transmission in human history. Chapter 10, Section "Paradigmatic Shifts in Geometry, Philosophy, and Technology" is concerned with how diagrams have been affected by major shifts in philosophy, geometry, and technologies of image production. The analysis focuses on how new diagrammatic forms have developed in interplay with preexisting historical conventions around diagrams and in conformity with cultural, primary, and innate principles of spatial cognition. Chapter 10, Section "History, Science, and Cognition" reflects on what historical studies can contribute to the cognitive sciences and a discussion on the relationship between diagrammatic thinking and historical change. This final chapter ends with a discussion of how our preference for certain spatial models affects our worldviews, and why we can benefit from awareness of this fact.

J. Holmén, *A History of Diagrams*, Palgrave Studies in the History of Science and Technology,
https://doi.org/10.1007/978-3-031-99691-7_10

The History of Diagrams and the Waves of Knowledge

This book has followed the historical development of diagrams, in the sense of spatial representations of non-spatial relationships. When such diagrams first appeared in their fully developed form, more than two millennia ago, they already built on a long development. Their oldest roots are found in evolutionary history, which developed the biological foundations of the spatial cognition that enables diagrammatic thinking. It has been suggested that the capacity to orient oneself spatially in the landscape developed first, where after the same neural system for spatial reference frames was first reused to make sense of the locations of parts of objects, and later, through analogy, to represent abstract concepts.

Thereafter followed a period, which we can call the prehistory of diagrams, during which humans utilized various spatial aids in order to enhance their cognition. Although gestures have left no archaeological evidence, it is clear that they have a role in speech similar to that of diagrams in text, and that they are probably older than vocal speech. After the appearance of speech, oral cultures used physical locations in terrain and architecture as memory loci that indexed songs containing their entire cultural knowledge. Similar mnemonic functions could also be performed by portable objects. The most advanced of them, such as the Inca quipus, can be described as a mix of proto-writing and proto-infographics. With the invention of writing, texts became spatially structured in ways that aided cognition, for example, in lists and tables. The invention of complete systems of writing also meant that the figurative, diagrammatic, and symbolic aspects of communication became separated, after having been closely intertwined in proto-writing systems. This paved the way for the emergence of diagrams as a distinct form of illustrations.

The history of diagrams is characterized by several periods of intensive diagram construction, followed by long periods dominated by reuse of previously established forms. Golden Ages of diagram creation seem to occur when large amounts of new knowledge need to be systematized and explained, or when old knowledge needs to be explained to large new groups. Since diagrams help humans structure information, it is not surprising that during periods where the rate of accumulation of new information has been particularly high, new forms of diagrams have also been invented. For example, during the twelfth and thirteenth centuries, the Latin scholars of Europe were inundated with translations of rediscovered

Greek texts, which forced them to develop diagrammatic tools for structuring textual information.

However, this was only part of a larger pattern, since both the western and eastern parts of the Eurasian landmass saw a new great flowering of diagrams in the period from approximately 800 to 1300 CE, characterized by an increase in the use of diagrams and in the development of new forms.

Another wave followed from the 1700s, with the introduction of data visualizations. Although scholars such as Nicholas Oresme in the 1300s made theoretical advances that could have been used to represent data in diagrams, they were not used for this purpose until the 1700s. Then, data visualization took off as the data amassed by modern societies became the object of human spatial thinking. This wave culminated in the second half of the 1800s. It has been described as a golden age of data visualization, when nation-building statistical agencies experimented with new diagrammatic forms. The first half of the 1900s was a period with fewer inventions, but for the first time, statistical graphics now reached the majority of the population through popular media, such as newspapers and school textbooks. A revival for development of diagrammatic forms began in the 1960s, this time accompanied by computerized methods for visualization and data analysis.

In order to fully understand such periods of intense diagram creation we need to reflect on the relationship between the cognitive and historical aspects of diagramming. Cognitive anthropology has shown that humans everywhere tend to classify knowledge into hierarchical systems, folk taxonomies. These have a limited number of levels and branches at each node, a structure adapted to human working memory capacity. Undoubtedly, this phenomenon also underlies the ubiquitous attempts by scholars to construct, for example, tree diagrams. However, while universal principles of cognition can explain these similarities, they cannot explain the differences and changes that appear. Although the tendency to classify is found everywhere, it is not always made the center of science and philosophy, such as in the European High Middle Ages, when classificatory subdivisions permeated scholasticism. Instead, we can understand such periods as a response to a particularly severe overload of new knowledge and information, such as translations of texts in twelfth-century Europe or some centuries earlier in the Arab world or of newly described plants from the sixteenth through the eighteenth centuries.

Bar charts, time series diagrams, and other data visualizations rest on cognitive principles such as the analogy between line length and number, that are either innate or primary metaphors, acquired by everyone early in life. However, it was only in the late 1700s, when waves of numerical data flooded the world as a result of improved measurement techniques and changed scientific ideals, that the need for externalization and oversight spurred the invention of diagrams for data visualization.

An intriguing question is why golden ages for diagram creation in geographically distant civilizations have coincided in time, such as the 800s to 1300s in the Arab, Latin, and Chinese worlds. That these periods were characterized by information overload does not suffice as an explanation, since we still need to explain why this overload would take place across the globe simultaneously. Its primary cause might be the long cycles in traditional agricultural societies. Periods of good harvest increased the population and enabled a larger share of it to engage in activities other than food production. A new larger class of literate people became educated in complicated theories earlier reserved for a smaller intellectual elite, and this called for diagrammatic simplification. The influx of information in the form of translations was also enabled by growing literate groups; texts do not just flow between cultures unless someone is there to read and translate them.

The periods of diagram development in Europe in the Middle Age and the late 1700s to 1800s took place after inflection points in population increase around the years 1100 and 1700. In Asia, the medieval inflection point took place earlier, as did the period of diagram development.[1] Although recent research indicates that historic climate trends have not coincided all over the world, both Europe and China experienced a medieval warm period, which through increased harvests might explain their simultaneous population growth.[2] Of course, this synchronicity is not unique to diagrams, since they merely form part of a general strong development of culture and science during the periods in question, when growing populations stimulated trade, commerce, and industry.

Compared to the libraries of modern and even early modern times, the absolute number of books that entered Europe in the high Middle Ages was not impressive. However, considering the severe limitations of the knowledge infrastructure and information technology of the day, the relative overload that the system experienced no doubt rivaled that of the Enlightenment and the present. As a reaction to this, and to the pressure

[1] See, for example, Livi Bacci (2017, 25), and Our World in Data.

[2] Mann et al. (2009); Hao et al. (2020).

from rising numbers of students, medieval Europe invented, developed, or intensified the use of institutions such as universities, methods such as scholasticism, practices such as diagramming, materials such as paper, and, eventually, technologies such as printing.

The number of diagrammatic forms is the highest in the beginning of the cycles of diagram innovation, where after they begin to plateau and decrease. There is an initial period of experimentation, during which many new diagrams are created, but some of them eventually come to dominate and suppress other forms. Diagrams have an advantage in this "survival of the fittest" if they can function as efficient analogies for concepts and information by being well adapted to cognitive principles, and if they are easy to reproduce given the technologies of the day. Thereafter follows a period of very little or no innovation in abstract diagrams. For example, following the intense Middle Age innovation of diagrams, in the European Renaissance and the scientific revolution, scientific texts were either accompanied by figurative illustrations, such as images of naturalia and scientific equipment, or tabulated measurement data. The invention of print was not associated with any new diagrams, but reproduced and perfected some of the medieval types with new technical means, primarily tree diagrams printed with bracket types.

However, successful diagrams often also connect to preexisting systems of knowledge, values, and conventions. Entirely new diagrammatic forms might be difficult for the public to interpret, and are often accompanied by long explanatory texts at their first appearances, such as was the case with Priestley's timelines. When some new diagrams become more common and recognized than others, they form schemas that can be reused with limited cognitive load. Thereby, they get an additional reproductive advantage over others forms, which eventually fade into the background.

Paradigmatic Shifts in Geometry, Philosophy, and Technology

The development of diagrams has been closely intertwined with the general development of philosophy and geometry. For most of human history, knowledge production has relied on direct observations through the senses, inference from these observations and on deduction. Beside exceptions such as astronomy, the core of science was not analysis of measurements but of language. It was mainly in logical analysis of language and texts that diagrams played a role in the premodern world. For millennia, the overwhelming majority of diagrams[3] were constructed in order to illustrate logical and linguistic operations, relations, distinctions, and classifications. The interpretation of the sources to the first diagrams is complicated, since it is generally difficult to know whether diagrams in later copies existed in the original. However, there are strong indications that diagrams were used in classical Greece, and likely in India and China around the same time as well.

In classical Greece and in the later cultures that it inspired—the Byzantine, the Latin in Western Europe, and the Arab—the deductive methods of geometry became an ideal for philosophy in general, and the diagrams of Euclidean geometry became blueprints for diagrams in all fields. Ancient Indian and Chinese philosophy had less of a distrust for sense perception, which was also reflected in the diagrams produced in these cultures. For example, the Buddhist philosophical diagram Hetucakra was concerned with the conditions for correct inference from perception, rather than the deductive logic of the similar Western square of opposition. Likewise, Chinese tree diagrams analyzed particular texts, and not general abstract principles as medieval European tree diagrams often did. This difference is also reflected in the fact that Greek geometry worked with lettered diagrams that illustrated general principles, while geometrical diagrams in most other cultural areas represented specific examples. Still, that the spatial forms of the diagrams were similar across all cultural areas, such as the tree diagram, reflects that they were founded on the same universal cognitive principles.

It is also important not to overemphasize the philosophical differences between ancient cultures, and be aware of the bias in what has been preserved. There is reason to believe that many ancient Greek and Indian diagrams have been lost, and that what remains is not a representative selection. For example, the successors of Ancient Greece were highly interested in preserving the deductive logic tradition of Plato, but less so regarding the writings of skeptical philosophers. In addition, diagrams from medieval Europe have seen much more scholarly interest in the last decades than what is the case with India and China. Still, even for a single Indian logician like Dignaga, who has attracted a fair amount of research, it has been possible to interpret his primary interests, expressed, for

[3] As defined in this volume: nonfigurative and not illustrating spatial relations, thereby excluding, for example, maps.

example, in the Hetucakra, as being either metaphysical, epistemological, linguistic, or rooted in skeptical philosophy.[4] Given the multitude of interpretations regarding a single individual, generalization about century-long trends in large cultural areas has to be done with great caution.

The next large revolution in the history of diagrams, the development of statistical graphics in the 1700s, was not merely a result of improved measurement techniques that created increased amounts of data, but also of an intellectual revolution: analytical geometry created the conception of a quantified space that could be used to illustrate measurements. The development of mathematical functions and the coordinate system endowed space with mathematical properties. Today, the understanding of space as a cubic lattice is so intuitive that we have difficulties grasping that it is a relatively recent invention. However, this conception of space has managed to achieve the status of common sense since it resonates with fundamental cognitive principles. The preference for horizontal or vertical lines in human perception, as well as our grid cell-based neurological system of orientation, favors the construction of grid structures, and the innate association of line length and number facilitates that we ascribe mathematical properties to the lines in such structures.

David Dunér has described how, during the early modern period and the scientific revolution, techniques were invented for distributed cognition that extended the human mind. Among them he includes analytical geometry.[5] Dunér's analysis focuses on tools for deductive reasoning, and the period's strong deductive ideals might indeed initially have counteracted the development of such an empirical field as data visualization, which remained a marginal phenomenon throughout the eighteenth century. However, analytical geometry also offered a template for how to visualize data, which facilitated a revolution in inductive reasoning in the nineteenth and twentieth centuries.

The two persons most responsible for enabling this development were Leonard Euler and Johann Heinrich Lambert, both attached to the Berlin Academy in the mid-1700s. While Euler completed the mathematical development of the function concept, Lambert was the first to apply it to scientific data by creating time series diagrams of evaporation. Another inspiration for Lambert was musical notation, a forerunner of time series diagrams that he discussed at length in his writings on semiotics.

[4] Hayes (1988, 32–35).
[5] Dunér (2016, 2019).

The emergence of a distinction between quantitative and qualitative properties was another intellectual development important for the introduction of data visualizations. This originated in the seventeenth-century discussion on primary and secondary qualities, which was influenced by Greek atomism. The primary qualities, such as size and shape, were considered more fundamental than the secondary qualities, which were perceived through the senses and were considered an effect of the primary qualities. When a property such as speed could now be considered quantitative, and thereby measurable, it made sense to plot it in the quantified space provided by analytical geometry.

The field of music was early in developing forerunners of modern diagrams for data visualization. This might be due to the fact that in music the relationships between primary and secondary qualities, such as the length of a string and pitch, were relatively easy to untangle. What we today would describe as the quantitative aspects of music was therefore of direct relevance for ancient, medieval, and Renaissance thinkers in their search for its qualities, and they were represented in diagrams that anticipated the quantitative turn in diagramming that was to take place in the 1700s.

Marshall McLuhan claimed that the invention of printing with movable types in Europe created the foundation for three-dimensional spatial cognition by exposing readers to the grid pattern of the printed page. That idea has been thoroughly refuted in this volume. There is ample evidence that spatial cognition is much older than printing, in all likelihood as old as humanity. However, there were some important differences between the printing technologies in Europe and China that affected the ability for scholars to include and update diagrams in their works. The earlier invention of printing in China also made use of movable type, but only for large editions. The larger number of characters made movable type printing unfeasible for low-volume print, such as most scientific texts, which were instead produced using block printing. Using movable type, European printers could produce not only text but also diagrams, allowing for inexpensive production and easy editing of low- and medium-volume scientific texts. Tree diagrams printed with bracket types provided a flexible means of accommodating and analyzing the growing amounts of observations and data collected during the scientific revolution, for example, in branches of science such as botany.

It is not difficult to find other instances where technologies of image production have had an influence on what diagrams have been produced.

For example, it is possible that the shape of the bamboo strips used in Chinese chronological tables inspired their tabular design. It has also been claimed that codex pages invited to a conscious spatial layout, which included diagrams, in a way that was not the case with earlier papyrus scrolls. From the late eighteenth century, advances in copperplate printing and a general proliferation of the technology made it easier to reproduce high-quality diagrams. In the twentieth century, computerization facilitated new forms of diagrams and simplified moving graphics.

However, regardless of these changes in image production, the basic structure of diagrams has developed remarkably little. Good diagrams need to be adapted to the principles of human cognition, and those have remained unchanged. Clearly delineated diagrams with a structure simple enough to be grasped by human consciousness look similar throughout the ages.

History, Science, and Cognition

This book has used theories from the cognitive sciences in order to analyze the history of diagrams. Historians are generally users and interpreters rather than producers of theories. Hopefully, cognitive scientists can find inspiration from the history of diagrams, and discover fruitful parallels to the problems in their own field from the examples from other disciplines and time periods presented in this volume. However, historical inquiry might also more directly contribute to the cognitive sciences by testing theories against a different body of empirical evidence.

For example, it has been debated whether spatial metaphors in diagrams are mere visualizations of verbal metaphors. The history of the left–right political spectrum in Sweden suggests that this is not the case. There, for most of the 1900s, the spatial model of the political system in tables and diagrams, with the conservative parties placed to the left and the radical to the right, was contrary to the verbal metaphor where the radicals were called left and the conservatives right. Since these metaphors were the absolute opposites of each other, one cannot possibly be considered a visualization of the other.

This study has also confirmed the ubiquity of what cognitive anthropologists call folk taxonomies, hierarchical structures of knowledge with a limited number of branches at each node. We can find them not only in different tree diagrams and tables, but also, for example, in the quarterings of heraldry and the layout of keyboards. The cognitive load theory assumes that the limitation of the human consciousness to a maximum of seven items of information is related to the exponential growth in possible permutations. However, there is no evidence that the brain actually follows this random generate and test hypothesis,[6] and cognitive anthropology can give us another explanation, also related to exponential growth, for why the capacity of working memory is so limited.

A system of classification with five levels, and up to five branches at each node—which cognitive anthropology suggests is typical for folk taxonomies, and which this study has confirmed is applicable also to historical sources—can contain 3905 nodes, which seems to be sufficient for most purposes. If we raise the number of levels and branches to the magical number seven, close to the limit of working memory capacity, we can squeeze 960,799 nodes into the system. This might help explain why human working memory has never evolved beyond the seven-item limit: it is simply more than enough for all practical purposes, as long as the information is stored in the hierarchical structure of a tree diagram. Rather than the magical number seven being dictated by more being too much, it might stem from the fact that less is good enough. All human traits, whether physical such as body length or weight, or mental such as working memory capacity, are trade-offs between benefits and drawbacks. Since five bits of working memory is good enough in a hierarchical structure of knowledge, the selection pressure for more capacity is limited. In addition, the process of creating hierarchical categories constitute an important part of cognition, since it facilitates associations and analogies between related items.

Thus, there are instances where this study can shed light on debates within the cognitive sciences, but more important than their direct contribution is that they confirm the general principle that historical sources can provide clues to the workings of human cognition. This has huge implications in an age where more and more historical sources are getting digitized, and the development of AI and machine learning rapidly advances the possibilities for large-scale advanced analyses of them. The costs in time and money for asking questions regarding, for example, spatial cognition, to the entire corpus of preserved texts and images might soon be lower than that of conducting laboratory experiments with a handful of test subjects. In addition, a history-based approach to cognitive science, drawing from the total experience of global history, might also provide

[6] Rather, it seems as if we are unable to randomize: Wagenaar (1972); Figurska et al. (2008).

more generalizable answers than experimental studies with test subjects representing a demography that is extremely narrow in time and space.[7]

Historians have been slower than archaeologists and anthropologists, for example, in applying a cognitive science perspective to their field. One contributing factor for this might be that historians generally seek to explain social, cultural, and political change. Since the basic principles of human cognition have changed little in historical time, it may seem unlikely that they can underlie processes of change. However, it can be argued that understanding the continuous factors in human history is also important in explaining change. After all, change is a pervasive, or at least frequently recurring, component of history. Cognitive history can help explain both why historical change occurs and the recurring patterns we see in these processes of change. For example, that stress induces us to think in terms of simple dichotomies and borderlines might help us understand conflicts and revolutions, which are important drivers of historical change.

The interaction between cognition and culture is also crucial in understanding processes of change. Although there is a universal cognitive foundation for why space is used as an analogy for abstract relations in all societies, how space is understood and which diagrammatic models are used differ between cultures and historical periods. For example, the turning point in the 1700s is associated with a new quantified approach to space and a functional understanding of geometry, which was accompanied by diagrammatic inventions such as timelines and statistical graphics. With time, this enabled new ways of viewing society and the universe, such as the idea of progress as a function of time, the quantification of social hierarchies, time as a fourth dimension, and time travel. While some of these changes might be dead ends or curiosities, others have become backbones in our worldviews and thereby catalyzers for scientific development and social transformation.

In addition, historians need to understand spatial cognition in order to investigate how historical agents, intent on bringing about or impeding change, have used diagrams as tools of persuasion. Even the most neutral-looking diagram will always contain metaphorical elements, since the spatial dimensions themselves are associated with values. For example, cognitive principles such as the primary metaphor "Control is Up" and

[7] Although psychologists aim for generalizable truths, they often do so by investigating a selective population of psychology students at major research universities. Medin and Atran (1999).

spatial agency bias lead us to perceive items that are placed on top or to the left in a diagram as more powerful than those in the bottom or to the right, for example, in social pyramids or charts of political systems.

Another strategy of persuasion is to create overly complex diagrams, in this volume labeled "labyrinthic chartjunk," that illustrate that everything is connected to everything. These figures function less as diagrams, tools for detecting and presenting relevant traits in a seemingly chaotic universe, and more as labyrinths that metaphorically represent this chaos. They do so by turning the principles of diagramming on their head, and consciously overloading the figures with more information than we can process. This creates a feeling of something being wrong, which can be used to discredit the portrayed system.

An improved understanding of these and other principles of spatial cognition can help us analyze diagrams, and also the diagrammatic aspects and spatial configurations in figurative images. This will enable historians to use new sources, or find new interpretations based on old ones. Today, the historian's toolbox of source evaluation is heavily skewed toward written sources, and instruments for analyzing spatial representations will be a valuable complement in the study of historical development.

With the help of diagrams, we can persuade not only others but also ourselves. Diagrams are interfaces between a complex world and our limited cognitive ability to understand this world. Characteristics of our cognition, such as limitations of working memory and a preference for symmetry, recur in spatial models from widely disparate domains. This has led many to search for connections between, for example, musical octaves and octaves in other fields, or between different forms of cyclicality, such as in Elliott wave theory. As he expanded his wave patterns from finance to all of nature and culture, Hermann Weyl did the same with his ideas of symmetry, which originated from group theory. The realization that such patterns and similarities might come from the cognitive lens through which we observe the world rather than from the world itself lends perspective on such attempts at grand theory building.

Diagram helps our understanding by allowing us to draw analogies to a new domain from one that is familiar to us, space. However, analogies might be false and misleading. One example of this is the invention of the timeline. Since you can move forward or backward on a timeline, either with your eyes or, as Mark Twain's children, physically, the analogy creates the impression that we can move backward in time. Although this idea has undoubtedly renewed fiction and perhaps stimulated thinking in theoretical physics, it is still not an analogy that helps us understand the unique characteristics of time.

The most successful scientific diagrams are those that, once constructed, can reveal relations that were not known to the original diagram constructors. Maybe the most prominent is the periodic table of elements, which has survived as a tool and an inspiration through several shifts of scientific paradigms. The periodic table of elements illustrates how a spatial reference frame, gradually improved through empirical work, can help transform the labyrinthic and mysterious nature into a comprehensible diagram.

An opposite example is provided by the string-diagrams that form the foundation of string theory. It was intended to replace the empirically derived standard model of particle physics, which seemed too complicated and was described as a zoo of particles. The idea that a single string could account for all this diversity by vibrating in different ways was alluring in its diagrammatic simplicity. String theory has been impossible to prove empirically, but in the absence of experiments or observations, defenders of the theory have resorted to arguments such as the beauty and symmetry of the theory and the diagrams associated with it. In this paradigm, one characteristic of a beautiful theory is that it should not contain very large numbers. However, it seems highly likely that this criterion is derived from the cognitive limitations of the human brain rather than from any fundamental principle of the universe. Nature might be more complicated than we would like it to be, and getting seduced by the beauty of simple diagrams might hinder us from trying to construct more complex but in other ways more appropriate models of the universe.

When necessary, researchers might therefore be entirely motivated to construct labyrinths rather than diagrams. However, to not accidentally step over the line between these two forms of visual representations one needs to be aware of what differentiates them. For this purpose, it is advisable to familiarize oneself with the limitations of human cognition, as they constitute the demarcation line between the diagram and the labyrinth. An important role of science and education is to find relevant spatial analogies that can help us turn labyrinths into diagrams. However, a second and equally important role is to raise awareness of the fact that these diagrams are simplifications imposed on us by our cognitive limitations. By being aware that our view of the world is influenced by a preference for certain spatial models, we can hopefully avoid some of the pitfalls that our appetite for diagrammatic simplicity might lure us into. The best way of creating such an awareness is to study the cognitive history of humanity.

References

Dunér, David. 2016. Science: The structure of scientific evolutions. In *Human lifeworlds: The cognitive semiotics of cultural evolution*, ed. David Dunér and Göran Sonesson, 229–266. Frankfurt am Main: Peter Lang.

Dunér, David. 2019. The axiomatic deductive ideal in early modern thinking. In *Cognitive history: Mind, space, and time*, ed. David Dunér and Christer Ahlberger, 3–34. Berlin: De Gruyter Oldenbourg.

Figurska, Małgorzata, Maciej Stańczyk, and Kamil Kulesza. 2008. Humans cannot consciously generate random numbers sequences: Polemic study. *Medical hypotheses* 70 (1): 182–185.

Hao, Zhixin, Wu Maowei, Yang Liu, Xuezhen Zhang, and Jingyun Zheng. 2020. Multi-scale temperature variations and their regional differences in China during the medieval climate anomaly. *Journal of Geographical Sciences* 30 (1): 119–130.

Hayes, Richard P. 1988. *Dignaga on the interpretation of signs*. Dordrecht: Kluwer.

Livi Bacci, Massimo. 2017. *A concise history of world population*. 6th ed. Hoboken, NJ: John Wiley & Sons.

Mann, Michael E., Zhihua Zhang, Scott Rutherford, Raymond S. Bradley, Malcolm K. Hughes, Drew Shindell, Caspar Ammann, Greg Faluvegi, and Fenbiao Ni. 2009. Global signatures and dynamical origins of the little ice age and medieval climate anomaly. *Science* 326 (5957): 1256–1260.

Medin, Douglas L., and Scot Atran. 1999. Introduction. In *Folkbiology*, ed. Douglas L. Medin and Scot Atran, 1–15. Cambridge, MA: MIT Press.

Wagenaar, Willem A. 1972. Generation of random sequences by human subjects: A critical survey of literature. *Psychological Bulletin* 77 (1): 65–72.

References

Alsted, Johann Heinrich. 1630. *Johannis-Henrici Alstedii Encyclopaedia Septem Tomis Distincta*. Herborn: Friedrich Hulsius.

Ambasciano, Leonardo. 2017. What is cognitive historiography, anyway? Method, theory, and a cross-disciplinary decalogue. *Journal of Cognitive Historiography* 4 (2): 136–150.

Ambasciano, Leonardo, Mathias Clasen, and Darryl Jones. 2020. Horror studies between humanistic interdisciplinarity and scientific consilience: A conversation with Darryl Jones and Mathias Clasen. *Journal of Cognitive Historiography* 5 (1–2): 159–179.

Ammon, Otto. 1895. *Die Gefellschaftsordnung und ihre natürlichen Grundlagen: Entwurf einer Sozial—Anthropologie zum Gebrauch für alle Gebildeten, die sich mit sozialen Fragen befassen* (The social order and its natural foundations: A sketch of a social anthropology for the use of all educated people who deal with social questions). Jena: Gustav Fischer.

Amos, Percy A. 1912. *Process of flour manufacture*. London: Longmans.

Anacker, Stefan. 2005. *Seven works of Vasubandhu: The Buddhist psychological doctor*. Delhi, India: Motilal Banarsidass.

Arai, Teishirō. 1950. *Sakata senjutsu shōkai*. (The Sakata method in detail). Tokyo: Kabushiki no kenkyū-sha.

Archibald, J. David. 2009. Edward Hitchcock's pre-darwinian (1840) 'tree of life'. *Journal of the History of Biology* 42 (3): 561–592.

Archibald, J. David. 2014. *Aristotle's ladder, Darwin's tree: The evolution of visual metaphors for biological order*. New York: Columbia University Press.

J. Holmén, *A History of Diagrams*, Palgrave Studies in the History of Science and Technology,
https://doi.org/10.1007/978-3-031-99691-7

Arethas of Caesarea and Michael Share. 1994. *Aretha Kaisareias Scholia Eis Tēn Porphyriou Eisagōgēn Kai Tas Aristotelous Katēgorias: Arethas of Caesarea's Scholia on Porphyry's isagoge and Aristotle's categories: (Codex Vaticanus Urbinas Graecus 35): A Critical Edition, vol. 1*. Athens: Akad. Athēnōn.

Atkinson, Ti-Grace. 1974. *Amazon Odyssey: The first collection of writings by the political pioneer of the women's movement*. New York: Links Books.

Atran, Scott. 1990. *Cognitive foundations of natural history: Towards an anthropology of science*. Cambridge: Cambridge University Press.

Ayala, María de los Ángeles. 1998. La obra narrativa de Enrique Gaspar: El Anacronópete (1887). In *Sociedad de Literatura Española del Siglo XIX. Coloquio, Del Romanticismo al Realismo*, ed. Luis F. Díaz Larios and Enrique Miralles, 403–411. Barcelona: Universitat de Barcelona.

Backhouse, Roger E., and Yann Giraud. 2010. Circular flow diagrams. In *Famous figures and diagrams in economics*, ed. Mark Blaug and Peter Lloyd, 221–229. Cheltenham: Edward Elgar.

Baetzhold, Howard G. 2018. 'Well, My Book Is Written—Let It Go…': The making of a Connecticut Yankee in King Arthur's court. In *A Connecticut Yankee in King Arthur's court: A Norton critical edition. An Authoritative Text. Contexts. Criticism*, ed. Henry B. Wonham, 445–477. New York: Norton.

Barnhart, Robert K., and Sol Steinmetz. 1988. *The Barnhart dictionary of etymology*. New York: H.W. Wilson Co.

Baron, Margaret E. 1969. A note on the historical development of logic diagrams: Leibniz, Euler and Venn. *Mathematical Gazette* 53 (384): 113–125.

Barsanti, Giulio. 1992. *La scala, la mappa, l'albero. Immagini e classificazioni della natura fra sei e ottocento*. Firenze: Sansoni.

Bates, D. R. 1961. *Quantum theory, vol. 1: Elements*. New York: Academic Press.

Bellezza, Francis S. 1996. Mnemonic methods to enhance storage and retrieval. In *Memory: Handbook of perception and cognition*, ed. Robert A. Bjork and Elisabet Ligon Bjork, 345–380. San Diego: Academic Press.

Bellhouse, David R. 2023. *The flawed genius of William Playfair: The story of the Father of Statistical Graphics*. Toronto: University of Toronto Press.

Bender, John B., and Michael Marrinan. 2010. *The culture of diagram*. Stanford, CA: Stanford University Press.

Bendix, Reinhard. 1978. *Kings or people: Power and the mandate to rule*. Berkeley: University of California Press.

Beniger, James R., and Dorothy L. Robyn. 1978. Quantitative graphics in statistics: A brief history. *The American Statistician* 32 (1): 1–11.

Benson, Elizabeth P. 1975. The quipu: 'Written' texts in ancient Peru. *The Princeton University Library Chronicle* 37 (1): 11–23.

Bent, Ian D., David W. Hughes, Robert C. Provine, Richard Rastall, Anne Kilmer, David Hiley, Janka Szendrei, Thomas B. Payne, Margaret Bent, and Geoffrey Chew. 2014. Notation. *Grove Music Online*. Oxford: Oxford University Press.

https://www.oxfordmusiconline.com/grovemusic/display/10.1093/gmo/9781561592630.001.0001/omo-9781561592630-e-0000020114?rskey=PZblv0&result=3. Accessed 20 February 2023.

Berger, Carol. 1981. The hand and the art of memory. *Musica disciplina* 35:87–120.

Bertin, Jacques. 1983. *Semiology of graphics: Diagrams, networks, maps*. Madison: University of Wisconsin Press.

Best, Phillip J., and Aaron M. White. 1999. Placing hippocampal single-unit studies in a historical context. *Hippocampus* 9 (4): 346–351.

Bickerton, Derek, and Eörs Szathmáry. 2011. Confrontational scavenging as a possible source for language and cooperation. *BMC Evolutionary Biology* 11 (1): 261.

Biggs, Norman E., Keith Lloyd, and Robin J. Wilson. 1986. *Graph theory 1736–1936*. Oxford: Oxford University Press.

Bigotti, Fabrizio, and David Taylor. 2017. The pulsilogium of Santorio: New light on technology and measurement in early modern medicine. *Societate şi politică* 11 (2): 53–113.

Bird-David, Nurit. 2019. Dis/working with diagrams: How genealogies and maps obscure nanoscale worlds (a hunter-gatherer case). *Social Analysis* 63 (4): 43–62.

Bissell, Chris C. 2007. The Moniac: A hydromechanical analog computer of the 1950s. *IEEE: Control Systems Magazine* 27 (1): 69–74.

Blasius, Jörg, and Andreas Schmitz. 2015. History of correspondence analysis. In *International encyclopedia of the social and behavioral sciences*, ed. James D. Wright, 2nd ed., 6–11. Amsterdam: Elsevier.

Blaug, Mark, and Peter Lloyd. 2010. Introduction. In *Famous figures and diagrams in economics*, ed. Mark Blaug and Peter Lloyd, 1–23. Cheltenham: Edward Elgar.

Bloom, Jonathan M. 2001. *Paper before print: The history and impact of paper in the Islamic world*. New Haven: Yale University Press.

Blount, Benjamin G. 2011. A history of cognitive anthropology. In *A companion to cognitive anthropology*, ed. David B. Kronenfeld, Giovanni Bennardo, and Victor C. Munck, 11–29. Chichester: John Wiley & Sons.

Boberg, Johan. 2020. *Scientifically minded: Science, the subject and Kant's critical philosophy*. Uppsala: Uppsala University.

Boiteau, Timothy W., Cameron Smith, and Amit Almor. 2020. Rightward directional bias in art produced by cultures without a written language. *Laterality* 25 (2): 165–176.

Bolton, G. M. 1983. *Testing in selection decisions*. London: NFER-Nelson.

Bonner, Anthony. 2007. *The art and logic of Ramon Llull: A User's Guide*. Leiden: Brill.

Börjesson, Mikael, Donald Broady, and Mikael Palme. 2020. 'In our school we have students of all sorts'. Mapping the space of elite education in a seemingly egalitarian system. In *Researching elites and power: Theory, methods, analyses*, ed. Francois Denord, Mikael Palme, and Bertrand Réau, 179–191. Cham: Springer Nature.

Boster, James S. 2005. Categories and cognitive anthropology. In *Handbook of categorization in cognitive science*, ed. Henri Cohen and Claire Lefebvre, 92–116. Amsterdam: Elsevier.

Boulding, Kenneth. 1973. Foreword. In *Image and environment: Cognitive mapping and spatial behavior*, ed. Roger M. Downs and David Stea, vii–xi. Chicago: Aldine.

Bouras-Vallianatos, Petros. 2022. Diagrams in Greek medical manuscripts. In *The diagram paradigm: Cross cultural approaches*, ed. Jefferey F. Hamburger, David J. Roxburgh, and Linda Safran, 287–329. Washington, DC: Dumbarton Oaks.

Bourdieu, Pierre. 2002. Habitus. In *Gotik. Arkitektur. Skolastik. Habitus*, ed. Vagn Lyhne, 143–172. Århus: Klim.

Boutell, Charles. 1900. *English Heraldry*. 6th ed. Boston: Page.

Bradley, John J. 2008. When a stone tool is a dingo: Country and relatedness in Australian Aboriginal notions of landscape. In *Handbook of landscape archaeology*, ed. Bruno David and Julian Thomas, 633–637. Walnut Creek, CA: Left Coast Press.

Brand, T. E., and A. J. Sherlock. 1972. VK'+ V'K. Venn diagrams or Karnaugh maps, but not both? *Mathematics in School* 1 (7): 4–7.

Braun, Marta. 1992. *Picturing time: The work of Etienne-Jules Marey (1830–1904)*. Chicago: University of Chicago Press.

Bray, Francesca. 2007. Introduction: The powers of *Tu*. In *Graphics and text in the production of technical knowledge in China: The Warp and the Weft*, ed. Francesca Bray, Vera Dorrofeeva-Lichtmann, and Georges Métailié, 1–79. Leiden: Brill.

Bray, Francesca, Vera Dorrofeeva-Lichtmann, and Georges Métailié, eds. 2007. *Graphics and text in the production of technical knowledge in China: The Warp and the Weft*. Leiden: Brill.

Brewer, Maura. 2023. Chartjunk. *X-traonline*. 13 March. https://www.x-traonline.org/online/chartjunk. Accessed 25 May 2023.

Broberg, Gunnar. 1983. Homo sapiens: Linnaeus's classification of man. In *Linnaeus, the man and his work*, ed. Tore Frängsmyr, Gunnar Broberg, and Sten Lindroth, 156–194. Berkeley: University of California Press.

Broberg, Gunnar. 1990. The broken circle. In *The quantifying spirit in the 18th century, vol. 7*, ed. Tore Frängsmyr, J. L. Heilbron, and Robin E. Rider, 46–71. Berkeley: University of California Press.

Bullynck, Maarten. 2010. Johann Lambert's scientific tool kit: Exemplified by the measurement of humidity. *Science in Context* 23 (1): 65–89.

Burch, George E., and Nicholas P. DePasquale. 1990. *A history of electrocardiography*. San Francisco: Norman.
Byrne, Patrick H. 1997. *Analysis and science in Aristotle*. New York: SUNY.
Campbell, Stephen R. 2004. On the concepts of quantity and quality in the history of Western thought. Online Submission. *The Annual Meeting of the American Educational Research Association*. https://eric.ed.gov/?id=ED495204
Candea, Matei. 2019. On visual coherence and visual excess: Writing, diagrams, and anthropological form. *Social Analysis* 63 (4): 63–88.
Carlgren, Fredrik. 2024. *Skattebaseffekter av värnskattens avskaffande*. Stockholm: Svenskt näringsliv.
Carothers, John Colin. 1959. Culture, psychiatry, and the written word. *Psychiatry* 22 (4): 307–320.
Carruthers, Mary J. 2008. *The book of memory: A study of memory in medieval culture*. 2nd ed. Cambridge: Cambridge University Press.
Cassirer, Ernst. 1955. *The philosophy of symbolic forms, vol. 2: Mythical thought*. New Haven, CT: Yale University Press.
Celaya, Juan de. 1525. *Expositio … in primum tractatum Summularum magistri Petri Hispani*. Brunello.
Chajes, J. H. 2020. The kabbalistic tree. In *The visualization of knowledge in medieval and early Europe*, ed. Marcia Kupfer, Adam S. Cohen, and J. H. Chajes, 449–473. Turnhout: Brepolis.
Charmantier, Isabelle. 2011. Carl Linnaeus and the visual representation of nature. *Historical Studies in the Natural Sciences* 41 (4): 365–404.
Châtelet, Gilles. 2000. *Figuring space: Philosophy, mathematics, and physics*. Dordrecht: Kluwer.
Chatterjee, Anjan. 2011. Directional asymmetries in cognition: What is left to write about? In *Spatial Dimensions of Social Thought*, ed. Anne Maas and Thomas W. Schubert, 189–210. Berlin: De Gruyter Mouton.
Christensen, J. Cornelius. 1912. *University business administration*. Topeka: Kansas state agricultural college.
Christensen, Thomas. 2014. *River of ink: Literature, history and art*. Berkeley: Counterpoint.
Cicero, Marcus Tullius, H. Rackham, and E. W. Sutton. 2014. *De Oratore*. Cambridge, MA: Harvard University Press.
Clagett, Marshall. 1959. *The science of mechanics in the Middle Ages*. Madison: University of Wisconsin Press.
Clagett, Marshall. 1968. *Nicolas Oresme and the geometry of qualities and motions*. Madison: University of Wisconsin Press.
Clark, Andy. 1997. *Being there: Putting brain, body, and world together again*. Cambridge, MA: MIT Press.
Clark, Andy, and David Chalmers. 1998. The extended mind. *Analysis* 58 (1): 7–19.

Coecke, Bob, and Aleks Kissinger. 2017. *Picturing quantum processes: A first course in quantum theory and diagrammatic reasoning*. Cambridge: Cambridge University Press.

Cohen, Henri, and Claire Lefebvre. 2005. Bridging the category divide. In *Handbook of categorization in cognitive science*, ed. Henri Cohen and Claire Lefebvre, 2–15. Amsterdam: Elsevier.

Colp, Ralph Jr. 1974. The contacts between Karl Marx and Charles Darwin. *Journal of the History of Ideas* 35 (2): 329–338.

Conlon, Joseph. 2016. *Why string theory?* Boca Raton, FL: CRC Press.

Conrad, Klaus. 1958. *Die beginnende Schizophrenie: Versuch einer Gestaltanalyse des Wahns*. Stuttgart: Georg Thieme Verlag.

Constantinescu, Alexandra O., Jill X. O'Reilly, and Timothy E. J. Behrens. 2016. Organizing conceptual knowledge in humans with a gridlike code. *Science* 352 (6292): 1464–1468.

Cooperrider, Kensy, Dedre Gentner, and Susan Goldin-Meadow. 2016. Spatial analogies pervade complex relational reasoning: Evidence from spontaneous gestures. *Cognitive Research: Principles and Implications* 1 (1): 1–17.

Costigan-Eaves, Patricia, and Michael Macdonald-Ross. 1990. William Playfair (1759–1823). *Statistical Science* 5 (3): 318–326.

Couzin, Robert. 2021. *Right and left in early Christian and medieval art*. Leiden: Brill.

Creese, David E. 2010. *The monochord in ancient Greek harmonic science*. Cambridge: Cambridge University Press.

Crisp, Ryan Patrick. 2003. Consanguinity and the Saint-Aubin genealogies. *Haskins Society Journal: Studies in Medieval History* 14:105–115.

D'Andrade, Roy G. 1995. *The development of cognitive anthropology*. Cambridge: Cambridge University Press.

Darvas, György. 2007. *Symmetry: Cultural-historical and ontological aspects of science-arts relations: The natural and man-made world in an interdisciplinary approach*. Basel: Birkhäuser.

Darwin, Charles. 1859. *On the origin of species by means of natural selection, or, the preservation of favoured races in the struggle for life*. London: John Murray.

Darwin, Charles. 1866. *On the origin of species by means of natural selection, or, the preservation of favoured races in the struggle for life*. 4th ed. London: John Murray.

Dau, Frithjof. 2009. The advent of formal diagrammatic reasoning systems. In *International Conference on Formal Concept Analysis*, 38–56. Berlin: Springer.

Dawid, Richard. 2015. *String theory and the scientific method*. Cambridge: Cambridge University Press.

Deane, Paul D. 1995. Metaphors of center and periphery in Yeats's *The Second Coming*. *Journal of Pragmatics* 24 (6): 627–642.

Demos, Tim. 2013. Explanation of my ECG interpretation flow diagram. https://www.youtube.com/watch?v=Cy8RM4W8b_U. Accessed 3 January 2024.

Derdikman, Dori, Jonathan R. Whitlock, Albert Tsao, Marianne Fyhn, Torkel Hafting, May-Britt Moser, and Edvard I. Moser. 2009. Fragmentation of grid cell maps in a multicompartment environment. *Nature Neuroscience* 12 (10): 1325–1332.

Digitale, Jean C., Jeffrey N. Martin, and Medellena Maria Glymour. 2022. Tutorial on directed acyclic graphs. *Journal of Clinical Epidemiology* 142:264–267.

Dobrzański, Michal, and Jens Lemanski. 2020. Schopenhauer diagrams for conceptual analysis. In *International conference on theory and application of diagrams*, ed. Ahti-Veikko Pietarinen, Peter Chapman, Leonie Bosveld-de Smet, Valeria Giardino, James Corter, and Sven Linker, 281–288. Cham: Springer.

Dodwell, Charles Reginald. 1993. *The pictorial arts of the West, 800–1200*. New Haven, CT: Yale University Press.

Donald, Merlin. 1991. *Origins of the modern mind: Three stages in the evolution of culture and cognition*. Cambridge, MA: Harvard University Press.

Donald, Merlin. 2001. *A mind so rare: The evolution of human consciousness*. New York: Norton.

Doob, Penelope Reed. 2019. *Idea of the labyrinth from classical antiquity through the Middle Ages*. Ithaca, NY: Cornell University Press.

Doolittle, W. Ford. 1999. Phylogenetic classification and the universal tree. *Science* 284 (5423): 2124–2128.

Doucet-Rosenstein, Diane. 1981. *Die Kombinatorik als Wissenschaftmethode Bei Raimund Lull und G.W. Leibnitz*. Dissertation. Münich: University of Münich.

Dove, Guy. 2015. How to go beyond the body: An introduction. *Frontiers in Psychology* 6:660.

Dreyer, John Louis, and Emil. 1953. *A history of astronomy from Thales to Kepler*. 2nd ed. New York: Dover.

Dunér, David. 2016. Science: The structure of scientific evolutions. In *Human lifeworlds: The cognitive semiotics of cultural evolution*, ed. David Dunér and Göran Sonesson, 229–266. Frankfurt am Main: Peter Lang.

Dunér, David. 2019a. Human mind in space and time: Prolegomena to a cognitive history. In *Cognitive history: Mind, space, and time*, ed. David Dunér and Christer Ahlberger, 3–34. Berlin: De Gruyter Oldenbourg.

Dunér, David. 2019b. The axiomatic deductive ideal in early modern thinking. In *Cognitive history: Mind, space, and time*, ed. David Dunér and Christer Ahlberger, 3–34. Berlin: De Gruyter Oldenbourg.

Eco, Umberto. 2014. *From the tree to the labyrinth: Historical studies on the sign and interpretation*. Boston: Harvard University Press.

Eddy, Matthew D. 2010. Tools for reordering: Commonplacing and the space of words in Linnaeus's philosophia botanica. *Intellectual History Review* 20 (2): 227–252.

Efficiency Society. 1912. The need of organization. *Bulletin of the Efficiency Society 1* (3): 5–6.

Eichenbaum, Howard. 2014. Time cells in the hippocampus: A new dimension for mapping memories. *Nature Reviews Neuroscience* 15 (11): 732–744.

Einthoven, Willem. 1896. *Onderzoekingen gedaan in het physiologisch laboratorium der universiteit te Leiden.* [Research conducted in the Physiological Laboratory of the University of Leiden] 2:2. Leiden: Van Doesburgh

Eisenstein, Elizabeth L. 1979. *The printing press as an agent of change: Communications and cultural transformations in early-modern Europe, vol. 1.* Cambridge: Cambridge University Press.

Eisenstein, Elizabeth L. 1983. *The printing revolution in early modern Europe.* Cambridge: Cambridge University Press.

Ekengren Oscarsson, Henrik. 2018. *Flytande väljare 2018.* Gothenburg: University of Gothenburg.

Ekstrom, Arne D., and Charan Ranganath. 2018. Space, time, and episodic memory: The hippocampus is all over the cognitive map. *Hippocampus* 28 (9): 680–687.

Elliott, Ralph Nelson. 1994. Natures law: The secret of the universe. In *R.N. Elliott's masterworks: The definitive collection*, ed. Robert Rougelot Pretcher Jr., 213–308. Gainesville, Georgia: New Classics Library.

Elngar, Ahmed A., Mohamed Arafa, Amar Fathy, Basma Moustafa, Omar Mahmoud, Mohamed Shaban, and Nehal Fawzy. 2021. Image classification based on CNN: A survey. *Journal of Cybersecurity and Information Management* 6 (1): 18–50.

Embach, Carolyn. 1969. *On the nature of things* (De Natura Rerum), by Isidore of Seville (ca. 560–636 CE) Trans. C. Embach. https://www.researchgate.net/publication/315663902_On_the_Nature_of_Things_De_Natura_Rerum_by_Isidore_of_Seville_ca_560-636_AD_Translated_by_Carolyn_Embach_1969.

Emmer, Michele. 1982. Art and mathematics: The Platonic solids. *Leonardo* 15 (4): 277–282.

Engelhardt, Yuri, and Clive Richards. 2021. A universal grammar for specifying visualization types. In *Diagrammatic representation and inference: 12th international conference, diagrams 2021*, ed. Amrita Basu, Gem Stapleton, Sven Linker, Catherine Legg, Emmanuel Manalo, and Petrucio Viana, 395–403. Cham: Springer.

Enger, Håkon. 2003. String theory and the scientific method. Håkon Enger homepage. https://henger.org/~henger/publ/mnvit-essay.pdf. Accessed 29 July 2024.

Englebretsen, George. 2020. *Figuring it out: Logic diagrams.* Berlin: De Gruyter.

Enriquez, León. 2012. A Peircean model for music and sound-based art: A pragmatist approach to experiences in the artistic use of sound. In *Proceedings of the electroacoustic music studies network conference 'meaning and meaningfulness in electroacoustic music.'* Stockholm: Electroacoustic Music Studies Network.

Epstein, Seymour, Rosemary Pacini, Veronika Denes-Raj, and Harriet Heier. 1996. Individual differences in intuitive–experiential and analytical–rational thinking styles. *Journal of Personality and Social Psychology* 71 (2): 390–405.

Euclid. 1301. *Euclidis elementa: interprete Adelardo, Bathoniens; accedit Campani commentarius.* Bibliothéque Nationale de France, MS Latin 7214. https://www.europeana.eu/sv/item/9200519/ark__12148_btv1b100352862. Accessed 29 July 2024.

Euler, Leonhard. 1748. *Introductio in analysin infinitorum.* Lausanne: Marcum-Michaelem Bousquet & socios.

Euler, Leonhard. 2000. *Foundations of differential calculus.* Trans. J.D. Blanton. Cham: Springer.

Evans, Michael W. 1980. The geometry of the mind. *Architectural Association Quarterly* 12 (4): 32–55.

Evans, Steven N. 2008. *Probability and Real Trees: École d'Été de Probabilités de Saint-Flour XXXV-2005.* Berlin: Springer.

Even-Ezra, Ayelet. 2021. *Lines of thought: Branching diagrams and the medieval mind.* Chicago: The University of Chicago Press.

Eves, Howard Whitley. 1990. *Foundations and fundamental concepts of mathematics.* 3rd ed. Boston: PWS-Kent.

Fauconnier, Gilles, and Mark Turner. 2003. Conceptual blending: Form and meaning. *Recherches en Communication* 19:57–86.

Felleman, Daniel J., and David C. Van Essen. 1991. Distributed hierarchical processing in the primate cerebral cortex. *Cerebral cortex* 1 (1): 1–47.

Ferede, Ergete. 2019. *The revenue effects of tax rate increases on high-income earners.* Vancouver: Fraser Institute.

Feynman, Richard P. 1949. Space-time approach to quantum electrodynamics. *Physical Review* 76 (6): 769–789.

Feynman, Richard. 1965. New textbooks for the "new" mathematics. *Engineering and Science* 28 (6): 9–15.

Figurska, Małgorzata, Maciej Stańczyk, and Kamil Kulesza. 2008. Humans cannot consciously generate random numbers sequences: Polemic study. *Medical hypotheses* 70 (1): 182–185.

Firestone, Shulamit. (1970) 2015. *The dialectic of sex: The case for a feminist revolution.* London: Verso.

Floyer, John. 1707. *The physicians pulse watch.* London: Smith & Walford.

Fraenkel, Aviezri S. 1970. Economic traversal of labyrinths. *Mathematics Magazine* 43 (3): 125–130.

Franklin, John Curtis. 2002. Diatonic music in Greece: A reassessment of its antiquity. *Mnemosyne* 55 (6): 669–702.

Friedrich, Trista E., Victoria L. Harms, and Lorin J. Elias. 2014. Dynamic stimuli: Accentuating aesthetic preference biases. *Laterality* 19 (5): 549–559.

Friendly, Michael. 2006. A brief history of data visualization. In *Handbook of computational statistics: Data visualization*, ed. Chun-houh Chen, Wolfgang Härdle, and Antony Unwin, vol. 3, 15–56. Heidelberg: Springer.

Friendly, Michael. 2008. The golden age of statistical graphics. *Statistical Science* 23 (4): 502–535.

Friendly, Michael, and Raymond Andrews. 2021. The radiant diagrams of Florence Nightingale. *Statistics and Operations Research Transactions* 45 (1): 1–18.

Friendly, Michael, and Howard Wainer. 2021. *A history of data visualization and graphic communication*. Cambridge, MA: Harvard University Press.

Funk, Holger. 2014. Describing plants in a new mode: The introduction of dichotomies into sixteenth-century botanical literature. *Archives of Natural History* 41 (1): 100–112.

Funkhouser, H. Gray. 1937. Historical development of the graphical representation of statistical data. *Osiris* 3:269–404.

Galilei, Vincentio. 1581. *Dialogo di Vincentio Galilei … della musica antica, et della moderna*. Florence: Giorgio Marescotti.

Galison, Peter. 2003. *Einstein's clocks, Poincaré's maps*. London: Sceptre.

Galton, John. 1869. *Hereditary genius: An inquiry into its laws and consequences*. London: Macmillan.

Ganeri, Jonardon. 2004. Indian logic. In *Handbook of the history of logic, vol. 1: Greek, Indian and Arabic Logic*, ed. Dov M. Gabbay and John Woods, 309–396. Amsterdam: Elsevier.

Gardner, Martin. 1982. *Logic machines and diagrams*. 2nd ed. Brighton: The Harvester Press.

Gardner, Howard. 1985. *The mind's new science: A history of the cognitive revolution*. New York: Basic books.

Gartley, Harold M. (1935) 1981. *Profits in the stock market*. Washington, DC: Lambert-Gann.

Gattis, Merideth. 2001a. Space as a basis for abstract thought. In *Spatial schemas and abstract thought*, ed. Merideth Gattis, 1–12. Cambridge, MA: MIT Press.

Gattis, Merideth. 2001b. Reading pictures. In *Spatial schemas and abstract thought*, ed. Merideth Gattis, 223–245. Cambridge, MA: MIT Press.

Gauchet, Marcel. 1996. Right and left. In *Realms of memory: The construction of the French past. I. Conflicts and divisions*, ed. Pierre Nora, 241–298. New York: Columbia University Press.

Georges, Karl Ernst, and Heinrich Georges. 1913. *Ausführliches lateinisch-deutsches Handwörterbuch: aus den Quellen zusammengetragen und mit besonderer bezugnahme auf Synonymik und Antiquitäten unter berücksichtigung der besten Hilfsmittel. 8., verb. und verm. Aufl. ed.* [Comprehensive Latin-German dictionary: compiled from the sources and with special reference to synonyms and antiquities, taking into account the best resources. 8th, revised and enlarged edition]. Hannover: Hahnsche Buchhandlung.

Giardino, Valeria. 2016. Behind the diagrams: Cognitive issues and open problems. In *Thinking with diagrams: The semiotic basis of human cognition*, ed. Sybille Krämer and Christina Ljungberg, 77–101. Boston: De Gruyter.

Giere, Ronald N., and Barton Moffatt. 2003. Distributed cognition: Where the cognitive and the social merge. *Social Studies of Science* 33 (2): 301–310.

Gilbreth, Frank Bunker, and Lillian Moller Gilbreth. 1921. *Process charts*. New York: American Society of Mechanical Engineers.

Giraud, Yann. 2014. Legitimizing napkin drawing: The curious dispersion of Laffer curves, 1978 – 2008. In *Representation in Scientific Practice Revisited*, ed. Catelijne Coopmans, Janet Vertesi, Michael Lynch, and Stephen William Woolgar, 269–290. Cambridge, MA: MIT Press.

Glasgow, Janice, N. Hari Narayanan, and B. Chandrasekaran, eds. 1995. *Diagrammatic reasoning: Cognitive and computational perspectives*. Menlo Park, MA: MIT Press.

Gobet, Fernand, and Gary Clarkson. 2004. Chunks in expert memory: Evidence for the magical number four … or is it two? *Memory* 12 (6): 732–747.

Gohary, Jocelyn. 1992. *Akhenaten's Sed-festival at Karnak*. London: Kegan Paul.

Gombrich, Ernst. 1976. *The heritage of Apelles*. Oxford: Phaidon.

Gombrich, Ernst. 1990. Pictorial instructions. In *Images and understanding: Thoughts about images. Ideas about understanding*, ed. Horace Barlow, Colin Blakemore, and Miranda Weston-Smith, 26–45. Cambridge: Cambridge University Press.

Gontier, Nathalie. 2011. Depicting the tree of life: The philosophical and historical roots of evolutionary tree diagrams. *Evolution Education & Outreach* 4 (3): 515–538.

Goodsell, Charles T. 1988. The architecture of parliaments: Legislative houses and political culture. *British Journal of Political Science* 18 (3): 287–302.

Gozza, Paolo. 2000a. 'Desiderio da Pavia' and renaissance musical theory. In *Number to sound: The musical way to the Scientific revolution*, ed. Paolo Gozza, 79–96. Dordrecht: Springer.

Gozza, Paolo. 2000b. Introduction. In *Number to sound: The musical way to the scientific revolution*, ed. Paolo Gozza, 1–63. Dordrecht: Springer.

Gozza, Paolo. 2000c. Preface. In *Number to sound: The musical way to the scientific revolution,* ed. Paolo Gozza, xi–xiii. Dordrecht: Springer.

Gragg, Gene B. 1995. Babylonian grammatical texts. In *Concise history of the language sciences from the Sumerians to the cognitivists*, ed. Ernst Frideryk Konrad Koerner and Ronald E. Asher, 19–21. Oxford: Pergamon.

Graham, Ben B. 2004. *Detail process charting: Speaking the language of process*. Hoboken, NJ: Wiley.

Granger, Herbert. 1980. Aristotle and the genus-species relation. *The Southern Journal of Philosophy* 18 (1): 37–50.
Greco, Pietro. 2018. *Galileo Galilei: The Tuscan artist.* Cham: Springer.
Gregoric, Pavel, and Jakob Leth Fink. 2022. Introduction: Sense perception in Aristotle and the Aristotelian tradition. In *Forms of representation in the Aristotelian tradition, vol. 1: Sense perception*, ed. Juhana Toivanen, 15–39. Leiden: Brill.
Griffiths, Antony. 2016. *The print before photography: An introduction to European printmaking 1550–1820.* London: The British Museum.
Grønfeldt Winter, Rasmus. 2020. *When maps become the world.* Chicago: University of Chicago Press.
Günther, Siegmund. 1877. *Die Anfänge und Entwickelungsstadien des Coordinatenprincipes.* Nürnberg: Naturhistorischen Gesellschaft.
Günther, Hans. 1924. *Rassenkunde des deutschen Volkes.* 6, Umgearb. Aufl. München: Lehmann.
Günther, Hans. 1928. *Rassenkunde des deutschen Volkes.* 12., Umgearb. Aufl. München: Lehmann.
Güntürkün, Onur, Felix Ströckens, and Sebastian Ocklenburg. 2020. Brain lateralization: A comparative perspective. *Physiological Reviews* 100 (3): 1019–1063.
Gurnee, Wes, and Max Tegmark. 2023. Language models represent space and time. https://arxiv.org/abs/2310.02207. Accessed 29 July 2024.
Hacking, Ian. 2007. Trees of logic, trees of Porphyry. In *Advancements of learning: Essays in honour of Paolo Rossi*, ed. John L. Heilbron, 219–261. Firenze: Leo S. Olschki.
Hadamard, Jacques. (1945) 1996. *The mathematician's mind: The psychology of invention in the mathematical field.* Princeton, NJ: Princeton University Press.
Haemer, Kenneth W. 1948. Range-bar charts. *The American Statistician* 2 (2): 23–23.
Hägele, Günter, and Friedrich Pukelsheim. 2001. Llull's writings on electoral systems. *Studia Lulliana* 41:3–38.
Haines, John. 2020. The visualization of music in the early Middle Ages: Three case studies. In *The visualization of knowledge in medieval and early Europe*, ed. Marcia Kupfer, Adam S. Cohen, and J. H. Chajes, 327–340. Turnhout: Brepolis.
Hamburger, Jeffrey F. 2020. Mindmapping: The diagram paradigm in medieval art—And beyond. In *The visualization of knowledge in medieval and early Europe*, ed. Marcia Kupfer, Adam S. Cohen, and J. H. Chajes, 61–86. Turnhout: Brepolis.
Hamburger, Jefferey F., David J. Roxburgh, and Linda Safran, eds. 2022. *The diagram paradigm: Cross cultural approaches.* Washington, DC: Dumbarton Oaks.

Hannah, Matthew N. 2021. A conspiracy of data: Qanon, social media, and information visualization. *Social Media + Society* 7(3). https://doi.org/10.1177/20563051211036064.

Hao, Zhixin, Wu Maowei, Yang Liu, Xuezhen Zhang, and Jingyun Zheng. 2020. Multi-scale temperature variations and their regional differences in China during the medieval climate anomaly. *Journal of Geographical Sciences* 30 (1): 119–130.

Harley, John Brian, and David Woodward. 1987. *The history of cartography, vol. 1: Cartography in prehistoric, ancient, and medieval Europe and the Mediterranean*. Chicago: University of Chicago Press.

Harnad, Stevan. 2005. To cognize is to categorize: Cognition is categorization. In *Handbook of categorization in cognitive science*, ed. Henri Cohen and Claire Lefebvre, 19–43. Amsterdam: Elsevier.

Hawkins, Jeff. 2005. *On intelligence*. New York: Henry Holt.

Hawkins, Jeff. 2021. *A thousand brains: A new theory of intelligence*. New York: Basic Books.

Hawkins, Jeff, Subutai Ahmad, and Yuwei Cui. 2017. A theory of how columns in the neocortex enable learning the structure of the world. *Frontiers in Neural Circuits* 11:81.

Hawkins, Jeff, Marcus Lewis, Mirko Klukas, Scott Purdy, and Subutai Ahmad. 2019. A framework for intelligence and cortical function based on grid cells in the neocortex. *Frontiers in Neural Circuits* 12:121.

Hayes, Richard P. 1988. *Dignaga on the interpretation of signs*. Dordrecht: Kluwer.

He, Xi. 2020. *Lineage and community in China, 1100–1500: Genealogical innovation in Jiangxi*. Abingdon: Routledge.

He, Qiliang, and Thackery I. Brown. 2019. Environmental barriers disrupt grid-like representations in humans during navigation. *Current Biology* 29 (6): 2718–2722.

He, Huanhuan, and Leonard W. J. van der Kuijp. 2016. Once again on the "Hetucakraḍamaru": Rotating the wheels. *Journal of Indian Philosophy* 44 (2): 267–302.

Heeffer, Albrecht. 2010. The body in Renaissance arithmetic: From mnemonics to embodied cognition. In *AISB 2010 convention*, 21–22. The Society for the Study of Artificial Intelligence and the Simulation of Behaviour.

Heijman, Wim J. M., and Johan A. C. van Ophem. 2005. Willingness to pay tax: The Laffer curve revisited for 12 OECD countries. *The Journal of Socio-Economics* 34 (5): 714–723.

Hellström, Petter. 2012. Darwin and the tree of life: The roots of the evolutionary tree. *Archives of Natural History* 39 (2): 234–252.

Hellström, Petter. 2019. *Trees of knowledge: Science and the shape of genealogy*. Uppsala: Uppsala universitet.

Hernstein, Richard J., and Charles Murray. 1994. *The bell curve: Intelligence and class structure in American life*. New York: Free Press.

Heuer, Kurt, Lothar Schimmelpfennig, and Erwin Swillus. 1973. *Staatsbürgerkunde: Volk und Wissen. Klasse 9*. Berlin: Volkseigener Verlag.

Heuer, Kurt, Ruth Swillus-Baumann, Lothar Diersch, Christina Muller, and Erwin Swillus. 1979. *Staatsbürgerkunde. Volk und Wissen. Lehrbuch für Klasse 9*. Berlin: Volkseigener Verlag.

Hintze, Jerry L., and Ray D. Nelson. 1998. Violin Plots: A Box Plot-Density Trace Synergism. *The American Statistician* 52 (2): 181–184.

Hitchcock, Edward. 1840. *Elementary geology*. Amherst, MA: Adams.

Hobbs, Robert. 2004. *Mark Lombardi: Global networks*. In *Mark Lombardi: Global networks*, ed. Robert Carleton Hobbs, 11–49. New York: Independent Curators International.

Hodges, Wilfrid. 2020. Medieval Arabic notions of algorithm: Some further raw evidence. In *Fields of logic and computation III: Essays dedicated to Yuri Gurevich on the occasion of his 80th birthday*, ed. Andreas Blass, Patrick Cégielski, Nachum Dershowitz, Manfred Droste, and Bernd Finkbeiner, 133–146. Cham: Springer.

Hodgson, Derek. 2009. Evolution of the visual cortex and the emergence of symmetry in the Acheulean techno-complex. *Comptes Rendus Palevol* 8 (1): 93–97.

Hoff, Hebbel E., and L. A. Geddes. 1962. The beginnings of graphic recording. *Isis* 53 (3): 287–324.

Holdsworth, Ruth. 1982. *Psychology for careers counseling*. London: Macmillan.

Holladay, Joan H. 2019. *Genealogy and the politics of representation in the High and late Middle Ages*. Cambridge: Cambridge University Press.

Holmén, Janne. 2020. Time and space in time and space: Mapping the conceptual history of mental maps and historical consciousness. *Contributions to the History of Concepts* 15 (2): 105–129.

Holmén, Janne. 2024. Flipping the political scale: Spatial cognition and changes in diagrammatic representations of the Swedish party system in civics textbooks, 1900–2020. *Journal of Cognitive Historiography* 8 (1–2): 141–167.

Holmén, Janne, Ann-Sofie Jägerskog, Daniel Schumann, and Malin Tväråna. 2024. Connecting the nation: Diagrams of unity and diversity in civics textbooks. In *Diagrams in civic education: Visuospatial models of society in textbooks and teaching*, ed. Janne Holmén, Ann-Sofie Jägerskog, Daniel Schumann, and Malin Tväråna, 11–59. Palgrave Macmillan.

Honma, Munehisha. 1898. *Sō Uchimoto Kansōkyūō Isho Honmasōkyū Cho*. (The Last Will and Testament of Munehisa Honma of Shonai). Sakata: Hayasaka Toyozō.

Hoskin, Michael. 1997. Astronomy in antiquity. In *The Cambridge illustrated history of astronomy*, ed. Michel Hoskin, 22–47. Cambridge: Cambridge University Press.

Hossenfelder, Sabine. 2018. *Lost in math: How beauty leads physics astray.* New York: Basic Books.

Hotson, Howard. 2007. *Commonplace learning: Ramism and its German ramifications 1543–1630.* Oxford: Oxford University Press.

Hsu, Feng-Hsiung. 2002. *Behind Deep Blue: Building the computer that defeated the world chess champion.* Princeton, NJ: Princeton University Press.

Hullman, Jessica, and Nick Diakopoulos. 2011. Visualization rhetoric: Framing effects in narrative visualization. *IEEE Transactions on Visualization and Computer Graphics* 17 (12): 2231–2240.

Humphrey, Thomas M. 1992. Marshallian cross diagrams and their uses before Alfred Marshall. *Federal Reserve Bank of Richmond Economic Review* 78 (2): 3–23.

Hung, Wu. 2007. Picturing or diagramming the universe. In *Graphics and text in the production of technical knowledge in China: The warp and the weft*, ed. Francesca Bray, Vera Dorrofeeva-Lichtmann, and Georges Métailié, 191–214. Leiden: Brill.

Hutchins, Edwin. 2006. The distributed cognition perspective on human interaction. In *Roots of human sociality*, ed. Nicholas J. Enfield and Stephen C. Levinson, 375–398. London: Routledge.

Ibn Fariun. 1985. *Compendium of sciences.* Frankfurt am Main: Institute for the History of Arabic-Islamic Science at the Johann Wolfgang Goethe University.

Isacsson, Walter. 2017. *Leonardo da Vinci.* New York: Simon and Schuster.

Ishikawa, Kaoru. 1998. *Guide to quality control.* 2nd rev. ed., 14th printing. Tokyo: Asian Productivity Organization.

Izard, Véronique, Pierre Pica, Stanislas Dehaene, Danielle Hinchey, and Elizabeth Spelke. 2011. Geometry as a universal mental construction. In *Space, time and number in the brain: Searching for the foundations of mathematical thought*, ed. Stanislas Dehaene and Elizabeth Brannon, 319–332. London: Academic Press.

Jacobsen, Lyle E. 1983. Use of knotted string accounting records in old Hawaii and ancient China. *The Accounting Historians Journal* 10 (2): 53–61.

James, William. 1902. *The varieties of religious experience: A study in human nature.* New York: Longmans, Green, and Co.

Jha, Aditya Kumar. 2020. A mathematical model of Dignāga's hetu-cakra (within the framework of Simon's bounded rationality). *Journal of Indian Council of Philosophical Research* 37 (3): 471–479.

Jones, Matthew, and John Ormond. 2015. Introduction. In *Time travel in popular media: Essays on film, television, literature and video games*, ed. Matthew Jones and John Ormond, 5–18. Jefferson, NC: McFarland.

Jönsson, Ann-Mari. 2014. Linnés dikt till Giske. In *Svenska linnésällskapets årsskrift*, 109–122. Uppsala: Svenska linnésällskapet.

Jussieau, Antonii–Laurentii de. 1789. *Genera plantarum secundum ordines naturales disposita.* Paris: Herissant et Theophilum Barrois.

Kahneman, Daniel. 2003. A perspective on judgment and choice: Mapping bounded rationality. *American Psychologist* 58 (9): 697–720.

Kaji, Masanori. 2018. The origin of Mendeleev's discovery of the periodic system. In *Mendeleev to Oganesson: A multidisciplinary perspective on the periodic table*, ed. Eric Scerri and Guillermo Restrepo, 219–244. Oxford: Oxford University Press.

Kant, Immanuel. (1768) 1992. Concerning the ultimate ground of the differentiation of directions in space. In *The Cambridge edition of the works of Immanuel Kant: Theoretical philosophy, 1755–1770*, ed. David Walford and Ralf Meerbote. Cambridge: Cambridge University Press.

Katz, Victor J. 1998. *A history of mathematics: An introduction*. 2nd ed. Reading: Addison Wesley Longman.

Kelly, Lynne. 2012. *When knowledge was power*. Dissertation, LaTrobe University. https://opal.latrobe.edu.au/articles/thesis/When_knowledge_was_power/21858909.

Kelly, Lynne. 2015. *Knowledge and power in prehistoric societies: Orality, memory and the transmission of culture*. Cambridge: Cambridge University Press.

Kennedy, Helen, Rosemary Lucy Hill, Giorgia Aiello, and William Allen. 2016. The work that visualisation conventions do. *Information, Communication & Society* 19 (6): 715–735.

Kent, Charles. 1961. Introduction. In *Compendium of music (Compendium Musicae)*, René Descartes. Trans. W. Robert. Rome: American Institute of Musicology.

Kepler, Johannes. 1597. *Prodromus dissertationvm cosmographicarvm continens Mysterivm cosmographicvm*. Frankfurt: Godefrid Tampach.

Kepler, Johannes. 1619. *Harmonices Mundi Libri V*. Linz: Johann Planck.

Khoddari Naeeni, Saeed, and Elham Akbari. 2019. A study on diagonal writing in Islamic manuscript layout from 11th–15th centuries. *Honar-Ha-Ye-Ziba: Honar-Ha-Ye-Tajassomi* 24 (3): 65–76.

Kick, Friedrich. 1878. *Die Mehlfabrikation: Ein Lehrbuch des Mühlenbetriebes*. 2. ergänzte und revidirte Aufl. Leipzig: Felix.

Kick, Freidrich, and Henry Handley Pridham Powles. 1888. *Flour manufacture: With supplement, recent progress in flour manufacture*. London: Lockwood.

Kilpatrick, Jeremy. 2012. The new math as an international phenomenon. *Zdm: The International Journal on Mathematics Education* 44(4): 563–571.

Kim-Renaud, Young-Key, ed. 1997. *The Korean alphabet: Its history and structure*. Honolulu: University of Hawai'i Press.

King, Daniel. 2013. Why were the Syrians interested in Greek philosophy? In *History and identity in the late antique Near East*, ed. Philip Wood, 61–82. New York: Oxford University Press.

Kinsella, Karl. 2016. *Edifice and education: Structuring thought in twelfth-century Europe, vol. 1: Text*. Oxford: Keble College.

Klapisch-Zuber, Christiane. 2000. *L'ombre des ancêtres: essai sur l'imaginaire médiéval de la parenté*. Paris: Fayard.

Klein, Jacob. (1968) 1992. *Greek mathematical thought and the origin of algebra*. New York: Dover.

Krämer, Sybille. 2016. Is there a diagrammatic impulse with Plato? "Quasi-diagrammatic-scenes" in Plato's philosophy? In *Thinking with diagrams: The semiotic basis of human cognition*, ed. Sybille Krämer and Christina Ljungberg, 163–177. Boston: De Gruyter.

Krementsov, Nikolai. 2018. *With and without Galton: Vasilii Florinskii and the fate of eugenics in Russia*. Cambridge: Open Book Publishers.

Kropff, Emilio, James E. Carmichael, May-Britt Moser, and Edvard I. Moser. 2015. Speed cells in the medial entorhinal cortex. *Nature* 523 (7561): 419–424.

Kuhn, Thomas. 1962. *The structure of scientific revolutions*. Chicago: University of Chicago Press.

Kühnel, Bianca. 2005. Carolingian Diagrams. In *Seeing the invisible in late antiquity and the early Middle Ages*, ed. Giselle de Nie, Karl F. Morrison, and Marco Mostert, 359–389. Turnhout, Belgium: Brepols.

Kupfer, Marcia. 2020. Introduction. In *The visualization of knowledge in medieval and early Europe*, ed. Marcia Kupfer, Adam S. Cohen, and J. H. Chajes, 361–382. Turnhout: Brepolis.

Kupfer, Marcia, Adam S. Cohen, and J. H. Chajes, eds. 2020. *The visualization of knowledge in medieval and early Europe*. Turnhout: Brepolis.

Lackner, Michael. 2007. Diagrams as an architecture by means of words: The *Yanji Tu*. In *Graphics and text in the production of technical knowledge in China: The warp and the weft*, ed. Francesca Bray, Vera Dorrofeeva-Lichtmann, and Georges Métailié, 341–377. Leiden: Brill.

Lakoff, George. 1990. The invariance hypothesis: Is abstract reason based on image schemas? *Cognitive Linguistics* 1 (1): 39–74.

Lakoff, George, and Mark Johnson. 1980. *Metaphors we live by*. Chicago: University of Chicago Press.

Lakoff, George, and Mark Johnson. 1999. *Philosophy in the flesh: The embodied mind and its challenge to Western thought*. New York: Basic Books.

Lakoff, George, and Rafael E. Núñez. 2000. *Where mathematics comes from: How the embodied mind brings mathematics into being*. New York: Basic Books.

Lamarck, Jean-Baptiste. 1778. *Flore françoise, ou, Description succincte de toutes les plantes qui croissent naturellement en France, vol. 1–3*. Paris: Imprimerie Royale.

Lambert, Johann Heinrich. 1764. *Neues Organon oder Gedanken über die Erforschung und Bezeichnung des Wahren und dessen Unterscheidung vom Jrrthum und Schein*. Bd. 2. Leipzig: Wendler.

Lambert, Johann Heinrich. 1771. Essai d'Hygrométrie ou sur la mesure de l'humidité. In *Histoire de l'Académie Royale des Sciences et Belles-Lettres de*

Berlin. Année MDCCLXIX, 68–127. Berlin: Libraires de la Cour & de l'Académie Royale.

Laponce, Jean A. 1975. Spatial archetypes and political perceptions. *The American Political Science Review* 69 (1): 11–20.

Laponce, Jean A. 1981. *Left and right: The topography of political perceptions*. Toronto: University of Toronto Press.

Larkin, Jill H., and Herbert A. Simon. 1987. Why a diagram is (sometimes) worth ten thousand words. *Cognitive Science* 1 (1): 65–100.

Larson, James L. 1971. *Reason and experience: The representation of natural order in the work of Carl von Linné*. Berkeley: University of California Press.

Latour, Bruno. 1986. Visualization and cognition: Thinking with eyes and hands. In *Knowledge and society: Studies in the sociology of science, past and present, vol. 6*, ed. Henrika Kuklick and Elizabeth Long, 1–40. New York: JAI Press.

Lay Yong, Lam. 1994. Nine chapters on the mathematical art: An overview. *Archive for History of Exact Sciences* 47 (1): 1–51.

Leatherdale, William Hilton. 1974. *The role of analogy, model and metaphor in science*. Amsterdam: North Holland.

Leeuwen, Cees van. 1998. Perception. In *A companion to cognitive science*, ed. William Bechtel and George Graham, 265–281. Malden, MA: Blackwell.

Lemanski, Jens. 2017. Periods in the use of Euler-type diagrams. *Acta Baltica Historiae Et Philosophiae Scientiarum* 5 (1): 50–69.

Lemanski, Jens, and Amirouche Moktefi. 2018. Making sense of Schopenhauer's diagram of good and evil. In *Diagrammatic representation and inference: Diagrams 2018*, ed. Peter Chapman, Gem Stapleton, Amirouche Moktefi, Sarah Perez-Kriz, and Francesco Bellucci, 721–724. Cham: Springer Nature.

Li, Yingli, and Michael W. O'Boyle. 2011. Differences in mental rotation strategies for native speakers of Chinese and English and how they vary as a function of sex and college major. *The Psychological Record* 61:2–19.

Liljequist, Kurt, Svenåke Björkryd, Sture Andersson, and Lennart Walles. 1978. *Världen—vårt samhälle: Samhällskunskap för gymnasieskolans 3- och 4-åriga linjer*. Uppsala: Almqvist and Wiksell.

Liljequist, Kurt, Anders Altersten, Svenåke Björkryd, Hans Olof Hanselid, and Reidar Larsson. 1985. *Världen vårt samhälle: Samhällskunskap för gymnasieskolan*. Uppsala: Almqvist and Wiksell.

Lima, Manuel. 2011. *Visual complexity: Mapping patterns of information*. New York: Princeton Architectural Press.

Lima, Manuel. 2014. *The book of trees: Visualizing branches of knowledge*. New York: Princeton Architectural Press.

Lima, Manuel. 2017. *The book of circles: Visualizing spheres of knowledge*. New York: Princeton Architectural Press.

Link, David. 2010. Scrambling TRUTH: Rotating letters as a material form of thought. *Variantology* 4:215–266.

Linné, Carl von. (1735) 1960. *Caroli Linnæi Systema Naturæ: Sive Regna Tria Naturæ Systematice Proposita Per Classes, Ordines, Genera, & Species*. Uppsala: Bokgillet.

Livesey, Steven J. 1989. *Theology and science in the fourteenth century: Three questions on the unity and subalternation of the sciences from john of reading's commentary on the sentences*. Leiden: Brill.

Livi Bacci, Massimo. 2017. *A concise history of world population*. 6th ed. Hoboken, NJ: John Wiley & Sons.

Lloyd, Geoffrey. 2014. *The ideals of inquiry: An ancient history*. Oxford: Oxford Academic Press.

Lo, Andre W., and Jasmina Hasanhodzic. 2010. *The evolution of technical analysis: Financial prediction from Babylonian tablets to Bloomberg terminals*. Hoboken, NJ: Bloomberg Press.

Londey, David, and Carmen Johanson. 1984. Apuleius and the square of opposition. *Phronesis* 29 (2): 165–173.

Lovejoy, Arthur O. 1936. *The great chain of being: A study of the history of an idea*. Boston: Harvard University Press.

Luckombe, Philip. 1771. *The history and art of printing*. London: Adlard and Browne.

Lundberg, Jacob. 2017. *The Laffer curve for high incomes*. Uppsala: Uppsala University. Department of Economics Working Paper 2017:9. http://uu.diva-portal.org/smash/get/diva2:1148666/FULLTEXT01.pdf

Lundborg, Herman. 1922. *Degenerationsfaran och riktlinjer för dess förebyggande*. Stockholm: Norstedts.

Lusthaus, Dan. 2003. *Buddhist phenomenology: A philosophical investigation of Yogacara Buddhism and the Ch'eng Wei-Shih Lun*. London: Routledge.

Lyhne, Vagn. 2002. Pierre Bourdieu. In *Gotik. Arkitektur. Skolastik. Habitus*, ed. Vagn Lyhne, 131–141. Århus: Klim.

Maass, Anne, and Thomas W. Schubert. 2011. Introduction: The interrelation of spatial and social cognition. In *Spatial dimensions of social thought*, ed. Anne Maas and Thomas W. Schubert, 1–14. Berlin: De Gruyter Mouton.

Maclean, Ian. 2006. Diagrams in the defence of Galen: Medical uses of tables, squares, dichotomies, wheels and latitudes. In *Transmitting knowledge: Words, images, and instruments in early modern Europe*, ed. Sachiko Kusukawa and Ian Maclean. Oxford: Oxford University Press.

MacMurdie, Meekyung. 2022. Proven recipes: Geometry and the art of Arabic medicine. In *The visualization of knowledge in medieval and early Europe*, ed. Marcia Kupfer, Adam S. Cohen, and J. H. Chajes, 331–358. Turnhout: Brepolis.

Madden, Samuel. 1733. *Memoirs of the twentieth century*. London: Osborn.

Maffei, Raffaele. 1511. *Commentariorum urbanorum octo et triginta libri cum duplici eorundem indice secundum tomos collecto*. Lyon: Sebastian Gryphium.

Magnello, M. Eileen. 2012. Victorian statistical graphics and the iconography of Florence Nightingale's polar area graph. *BSHM Bulletin: Journal of the British Society for the History of Mathematics* 27 (1): 13–37.

Manetti, Giovanni. 1993. *Theories of the sign in classical antiquity.* Bloomington: Indiana University Press.

Mann, Michael E., Zhihua Zhang, Scott Rutherford, Raymond S. Bradley, Malcolm K. Hughes, Drew Shindell, Caspar Ammann, Greg Faluvegi, and Fenbiao Ni. 2009. Global signatures and dynamical origins of the little ice age and medieval climate anomaly. *Science* 326 (5957): 1256–1260.

Marcus, Gary. 2023. Muddles about models: Why it's a stretch to say that large language models represent "literal world models." https://garymarcus.substack.com/p/muddles-about-models. Accessed 31 December 2023.

Martin, Luther H., and Jesper Sørensen, eds. 2011. *Past minds: Studies in cognitive historiography.* London: Equinox.

Maruyama, Magoroh. 1963. The second cybernetics: Deviation-amplifying mutual causal processes. *American scientist* 51 (2): 164–179.

May, Michael. 1999. Diagrammatic reasoning and levels of schematization. In *Iconicity: A fundamental problem in semiotics*, ed. Troels Degn Johansson, Martin Skov, and Bertil Brogaard, 175–194. Århus: NSU Press.

Mazurs, Edward G. 1974. *Graphic representation of the periodic system during one hundred years.* Tuscaloosa: University of Alabama Press.

McBean, Sam. 2021. Feminist diagrams. *Feminist Theory* 22 (2): 206–225.

McCaffrey, Joe. 2014. Ibn Khaldun: The forgotten father of economics. *ISOR Journal of Business and Management* 16 (3): 1–3.

McCulloch, Jock. 1995. *Colonial psychiatry and the "African mind."* Cambridge: Cambridge University Press.

McLuhan, Marshall. (1962) 1967. *The Gutenberg galaxy: The making of typographic man.* London: Routledge & Kegan Paul.

McLuhan, Marshall. (1964) 2005. *Understanding media: The extensions of man.* Abingdon: Routledge.

McNeill, David. 1992. *Hand and mind: What gestures reveal about thought.* Chicago: University of Chicago Press.

Medin, Douglas L., and Scot Atran. 1999. Introduction. In *Folkbiology*, ed. Douglas L. Medin and Scot Atran, 1–15. Cambridge, MA: MIT Press.

Mehra, Jagdish. 1994. *The beat of a different drum: The life and science of Richard Philips Feynman.* Oxford: Clarendon Press.

Meynell, Letitia. 2008. Why Feynman diagrams represent. *International Studies in the Philosophy of Science* 22 (1): 39–59.

Mialet, Hélène. 2012. *Hawking incorporated: Stephen Hawking and the anthropology of the knowing subject.* Chicago: University of Chicago Press.

Middleton, Roger. 2010. The Laffer curve. In *Famous figures and diagrams in economics*, ed. Mark Blaug and Peter Lloyd, 412–418. Cheltenham: Edward Elgar.

Miller, George A. 1956. The magical number seven, plus or minus two: Some limits on our capacity for processing information. *Psychological Review* 63 (2): 81–97.

Minami, Hideme. 1760. *Chaji-dan*. Kyoto: Nishimura Ichiroemon.

Mitchel, Edward Page. 1881. The clock that went backward. *The [New York] Sun*, 18 September 1881.

Mitchell, William John Thomas. 1980. Spatial form in literature: Toward a general theory. *Critical Inquiry* 6 (3): 539–567.

Mithen, Steven. 2024. *The language puzzle: How we talked our way out of the stone age*. London: Profile books.

Moktefi, Amirouche, and Sun-Joo Shin. 2012. In *A history of logic diagrams. In Logic: A history of its central concepts*, ed. Dov M. Gabbay, John Woods, and Francis Jeffry Pelletier, 611–682. Elsevier.: North-Holland.

Monroe, Dylan Louis. n.d. Deepstatemappingproject.com https://deepstatemappingproject.com/. Accessed 4 January 2024.

Morey-Holton, Emily R. 2003. The impact of gravity on life. In *Evolution on planet Earth*, ed. Lynn J. Rothschild and Adrian M. Lister, 143–159. Amsterdam: Academic Press.

Morris, Stephen J., and O. C. Z. Gotel. 2006. Flow diagrams: Rise and fall of the first software engineering notation. In *Diagrams 2006: Diagrammatic representation and inference*, ed. Dave Barker-Plummer, Richard Cox, and Nik Swoboda, 130–144. Berlin: Springer.

Moser, May-Britt, and Edvard I. Moser. 2015. Understanding the cortex through grid cells. In *The future of the brain: Essays by the world's leading neuroscientists*, ed. Gary Marcus and Jeremy Freeman, 67–77. Princeton, NJ: Princeton University Press.

Moxon, Joseph. (1683) 1958. *Mechanick exercises on the whole art of printing*, ed. Herbert Davis and Harry Carter. London: Oxford University Press.

Muirhead, James Patrick. 1854. *The origin and progress of the mechanical inventions of James Watt: Illustrated by his correspondence with his friends and the specifications of his patents*. Vol. 1. London: Murray.

Müller, Robert R. 1944. *Über die Enfaltungsordnung und den Stambaum der chemischen Grundstoffe: Eine morphologische Betrachtung. Gestalt 15*. [On the order of development and the family tree of the basic chemical substances: A morphological consideration. Figure 15.] Halle: Niemeyer.

Müller-Wille, Staffan, and Isabelle Charmantier. 2012. Natural history and information overload: The case of Linnaeus. *Studies in History and Philosophy of Science. Part C, Studies in History and Philosophy of Biological and Biomedical Sciences* 43(1): 4–15.

Murat, Zuleika. 2015. The compendium Historiae in Genealogia Christi and its iconographical tradition: Legacy of classical antiquity in a mediaeval biblical anthology. In *Actual problems of theory and history of art 5*, 404–411. St. Petersburg: St. Petersburg University.

Murdoch, John E. 1984. *Album of science: Antiquity and the Middle Ages*. New York: Charles Scribbner's Sons.

Murdoch, Stephen. 2007. *IQ: A smart history of a failed idea*. London: Duckworth.

Murphy, John J. 1999. *Technical analysis of the financial markets: A comprehensive guide to trading methods and applications*. New York: New York Institute of Finance/Penguin.

Muzzulini, Daniel. 2015. The geometry of musical logarithms. *Acta Musicologica* 87 (2): 193–216.

Nadaillac, Jean-François-Albert du Pouget, and W. H. Dall. 2005. *Prehistoric America*. Tuscaloosa: University of Alabama Press.

Nahin, Paul J. 2001. *Time machines: Time travel in physics, metaphysics, and science fiction*, 2nd ed., second corrected printing. New York: Springer.

Natali, Carlo, and Douglas S. Hutchinson. 2013. *Aristotle: His life and school*. Princeton, NJ: Princeton University Press.

Needham, Rodney. 1963. Introduction. In *Primitive classification*, ed. Émile Durkheim and Marcel Mauss, vii–xliii. Chicago: University of Chicago Press.

Nersessian, Nancy J. 1992. How do scientists think? Capturing the dynamics of conceptual change in science. In *Cognitive models of science*, ed. Ronald N. Giere, 3–47. Minneapolis: University of Minnesota Press.

Netz, Reviel. 1999. *The shaping of deduction in Greek mathematics: A study in cognitive history*. Cambridge: Cambridge University Press.

Nightingale, Florence. 1859. *A contribution to the sanitary history of the British Army during the late war with Russia*. London: John W. Parker and Son.

Nihon shōken shinbun-sha [Nihon Securities Journal]. 1969. *'Sakata gohō' wa fūrinkazan*. [The "five Sakata rules" are fūrinkazan (wind, forest fire, mountain, a quote from Zun Tzu's art of war)]. Tokyo: Nihon shōken shinbun-sha.

Nison, Steve. 1991. *Japanese candlestick charting techniques: A contemporary guide to the ancient investment technique of the Far East*. New York: New York Institute of Finance.

Nison, Steve. 1994. *Beyond candlesticks: New Japanese charting techniques revealed*. New York: Wiley Finance.

Nöth, Winfried. 2024. Peirce's iconicity and his image-diagram-metaphor triad revisited: Complements to Stjernfelt's sheets, diagrams, and realism. *Semiotica* 258:143–167.

Núñez, Rafael, and Kensy Cooperrider. 2013. The tangle of space and time in human cognition. *Trends in Cognitive Sciences* 17 (5): 220–229.

Núñez, Rafael, and Eve Sweetser. 2006. With the future behind them: Convergent evidence from Aymara language and gesture in the crosslinguistic comparison of spatial construals of time. *Cognitive Science* 30 (3): 401–450.

Núñez, Rafael, D. Kensy Cooperrider, and Doan, and Jürg Wassmann. 2012. Contours of time: Topographic construals of past, present, and future in the Yupno Valley of Papua New Guinea. *Cognition* 124 (1): 25–35.

O'Keefe, John, and Jonathan Dostrovsky. 1971. The hippocampus as a spatial map: Preliminary evidence from unit activity in the freely moving rat. *Brain Research* 34 (1): 171–175.

O'Meadhra, Uaininn. 2012. Medieval logic diagrams: In Bro church, Gotland, Sweden. *Acta Archaeologica* 83(1): 287–316.

Ong, Walter J. 1957. System, space, and intellect in renaissance symbolism. *Cross Currents* 7 (2): 123–136.

Ong, Walter J. 1958. *Ramus. Method, and the decay of dialogue: From the art of discourse to the art of reason.* Cambridge, MA: Harvard University Press.

Ong, Walter J. 1982. *Orality and literacy: The technologizing of the word.* London: Methuen.

Ortiz, Maria J. 2011. Primary metaphors and monomodal visual metaphors. *Journal of Pragmatics* 43 (6): 1568–1580.

Our World in Data. n.d. Population by country, available from 10,000 BCE to 2100, based on data and estimates from different sources. https://ourworldindata.org/grapher/population?time=-10000..latest. Accessed 30 July 2024.

Paff, Werner, Martin Döbler, Christina Müller, Horst Riechert, Günter Schippel, and Georg Warkotsch 1983. *Staatsbürgerkunde: Volk und Wissen. Klasse 9.* [Civics: People and Knowledge. Grade 9.] Berlin: Volkseigener Verlag.

Palisca, Claude V. 2003. Introduction. In Vincenzo Galileo, *Dialogue on Ancient and Modern Music*. Trans. C.V. Palisca, xv–lxix. New Haven, CT: Yale University Press.

Palmer, Richard E. 1969. *Hermeneutics: Interpretation theory in Schleiermacher, Dilthey, Heidegger, and Gadamer*. Evanston, IL: Northwestern University Press.

Panofsky, Erwin. 1951. *Gothic architecture and scholasticism.* Latrobe: Archabbey Press.

Park, A., Douglas S. Miller, Hamed Nili, Charan Ranganath, and Erie D. Boorman. 2020. Map making: Constructing, combining, and inferring on abstract cognitive maps. *Neuron* 107 (6): 1226–1238.

Pearl, Judea, and Dana Mackenzie. 2018. *The book of why: The new science of cause and effect.* New York: Basic books.

Pearson, Karl. 1901. On lines and planes of closest fit to systems of points in space. *Philosophical Magazine and Journal of Science* 6 (2): 559–572.

Penrose, Roger. 1971. Applications of negative dimensional tensors. In *Combinatorial mathematics and its applications: Proceedings of a conference held at the mathematical institute, Oxford, from 7–10 July, 1969*, ed. Dominic J. A. Welsh, 221–244. London: Academic Press.

Penrose, Roger. 1989. *The emperor's new mind: Concerning computers, minds and the laws of physics*. London: Vintage.

Perez, Daniela M., Siobhan J. Heatwole, Lesley J. Morrell, and Patricia R. Y. Backwell. 2015. Handedness in fiddler crab fights. *Animal Behaviour* 110:99–104.

Persaud, Navindra. 2005. Humans can consciously generate random number sequences: A possible test for artificial intelligence. *Medical Hypotheses* 65 (2): 211–214.

Peterson, Lloyd, and Margaret Jean Peterson. 1959. Short-term retention of individual verbal items. *Journal of Experimental Psychology* 58 (3): 193–198.

Pettegree, Andrew. 2010. *The book in the renaissance*. New Haven, CT: Yale University Press.

Pietarinen, Ahti-Veikko. 2016. Is there a general diagram concept? In *Thinking with diagrams: The semiotic basis of human cognition*, ed. Sybille Krämer and Christina Ljungberg, 121–137. Boston: De Gruyter.

Playfair, William. 1801. *The statistical breviary*. London: Bensley.

Plot, Robert. 1685. Observations of the wind, weather, and height of the mercury in the barometer, throughout the year 1684; taken in the musaeum ashmoleanum at Oxford, by Robert Plot, LLD. To which is prefixt a letter from him, to Dr. Martin Lister, F. off the R.S. concerning the use of this and such like historys of the weather. *Philosophical Transactions (1683–1775)* 15: 930–943.

Pratt, H. 1915. Long range reception with combined crystal detector and audion amplifier. *Proceedings of the Institute of Radio Engineers* 3 (2): 173–178.

Preda, Alex. 2001. The rise of the popular investor: Financial knowledge and investing in England and France, 1840–1880. *Sociological Quarterly* 42 (2): 205–232.

Priestley, Joseph. 1764. *A description of a chart of biography*. Warrington.

Pryer, Anthony. 2011. Notation. In *The Oxford companion to music*. Oxford: Oxford University Press., https://www.oxfordreference.com/display/10.1093/acref/9780199579037.001.0001/acref-9780199579037-e-4761?rskey=6GSQhH&result=5202. Accessed 8 August 2024.

Quetelet, Adolphe. 1846. *Lettres sur la théorie des probabilités, appliquée aux sciences morales et politiques*. Bruxelles: Hayez.

Ragen, Brian Abel. 1994. Semiotics and heraldry. *Semiotica* 100:5–34.

Ramus, Petrus. 1576. *Professio Regia: Hoc est septem artes liberales*. Basel: Sebastian Henric Petri.

Rankin, Susan. 2022. *Sounding the word of God: Carolingian books for singers*. Notre Dame: University of Notre Dame Press.

Read, Sara. 2018. Sir John Floyer's pulse-watch. *Early modern medicine: A blog about bodies and medicine c. 1500–1780*. https://earlymodernmedicine.com/sir-john-floyers-pulse-watch/. Accessed 3 January 2024.

Redlow, Götz, Helmut Frommknecht, Matthäus Klein, Eberhard Fromm, Herbert Hörz, Gottfried Stiehler, and Harald Wessel. 1971. *Einführung in den dialektischen und historischen Materialismus. Lehrbuch für Staatsbürgerkunde Klassen 11 und 12* [Introduction to dialectical and historical materialism: Textbook for civics classes 11 and 12]. Berlin: Dietz Verlag.

Reisch, Gregor. 1503. *Margarita Philosophica*. Strassbourg: Johann Schott.

Rendgen, Sandra. 2018. *The Minard system: The complete statistical graphics of Charles-Joseph Minard*. New York: Princeton Architectural Press.

Rendgen, Sandra, ed. 2019. *History of information graphics*. Cologne: Taschen.

Renfrew, Colin. 1982. *Towards an archaeology of mind: An inaugural lecture delivered before the University of Cambridge on 30 November 1982*. Cambridge: Cambridge University Press.

Richardson, George P. 1999. *Feedback thought in social science and systems theory*. Waltham, MA: Pegasus communications.

Risi, Vincenzo De. 2018. Analysis situs, the foundations of mathematics, and a geometry of space. In *The Oxford handbook of Leibniz*, ed. Maria Rosa Antognazza, 247–258. Oxford: Oxford University Press.

Robbins, Philip, and Murat Aydede, eds. 2009. *The Cambridge handbook of situated cognition*. Cambridge: Cambridge University Press.

Rosa, Mark. 2016. *Native writing systems in the Okinawan islands*. Dissertation. Tokyo: University of Tokyo. https://www.l.u-tokyo.ac.jp/postgraduate/database/2015/165.html.

Rosch, Eleanor. 1978. Principles of categorization. In *Cognition and categorization*, ed. Eleanor Rosch and Barbara B. Lloyd, 27–48. Hillsdale: Lawrence Erlbaum.

Rose, Richard, and Ian McAllister. 1986. *Voters begin to choose: From closed-class to open elections in Britain*. London: Sage.

Rosenberg, Daniel. 2013. The Pleasures of Memory. *Time OnLine*. https://timeonline.uoregon.edu/twain/pleasures.php#page-1. Accessed 3 January 2024.

Rosenberg, Daniel, and Anthony Grafton. 2010. *Cartographies of time: A history of the timeline*. New York: Princeton Architectural Press.

Roth, Wolff-Michael, and Alfredo Jornet. 2013. Situated cognition. *Wiley Interdisciplinary Reviews: Cognitive Science* 4 (5): 463–478.

Rothschild, Lynn J. 2003. The sun: The impetus for life. In *Evolution on planet Earth*, ed. Lynn J. Rothschild and Adrian M. Lister, 87–107. Amsterdam: Academic Press.

Rubin, David C. 1995. *Memory in oral traditions: The cognitive psychology of epic, ballads, and counting-out rhymes*. Oxford: Oxford University Press.

Rubin, Gayle. 2008 (1984). Thinking sex: Notes towards a radical theory of the politics of sexuality. In *The Routledge critical and cultural theory reader*, eds. Neil Badmington and Julia Thomas, 281–323. London: Routledge.

Safran, Linda. 2020. A prolegomenon to Byzantine diagrams. In *The visualization of knowledge in medieval and early Europe*, ed. Marcia Kupfer, Adam S. Cohen, and J. H. Chajes, 361–382. Turnhout: Brepolis.

Safran, Linda. 2022. Diagramming Byzantine orthodoxy. In *The diagram paradigm: Cross-cultural approaches*, ed. Jefferey F. Hamburger, David J. Roxburgh, and Linda Safran, 489–518. Washington, DC: Dumbarton Oaks.

Sahoo, Santanu, Asit Subudhi, Manasa Dash, and Sukanta Sabut. 2020. Automatic classification of cardiac arrhythmias based on hybrid features and decision tree algorithm. *International Journal of Automation and Computing* 17 (4): 551–561.

Salomon, Frank. 2004. *The cord keepers: Khipus and cultural life in a Peruvian village*. Durham, NC: Duke University Press.

Santiago, Julio, Antonio Román, and Marc Ouellet. 2011. Flexible foundations of abstract thought. In *Spatial dimensions of social thought*, ed. Anne Maas and Thomas W. Schubert, 39–108. Berlin: De Gruyter Mouton.

Sarfatti, Tamar. 2022. *The description of Egypt from Napoleon to Champollion*. Cham: Palgrave Macmillan.

Sarmiento, Ivan, Sergio Paredes-Solís, David Loutfi, Anna Dion, Anne Cockcroft, and Neil Andersson. 2020. Fuzzy cognitive mapping and soft models of indigenous knowledge on maternal health in Guerrero, Mexico. *BMC medical research methodology* 20(1): 1–16.

Savage, Patrick E., Steven Brown, Emi Sakai, and Thomas E. Currie. 2015. Statistical universals reveal the structures and functions of human music. *Proceedings of the National Academy of Sciences* 112 (9): 8987–8992.

Savage-Smith, Emilie. 2002. Galen's lost ophthalmology and the Summaria Alexandrinorum. *Bulletin of the Institute of Classical Studies* 77:121–138.

Saward, Jeff. 2003. *Labyrinths and mazes: The definitive guide to ancient and modern traditions*. London: Gaia Books.

Saxl, Franz. 1942. A spiritual encyclopaedia of the later Middle Ages. *Journal of the Warburg and Courtauld Institutes* 5:82–142.

Scerri, Eric R. 2006. *The periodic table: Its story and its significance*. New York: Oxford University Press.

Scerri, Eric. 2019. Can quantum ideas explain chemistry's greatest icon? *Nature* 565 (7741): 557–559.

Scerri, Eric R. 2022. Various forms of the periodic table including the left-step table, the regularization of atomic number triads and first-member anomalies. *ChemTexts* 8:1–13.

Schivelbusch, Wolfgang. 1978. Railroad space and railroad time. *New German Critique* 14 (14): 31–40.

Scholl, Brian J. 2005. Innateness and (Bayesian) visual perception: Reconciling nativism and development. In *The innate mind: Structure and contents*, ed. Peter Carruthers, Stephen Laurence, and Stephen Stich. New York: Oxford Academic.

Schopenhauer, Arthur. 1961. *Die Welt als Wille und Vorstellung. Erster Band* [The World as Will and Representation, vol. 1]. Wiesbaden: Brockhaus.

Schreiber, Peter. 2003. The beginnings of visualizing the invisible. *Prague Studies in the History of Science and Technology, new series* 7:157–168.

Schröder, Ulrike. 2014. Rediscovering the cognitive-semiotic and cognitive-pragmatic approaches to metaphor in the work of Johann Heinrich Lambert and Philipp Wegener. *Metaphoric.de* 25:79–102.

Schubert, Thomas W., Sven Waldzus, and Beate Seibt. 2011. More than a metaphor: How the understanding of power is grounded in experience. In *Spatial dimensions of social thought*, ed. Anne Maas and Thomas W. Schubert, 153–185. Berlin: De Gruyter Mouton.

Schultheis, Holger, and Laura A. Carlson. 2013. Spatial reasoning. In *The Oxford handbook of cognitive psychology*, ed. Daniel Reisberg. Oxford: Oxford University Press.

Scriba, Christoph J., Peter Schreiber, and Jana Schreiber. 2015. *5000 years of geometry: Mathematics in history and culture*. 15th ed. Basel: Springer.

Seligenstädter Lateinpädagogik. Uppsala University Library, Manuscript C 678. http://www.alvin-portal.org/alvin/view.jsf?pid=alvin-record%3A201587&dswid=-8746. Accessed 30 July 2024.

Shalev, Eran. 2018. Our federal sun: Planetary politics before the civil war. *American Political Thought* 7 (2): 189–215.

Singh, Manpreet, Levi Monteiro Martins, Patrick Joanis, and Vijay K. Mago. 2016. Building a cardiovascular disease predictive model using structural equation model and fuzzy cognitive map. In *2016 IEEE International Conference on Fuzzy Systems (FUZZ-IEEE)*, 1377–1382. IEEE.

Siskin, Clifford. 2016. *System: The shaping of modern knowledge*. Cambridge, MA: The MIT Press.

Siu, Man-Keung. 1993. Proof and pedagogy in ancient China. *Educational Studies in Mathematics* 24 (4): 345–357.

Slater, Steven, and Alfred Znamierowski. 2007. *The world encyclopedia of flags and heraldry*. London: Lorenz Books.

Slingerland, Edward. 2014. Toward a second wave of consilience in the cognitive scientific study of religion. *Journal of Cognitive Historiography* 1 (1): 121–130.

Sloman, Aaron. 1995. Musings on the roles of logical and nonlogical representations in intelligence. In *Diagrammatic reasoning: Cognitive and computational perspectives*, ed. Janice N. Glasgow, Hari Narayanan, and B. Chandrasekaran, 7–32. Menlo Park: MIT Press.

Smith, D. Eugene. 1932. *History of Mathematics.* Volume I. Boston: Ginn & Company.

Smith, D. Eugene. 1958. *History of mathematics.* Volume II. New York: Dover.

Smith, A. Mark. 2020. More than meets the eye: What made the printing revolution revolutionary. In *The visualization of knowledge in medieval and early Europe*, ed. Marcia Kupfer, Adam S. Cohen, and J. H. Chajes, 211–228. Turnhout: Brepolis.

Smolin, Lee. 2006. *The trouble with physics: The rise of string theory, the fall of a science, and what comes next.* Boston: Houghton Mifflin.

Snyder, Laura J. 2005. Consilience, confirmation and realism. In *Scientific evidence: Political theories and applications*, ed. Peter Achinstein, 129–148. Baltimore, MD: Johns Hopkins University Press.

Sohn, Pow-key. 1959. Early Korean printing. *Journal of the American Oriental Society* 79 (2): 96–103.

Spear, Mary Eleanor. 1952. *Charting statistics.* New York: McGraw-Hill.

Spear, Mary Eleanor. 1969. *Practical charting techniques.* New York: McGraw-Hill.

Spelke, Elizabeth S. 2011. Natural number and natural geometry. In *Space, time and number in the brain: Searching for the foundations of mathematical though*, ed. Stanislas Dehaene and Elizabeth Brannon, 287–317. London: Academic Press.

Spence, Ian. 2005. No humble pie: The origins and usage of a statistical chart. *Journal of Educational and Behavioral Statistics* 30 (4): 353–368.

Spence, Ian, and Howard Wainer. 2017. William Playfair and the invention of statistical graphs. In *Information design: Research and practice*, ed. Alison Black, Paul Luna, Ole Lund, and Sue Walker, 43–60. London: Routledge.

Spronsen, Johannes Willem van. 1969. *The periodic system of chemical elements: A history of the first hundred years.* Amsterdam: Elsevier.

Starcke, Katrin, and Matthias Brand. 2012. Decision-making under stress: A selective review. *Neuroscience & Biobehavioral Reviews* 36 (4): 1228–1248.

Stcherbatsky, Feodor Ippolitovic. 1984. *Buddhist logic, vol. 1.* New Delhi: Oriental Books.

Stefanou, Maria Ioanna. 2020. The footprints of neuroscience in Alexandria during the 3rd-century BC: Herophilus and Erasistratus. *Journal of Medical Biography* 28 (4): 186–194.

Stewart, Philip J. 2018. Amateurs and professionals in chemistry: The case of the periodic system. In *Mendeleev to Oganesson: A multidisciplinary perspective on the periodic table*, ed. Eric Scerri and Guillermo Restrepo, 66–79. Oxford: Oxford University Press.

Stjernfelt, Frederik. 2007. *Diagrammatology: An investigation on the borderlines of phenomenology, ontology, and semiotics.* London: Springer.

Striner, Richard. 1995. Political Newtonianism: The cosmic model of politics in Europe and America. *The William and Mary Quarterly* 52 (4): 583–608.

Strongman, Luke. 2010. Force field: Vitruvian Man and the physics of sensory perception. *International Journal of Arts and Sciences* 3 (9): 218–226.

Stückelberg, Ernst C. G. 1941. Remarque a propos de la création de paires de particules en théorie de relativité. *Helvetica Physica Acta* 14 (372): 588–594.

Stuessy, Tod F. 2009. *Plant taxonomy: The systematic evaluation of comparative data*. 2nd ed. New York: Columbia University Press.

Suedfeld, Peter, and A. Dennis Rank. 1976. Revolutionary leaders: Long-term success as a function of changes in conceptual complexity. *Journal of Personality and Social Psychology* 34 (2): 169–178.

Suitner, Caterina, and Chris McManus. 2011. Aesthetic asymmetries, spatial agency, and art history: A social psychological perspective. In *Spatial Dimensions of Social Thought*, ed. Anne Maas and Thomas W. Schubert, 277–301. Berlin: De Gruyter Mouton.

Suminoe, Saichirō. 1990. *Nihon Keisenshi: Nihon Tekunikaru Anarisuto Kyōkai* (History of technical analysis in Japan), 10th print. Tōkyō: Nihon Keizai Shinbunsha.

Sumner, William Graham. 1906. *Folkways: A study of the sociological importance of usages, manners, customs, mores, and morals*. Boston: Ginn.

Swade, Doron. 2000. The Phillips Machine and the history of computing. In *A.W.H. Phillips: Collected works in contemporary perspective*, ed. R. Leeson, 120–126. Cambridge: Cambridge University Press.

Sweller, John. 1988. Cognitive load during problem solving: Effects on learning. *Cognitive Science*. 12 (2): 257–285.

Sweller, John, Paul L. Ayres, and Slava Kalyuga. 2011. *Cognitive load theory*. New York: Springer.

Tabing, Felicia. 2018. Mathematics in drafting Japanese crest designs. In *Proceedings of bridges 2018: Mathematics, art, music, architecture, education, culture*, ed. Eve Torrence, Bruce Torrence, Carlo Séquin, and Kristóf Fenyvesi, 463–466. Phoenix: Tesselations Publishing.

Taleb, Nassim N. 2019. Fooled by correlation: Common misinterpretations in social 'science.' Academia Online. https://www.academia.edu/39797871/Fooled_by_Correlation_Common_Misinterpretations_in_Social_Science. Accessed 3 January 2024.

Tang, Anson C. Yan. 2012. Review of traditional Chinese medicine pulse diagnosis quantification: Complementary therapies for the contemporary healthcare. In *Complementary therapies for the contemporary healthcare*, ed. Marcelo Saadand Roberta De Medeiros, 61–80. Rijeka: InTech.

Tangian, Andranik. 2014. *Mathematical theory of democracy*. Berlin: Springer.

Taruskin, Richard. 2005. *The Oxford history of Western music, vol. 1: The earliest notations to the sixteenth century*. Oxford: Oxford University Press.

Taylor, Jon. 2013. Administrators and scholars: The first scribes. In *The Sumerian world*, ed. Harriet Crawford, 290–304. Abingdon: Routledge.

Theocharis, Reghinos D. 1983. *Early developments in mathematical economics.* 2nd ed. London: Macmillan.

Tiles, Mary, and Yuan Jinmei. 2004. Could the Aristotelian square of opposition be translated into Chinese? *Dao: A Journal of Comparative Philosophy* 4(1): 137–149.

Tilling, Laura. 1975. Early experimental graphs. *The British Journal for the History of Science* 8 (3): 193–213.

Timbro. n.d. Lafferkurvan. *Timbro.se.* https://timbro.se/skattekoll/lafferkurvan/. Accessed 15 February 2022.

Tōkyō kabushiki torihikisho. 1932. *Tōkyō kabushiki torihikisho* (Tokyo stock exchange). Tōkyō: Tōkyō kabushiki torihikisho chōsa-ka.

Tolman, Edward C. 1947. Cognitive maps in rats and men. *Psychological Review* 55 (4): 189–208.

Triantafillou, Eric. 2019. To make what is vertical horizontal: Picturing social domination. *Journal of Visual Culture* 18 (3): 265–288.

Trivedi, Kirti. 1989. Hindu temples: Models of a fractal universe. *The Visual Computer* 5:243–258.

Truesdell, Clifford. (1972) 2007. Leonhard Euler, supreme geometer. In *The genius of Euler: Reflections on his life and work,* ed. William Dunham, 13–41. Washington, DC: Mathematical Association of America.

Tsien, Tsuen-Hsuin. 1985. *Paper and printing: Science and civilization in China, vol. 5 part 1,* Series editor Joseph Needham. Cambridge: Cambridge University Press.

Tufte, Edward R. 2001. *The visual display of quantitative information.* 2nd ed. Cheshire: Graphics Press.

Tufte, Edward R. 2006. *Beautiful evidence.* Cheshire: Graphics Press.

Tukey, John W. 1977. *Exploratory data analysis.* Reading, MA: Addison-Wesley.

Tversky, Barbara. 2001. Spatial schemas in depictions. In *Spatial schemas and abstract thought*, ed. Merideth Gattis, 79–112. Cambridge, MA: MIT Press.

Tversky, Barbara. 2011. Spatial thought, social thought. In *Spatial dimensions of social thought*, ed. Anne Maas and Thomas W. Schubert, 17–38. Berlin: De Gruyter Mouton.

Tversky, Barbara. 2019. *Mind in motion: How action shapes thought.* New York: Basic.

Tversky, Barbara, Sol Kugelmass, and Atalia Winter. 1991. Cross-cultural and developmental trends in graphic productions. *Cognitive Psychology* 23 (4): 515–557.

Twain, Mark. 1889. *A Connecticut Yankee in King Arthur's court.* New York: Harper & Brothers.

Twain, Mark. 1914. How to make history dates stick. *Harper's Monthly Magazine,* December.

Underwood, Emily. 2016. Expanding our mental maps. *Science* 352 (6292): 1378.

Urton, Gary. 2017. *Inka history in knots: Reading khipus as primary sources*. Austin: University of Texas Press.

Vankeerberghen, Griet. 2007. The tables (*biao*) in Sima Qian's *shiji:* Rhetoric and remembrance. In *Graphics and text in the production of technical knowledge in China: The warp and the weft*, ed. Francesca Bray, Vera Dorrofeeva-Lichtmann, and Georges Métailié, 295–311. Leiden: Brill.

Varela, Francisco J., Evan Thompson, and Eleanor Rosch. 2016. *The embodied mind: Cognitive science and human experience*. Revised edition. Cambridge, MA: MIT Press.

Vartanian, Oshin. 2012. Dissociable neural systems for analogy and metaphor: Implications for the neuroscience of creativity. *The British Journal of Psychology* 103 (3): 302–316.

Vat. Syr. 158. 2024. Vatican Library, Manuscript Vat. Syr 158. https://digi.vatlib.it/view/MSS_Vat.sir.158. Accessed 20 July 2024.

Verboon, Annemieke R. 2010. *Lines of thought: Diagrammatic representation and the scientific texts of the Arts Faculty, 1200–1500*. Dissertation. Lieden: Leiden University Scholarly Publications.

Verboon, Annemieke R. 2014. The medieval tree of Porphyry: An organic structure of logic. In *The tree: Symbol, allegory, and mnemonic device in medieval art and thought*, ed. Pippa Salonius and Andrea Worm. Turnhout: Brepols.

Von Petzinger, Genevieve. 2017. *The first signs: Unlocking the mysteries of the world's oldest symbols*. New York: Atria.

Voss, Julia. 2010. *Darwin's pictures: Views of evolutionary theory, 1837–1874*. New Haven, CT: Yale University Press.

Wagenaar, Willem A. 1972. Generation of random sequences by human subjects: A critical survey of literature. *Psychological Bulletin* 77 (1): 65–72.

Waley, Arthur. (1952) 2005. *The real Tripitaka: And other pieces*. Oxfordshire: Routledge

Wallace, Anthony F. C. 1961. On being just complicated enough. *Proceedings of the National Academy of Science* 47 (4): 458–464.

Wallis Budge, Earnest A. 1920. *An Egyptian hieroglyphic dictionary: With an index of English words, king list and geographical list with indexes, list of hieroglyphic characters, Coptic and Semitic alphabets*. London: Murray.

Wammes, Jeffrey D., Melissa E. Meade, and Myra A. Fernandes. 2016. The drawing effect: Evidence for reliable and robust memory benefits in free recall. *Quarterly Journal of Experimental Psychology* 69 (9): 1752–1776.

Wardhaugh, Benjamin. 2008. Musical logarithms in the seventeenth century: Descartes, Mercator, Newton. *Historia Mathematica* 35(1): 19–36.

Watson, Arthur. 1934. *The early iconography of the Tree of Jesse*. Oxford: Oxford University Press.

Wells, Herbert George. 1888. *The chronic argonauts*. Telelib.com https://www.telelib.com/authors/W/WellsHerbertGeorge/prose/chronicargonauts/

index.html. Accessed 30 July 2024. Originally published in *The Science Schools Journal*, April, May, and June 1888.
Wentzel, Gregor. 1938. Schwere Elektronen und Theorien der Kernvorgänge [Heavy electrons and theories of nuclear processes]. *Die Naturwissenschaften* 26 (18): 273–279.
Wertheimer, Max. 1945. *Productive thinking*. New York: Harper.
Weyl, Hermann. 1928. *Gruppentheorie und Quantenmechanik*. Leipzig: Hirzel.
Weyl, Hermann. 1952. *Symmetry*. Princeton, NJ: Princeton University Press.
Wheat, Elizabeth Ruth Josie. 2013. *Terrestrial cartography in ancient Mesopotamia*. Birmingham, AL: University of Birmingham.
Wickham, Hadley, and Lisa Stryjewski. 2011. 40 years of boxplots. Hadley Wickham's academic portfolio. https://vita.had.co.nz/papers/boxplots.pdf. Accessed 3 January 2024.
Wilkins, John S. 2009. *Species: A history of the idea*. Berkeley: University of California Press.
Williams, Jean Anne Hayes. 2000. The earliest dated Tree of Jesse image: thematically reconsidered. *Athanor* 18:17–23.
Willman, Marshall. 2021. Logic and language in early Chinese philosophy. In *The Stanford encyclopedia of philosophy*, ed. Edward N. Zalta. Stanford: Stanford University. https://plato.stanford.edu/archives/spr2021/entries/chinese-logic-language/. Accessed 30 July 2024.
Woelert, Peter. 2011. Human cognition, space, and the sedimentation of meaning. *Phenomenology and the Cognitive Sciences* 10 (1): 113–137.
Woit, Peter. 2006. *Not even wrong: The failure of string theory and the continuing challenge to unify the laws of physics*. London: Jonathan Cape.
Wongpakaran, Tinakon, Nahathai Wongpakaran, Sitthinant Tanchakvaranont, Putipong Bookkamana, Manee Pinyopornpanish, Kamonporn Wannarit, Sirina Satthapisit, Daochompu Nakawiro, Thanita Hiranyatheb, and Kulvadee Thongpibul. 2016. Depression and pain: Testing of serial multiple mediators. *Neuropsychiatric disease and treatment* 12:1849–1860.
Worthing, Daniel. 2016. *Computational conceptual blending as a model of spatial transfer in natural language*. Thesis. Edinburgh: University of Edinburgh. https://project-archive.inf.ed.ac.uk/msc/20162033/msc_proj.pdf. Accessed 30 July 2024.
Wujastyk, Dominik. 2018. The elephant's footprint: An ancient Indian logic diagram. *Studia Orientalia Electronica* 6:56–61.
Wüthrich, Adrian. 2010. *The genesis of Feynman diagrams*. Dordrecht: Springer.
Wynn, Thomas Grant, and Frederick L. Coolidge. 2022. *An introduction to evolutionary cognitive archaeology*. New York: Routledge.
Yates, Frances A. (1966) 2014. *The art of memory*. London: The Bodley Head.
Yau, Shing-Tung, and Steven J. Nadis. 2010. *The shape of inner space: String theory and the geometry of the universe's hidden dimensions*. New York: Basic Books.

Yi, Hwang. 1988. *To become a sage: The ten diagrams on sage learning by Yi T'oegye*. Trans. M.C. Kalton. New York: Columbia University Press.
Youschkevitch, Adolph P. 1976. The concept of function up to the middle of the 19th century. *Archive for history of exact sciences* 16 (1): 37–85.
Zador, Anthony, Sean Escola, Blake Richards, Bence Ölveczky, Yoshua Bengio, Kwabena Boahen, Matthew Botvinick, et al. 2023. Catalyzing next-generation artificial intelligence through neuroai. *Nature Communications* 14 (1): 1597.
Zarlino, Gioseffi. 1588. *Sopplimenti musicali*. Venice: Franceschi.
Zenderland, Leila. 1998. *Measuring minds: Henry Herbert Goddard and the origins of American intelligence testing*. Cambridge: Cambridge University Press.
Zitser, Ernest A. 2018. Raphael Lemkin and the Soviet propaganda poster collection at Duke university library. *Slavic and East European Information Resources* 19 (3–4): 242–262.
Zlatev, Jordan. 2012. Cognitive semiotics: An emerging field for the transdisciplinary study of meaning. *Public Journal of Semiotics* 4 (1): 2–24.
Zobel, Justin. 2004. *Writing for computer science*. 2nd ed. London: Springer.

Index[1]

[1] Note: Page numbers followed by 'n' refer to notes.

J. Holmén, *A History of Diagrams*, Palgrave Studies in the History of Science and Technology,
https://doi.org/10.1007/978-3-031-99691-7

FSC
www.fsc.org
MIX
Papier | Fördert
gute Waldnutzung
FSC® C083411

Kolibri 360 GmbH
Ferdinand-Jühlke-Straße 7
99095 Erfurt, Deutschland
produktsicherheit@kolibri360.de